T0320615

QUANTUM MONTE CARLO APPROACHES FOR CORRELATED SYSTEMS

Over the past several decades, computational approaches to studying strongly interacting systems have become increasingly varied and sophisticated. This book provides a comprehensive introduction to state-of-the-art quantum Monte Carlo techniques relevant for applications in correlated systems. Starting with an overview of variational wave functions, it features a detailed presentation of stochastic samplings including Markov chains and Langevin dynamics, which are developed into a discussion of Monte Carlo methods. The variational technique is also described, from foundations to a detailed description of its algorithms. Other topics discussed include optimization techniques, real-time dynamics, and projection methods, including Green's function, reptation, and auxiliary-field Monte Carlo, from basic definitions to advanced algorithms for efficient codes, and concluding with recent developments on the continuum space. This book provides an extensive reference for students and researchers working in condensed matter theory or those interested in advanced numerical methods for electronic simulation.

FEDERICO BECCA is a researcher at the National Research Council (CNR) working in the theoretical group of the Condensed Matter section of the International School for Advanced Studies (SISSA) in Trieste. His research focuses on different aspects of correlated systems on the lattice. His major scientific contributions include advances in frustrated magnets, superconductivity from strong electronic correlation, disordered fermionic and bosonic models, and Mott metal-insulator transitions.

SANDRO SORELLA is Professor of Condensed Matter Physics at the International School of Advanced Studies (SISSA) in Trieste. His focus is on the study of strongly correlated electron systems by advanced numerical simulation techniques based on quantum Monte Carlo. He has developed novel Monte Carlo algorithms that are now widely used and considered state-of-the-art in the field.

QUANTUM MONTE CARLO APPROACHES FOR CORRELATED SYSTEMS

FEDERICO BECCA

National Research Council

SANDRO SORELLA

International School for Advanced Studies, Trieste

CAMBRIDGE
UNIVERSITY PRESS

Shaftesbury Road, Cambridge CB2 8EA, United Kingdom

One Liberty Plaza, 20th Floor, New York, NY 10006, USA

477 Williamstown Road, Port Melbourne, VIC 3207, Australia

314–321, 3rd Floor, Plot 3, Splendor Forum, Jasola District Centre, New Delhi – 110025, India

103 Penang Road, #05–06/07, Visioncrest Commercial, Singapore 238467

Cambridge University Press is part of Cambridge University Press & Assessment,
a department of the University of Cambridge.

We share the University's mission to contribute to society through the pursuit of
education, learning and research at the highest international levels of excellence.

www.cambridge.org
Information on this title: www.cambridge.org/9781107129931

DOI: 10.1017/9781316417041

First published 2017

A catalogue record for this publication is available from the British Library

Library of Congress Cataloging-in-Publication data
Names: Becca, Federico, 1972– author. | Sorella, Sandro, 1960– author.
Title: Quantum Monte Carlo approaches for correlated systems / Federico Becca
(Scuola Internazionale Superiore di Studi Avanzati), Sandro Sorella.
Description: Cambridge, United Kingdom ; New York, NY : Cambridge University
Press, 2017. | Includes bibliographical references and index.
Identifiers: LCCN 2017025043 | ISBN 9781107129931 (hardback ; alk. paper) |
ISBN 1107129931 (hardback ; alk. paper)
Subjects: LCSH: Monte Carlo method. | Variational principles.
Classification: LCC QC174.85.M64 B43 2017 | DDC 530.1201/518282–dc23
LC record available at https://lccn.loc.gov/2017025043

ISBN   978-1-107-12993-1   Hardback

Contents

Preface   *page* ix
Acknowledgements   xii

Part I   Introduction   1

1   Correlated Models and Wave Functions   3
    1.1   Introduction   3
    1.2   The Matrix Formulation   6
    1.3   Effective Lattice Models   7
    1.4   The Variational Principle   13
    1.5   Variational Wave Functions   14
    1.6   Size Extensivity   31
    1.7   Projection Techniques   34

Part II   Probability and Sampling   37

2   Probability Theory   39
    2.1   Introduction   39
    2.2   Events and Probability   41
    2.3   Moments of the Distribution: Mean Value and Variance   44
    2.4   Changing Random Variables   47
    2.5   The Chebyshev's Inequality   48
    2.6   Summing Independent Random Variables   48
    2.7   The Central Limit Theorem   53

3   Monte Carlo Sampling and Markov Chains   56
    3.1   Introduction   56
    3.2   Reweighting Technique and Correlated Sampling   59
    3.3   Direct Sampling   60
    3.4   Importance Sampling   61

3.5   Sampling a Discrete Distribution Probability   62
3.6   Sampling a Continuous Density Probability   64
3.7   Markov Chains   66
3.8   Detailed Balance and Approach to Equilibrium   69
3.9   Metropolis Algorithm   74
3.10 How to Estimate Errorbars   76
3.11 Errorbars in Correlated Samplings   82

4   Langevin Molecular Dynamics   85
    4.1   Introduction   85
    4.2   Discrete-Time Langevin Dynamics   87
    4.3   From the Langevin to the Fokker-Planck Equation   89
    4.4   Fokker-Planck Equation and Quantum Mechanics   91
    4.5   Accelerated Langevin Dynamics   96

    Part III   Variational Monte Carlo   101

5   Variational Monte Carlo   103
    5.1   Quantum Averages and Statistical Samplings   103
    5.2   The Zero-Variance Property   105
    5.3   Jastrow and Jastrow-Slater Wave Functions   106
    5.4   The Choice of the Basis Sets   108
    5.5   Bosonic Systems   109
    5.6   Fermionic Systems with Determinants   112
    5.7   Fermionic Systems with Pfaffians   123
    5.8   Energy and Correlation Functions   129
    5.9   Practical Implementation   129

6   Optimization of Variational Wave Functions   131
    6.1   Introduction   131
    6.2   Reweighting Techniques for the Optimization of Wave Functions   132
    6.3   Energy Derivatives   134
    6.4   The Stochastic Reconfiguration   137
    6.5   Stochastic Reconfiguration as a Projection Technique   146
    6.6   The Linear Method   147
    6.7   Calculations of Derivatives in the Jastrow-Slater Case 152

7   Time-Dependent Variational Monte Carlo   156
    7.1   Introduction   156
    7.2   Real-Time Evolution of the Variational Parameters   157
    7.3   An Example for the Quantum Quench in One Dimension   161

Part IV   Projection Techniques 165

8   Green's Function Monte Carlo 167
    8.1   Basic Notions and Formal Derivations 167
    8.2   Single Walker Technique 170
    8.3   Importance Sampling 173
    8.4   The Continuous-Time Limit 178
    8.5   Many Walkers Formulation 180
    8.6   Practical Implementation 188

9   Reptation Quantum Monte Carlo 189
    9.1   A Simple Path Integral Technique 189
    9.2   A Simple Way to Sample Configurations 191
    9.3   The Bounce Algorithm 195
    9.4   The Continuous-Time Limit 196
    9.5   Practical Implementation 197

10   Fixed-Node Approximation 199
    10.1   The Sign Problem 199
    10.2   A Simple Example on the Continuum 204
    10.3   A Simple Example on the Lattice 208
    10.4   The Fixed-Node Approximation on the Lattice 209
    10.5   Practical Implementation 213

11   Auxiliary Field Quantum Monte Carlo 214
    11.1   Introduction 214
    11.2   Trotter Approximation 216
    11.3   Hubbard-Stratonovich Transformation 217
    11.4   The Path-Integral Representation 219
    11.5   Sequential Updates 223
    11.6   Ground-State Energy and Correlation Functions 228
    11.7   Simple Cases without Sign Problem 228
    11.8   Practical Implementation 231

Part V   Advanced Topics 233

12   Realistic Simulations on the Continuum 235
    12.1   Introduction and Motivations 235
    12.2   Variational Wave Function with Localized Orbitals 237
    12.3   Size Consistency of the Variational Wave Functions 244

12.4 Optimization of the Variational Wave Functions   245
12.5 Lattice-Regularized Diffusion Monte Carlo   252
12.6 An Improved Scheme for the Lattice Regularization   256

Appendix   Pseudo-Random Numbers Generated by Computers   261
References   264
Index   272

Preface

This book originates from the lecture notes of the Ph.D. course on numerical simulations for strongly-correlated systems at the International School for Advanced Studies (S.I.S.S.A.) in Trieste. Even though the backbone of this book has been formed over the last fifteen years, almost all chapters have been completely rewritten, reshaped, or heavily modified in order to improve the clarity of the presentation. In addition, the second part of this book (i.e., Chapters 2, 3, and 4) has been taught during the C.E.C.A.M. Summer School on "Atomistic Simulation Techniques for Material Science, Nanotechnology, and Biophysics", held in Trieste from 2011 to 2016. This school was addressed to graduate and undergraduate students who had little experience in numerical simulations and part of the actual presentation has benefited from their suggestions. The book is envisaged for students and young researches, who do not know much on Monte Carlo methods and want to understand how to implement ground-state algorithms on interacting models. After a first chapter dealing with a bird's-eye review on correlated wave functions that have been used in the past, the second part is intended to give a rather pedagogical introduction to simple probability theory (just the concepts that are necessary for the further developments) and methods for statistical samplings, not only based on Monte Carlo techniques but also including molecular-dynamics approaches. The central part of the book deals with variational and projection Monte Carlo approaches that are well established and widely used to treat lattice models (for both fermions and bosons). Finally, the last part provides an introduction to Monte Carlo methods that have been recently developed for electron systems on the continuum.

Our main motivation to convert the preliminary lecture notes into a structured book is to put in a sequential order all the concepts and machineries that are needed for writing efficient codes based upon variational and projection Monte Carlo techniques. Moreover, a modern textbook describing how to treat wave functions within quantum Monte Carlo algorithms was lacking in the portfolio of books treating correlated systems. The methods chosen here are influenced by our own research and do not cover all the numerical approaches that are currently used in the field. For example, we do not discuss the algorithms that have been introduced to deal with impurity models, also employed in dynamical mean-field theory (DMFT), and wave functions that are defined by tensor networks, which, in our opinion, although representing a promising route, they are still in their infancy, especially for what concerns the optimization procedure. Instead, we preferred to focus on a selected number of well established methods that have been developed in the last twenty years and have been demonstrated to be very powerful in the description of several lattice models (e.g., Hubbard, $t-J$, Heisenberg, and many others). Excellent books treating various aspects, which are not considered here, are the ones written by Krauth (2006), Gubernatis et al. (2016), and Martin et al. (2016).

In this book, we have followed the scientific development of one of the authors. In fact, a few years ago, it became important to extend all the knowledge acquired in the construction of variational wave functions on the lattice to realistic simulations and perform direct comparisons with experiments. Here, we have introduced the field starting from lattice models, which is clearly original and quite unconventional; nonetheless, we believe that this approach is pedagogical, as any numerical method is more transparent and clearly defined when first applied to lattice models and then extended to more realistic situations. This way of thinking should be intriguing also for expert coming from electronic systems on the continuum, since a different perspective of a conventional method may clarify and make more transparent the original formulation (e.g., the lattice-regularized diffusion Monte Carlo approach for the standard diffusion Monte Carlo).

In the past two decades, there has been a substantial development in Monte Carlo methods for the optimization of variational wave functions and the definition of stable projection techniques, especially for fermionic models. Even though a final solution of correlated models in more than one spatial dimension is still far away elusive, variational wave functions have been proven to provide very accurate approximations to ground states, when compared to exact diagonalizations on small clusters or in particular regimes where exact solutions are known (e.g., when the sign problem is not present or for particularly designed cases where the ground state can be exactly evaluated). In particular, a deep investigation has been dedicated to frustrated Heisenberg models on various lattices (e.g., on square, triangular, honeycomb, and kagome lattices), to pursue the possibility of stabilizing so-called quantum spin liquids. Here, we would like to mention the case of the Heisenberg model on the kagome lattice, for which a variational wave function, constructed from applying the Gutzwiller projector to a non-correlated fermionic state, suggested the possibility that the ground state can be a spin liquid with Dirac cones in the spinon spectrum (i.e., a gapless spin liquid). This approach competes with other numerical techniques, like density-matrix renormalization group (DMRG) or its extensions based upon tensor networks. In our opinion, the final understanding of this kind of problem will require both unbiased methods, like DMRG in which the wave function is obtained without any initial guess, and more guided (but more transparent) approaches, like the ones based upon a suitable variational state, whose properties can be estimated by Monte Carlo sampling.

Correlated wave functions (and Monte Carlo methods to assess their physical properties) have been largely employed also in relation to the discovery of high-temperature superconductors. For sure, part of the motivation was given by the success of the Bardeen-Cooper-Schrieffer (BCS) theory for standard (low-temperature) superconductors. However, a crucial contribution in this direction was given by Philip Anderson, who suggested that superconductivity may naturally emerge from doping a resonating-valence bond (RVB) insulator, with a mechanism that does not involve phonons like in the BCS theory. Also in this case, suitably optimized RVB states, constructed from Gutzwiller-projected BCS wave functions, have been shown to give remarkably accurate energies, predicting a uniform RVB superconductivity in hole-doped Hubbard and $t-J$ models. Connected to this issue, we mention that a rather accurate description of a *bona fide* Mott insulator, with preformed Cooper pairs, has been accomplished by a full optimization of the Jastrow factor, which generalizes the Gutzwiller factor to include long-range density-density correlations. Then, within Jastrow-Slater wave functions, it has been possible to describe a continuous metal-insulator transition in the paramagnetic sector of the single-band Hubbard model, similarly to what has been obtained within DMFT.

In recent years, simple generalizations of the previous variational states have been used to study several other models, approaching more realistic models with few (or several) orbitals, disorder, and electron-phonon interactions. In all these cases, the enlarged Hilbert space becomes prohibitive for most of the other numerical methods, while the Monte Carlo sampling is relatively unaffected, allowing us to explore problems that are at the frontier of the present research in solid-state physics.

We would like to remark that very few investigations have been performed in the lattice versions of the quantum Hall effect. Indeed, even though, on the continuum, the Laughlin wave function has been pivotal to describe the fractional case with its exotic properties and several generalizations have been proposed over the years (like the Moore-Read state for the 5/2 filling factor), very few attempts have been done on the lattice (i.e., for the Hubbard-Hofstadter model). We hope that this book will give motivations to pursue this line of research.

Acknowledgements

We acknowledge all the collaborators over the last twenty years with whom we have worked and discussed. Among them, A. Parola holds a place of honor, even though he is proud of not using random numbers in all his codes. His constant presence over the years was fundamental for our research in frustrated magnetic systems and correlated superconductors. He also carefully read a few chapters of this book, correcting many inaccuracies and vagueness. L. Capriotti, who took part in the early and "heroic" days of variational Monte Carlo approaches and left physics for a successful career in finance. Then, many students and postdocs in S.I.S.S.A. took part of our adventure in Monte Carlo methods. We would particularly like to thank our long-lasting collaborators C. Attaccalite, G. Carleo, M. Casula, R. Hlubina, Y. Iqbal, G. Mazzola, L.F. Tocchio, S. Yunoki, and A. Zen. Heartfelt thanks for the data shown in the figures of Chapters 4 and 7 go to G. Carleo and G. Mazzola. We are indebted to L.F. Tocchio, C. Genovese, and C. de Franco for having carefully read the manuscript and for their precise and accurate (sometimes too precise and accurate!) corrections to improve all our errors and obscurities. Finally, we wish to thank S. Capelin, our editor from Cambridge University Press, who proposed and persuaded us in writing this book. Moreover, we thank R. Munnelly and people at Cambridge University Press for working with us throughout this project.

Part I

Introduction

1

Correlated Models and Wave Functions

1.1 Introduction

The aim of this chapter is to give a concise overview of the variational approach for strongly correlated systems on the lattice. In particular, we focus our attention on well-known wave functions that have been proven to describe a huge variety of stable phases of matter, e.g., metals, superconductors, band and topological insulators, and quantum Hall states. In the following (and throughout the entire book), we will assume that the reader is familiar with the second-quantization formalism; a simple and complete review of it can be found in the book by Fetter and Walecka (2003).

The use of variational wave functions is rooted in the early days of quantum mechanics. Indeed, soon after its development in 1925−26, Walter Heitler and Fritz London proposed a simplified treatment of the H_2 molecule and opened the way to a theoretical understanding of the chemical bond (Heitler and London, 1927). The Heitler-London *Ansatz* for the ground-state wave function of H_2 treats the two electrons as being highly correlated, since ionic configurations with both electrons on the same Hydrogen atom are excluded. By using the modern formalism of second quantization, the Heitler-London (singlet) state has the following form:

$$|\Psi_{HL}\rangle = \frac{1}{\sqrt{2}} \left(c^\dagger_{1,\uparrow} c^\dagger_{2,\downarrow} + c^\dagger_{2,\uparrow} c^\dagger_{1,\downarrow} \right) |0\rangle; \tag{1.1}$$

here $|0\rangle$ is the vacuum and $c^\dagger_{i,\sigma}$ creates an electron with spin $\sigma = \uparrow, \downarrow$ in a given orbital $\phi(\mathbf{r} - \mathbf{R}_i)$ centered around the atom $i = 1, 2$ at position $\mathbf{R}_i$:

$$c^\dagger_{i,\sigma} = \int \mathbf{dr} \, \phi(\mathbf{r} - \mathbf{R}_i) \psi^\dagger_\sigma(\mathbf{r}), \tag{1.2}$$

where $\psi_\sigma^\dagger(\mathbf{r})$ is the fermionic field operator. In the following, we consider the case in which the orbitals on different atoms are orthogonal:

$$\int d\mathbf{r}\ \phi^*(\mathbf{r} - \mathbf{R}_i)\phi(\mathbf{r} - \mathbf{R}_j) = \delta_{i,j}, \tag{1.3}$$

thus leading to the anti-commutation relations among fermion operators:

$$\left\{c_{i,\sigma}, c_{j,\tau}^\dagger\right\} = \delta_{i,j}\delta_{\sigma,\tau}, \tag{1.4}$$

$$\left\{c_{i,\sigma}^\dagger, c_{j,\tau}^\dagger\right\} = 0. \tag{1.5}$$

In the original work by Heitler and London, the orbitals were not taken to be orthogonal (e.g., they considered hydrogenoic orbitals centered around the two atoms), implying a different normalization of $|\Psi_{HL}\rangle$. The wave function (1.1) becomes very accurate when the Hydrogen molecule is "stretched out," namely when $|\mathbf{R}_1 - \mathbf{R}_2| \to \infty$, and becomes exact in the atomic limit, where the distance between the two atoms is infinite (here, $\phi(\mathbf{r})$ can be taken to be the $1s$ hydrogenoic orbital). In fact, when their relative distance becomes very large, it is energetically favorable to have one electron "localized" on each atom, while the ionic configurations with both electrons on the same Hydrogen have a large energy originating from the electron-electron repulsion.

A distinctly different approach for the same problem was taken soon thereafter by Douglas Hartree, Vladimir Fock, and John Slater, who treated the electrons as being independent from each other (Slater, 1930). Within the independent-electron (Hartree-Fock) approximation, the ground-state wave function is given by a Slater determinant constructed from filling a molecular orbital with up and down spins:

$$|\Psi_{HF}\rangle = \Phi_\uparrow^\dagger \Phi_\downarrow^\dagger |0\rangle, \tag{1.6}$$

where $\Phi_\sigma^\dagger$ creates an electron on a given molecular orbital. The simplest case is obtained by taking a linear combination of orbitals localized around each atom:

$$\Phi_\sigma^\dagger = \frac{1}{\sqrt{2}}(c_{1,\sigma}^\dagger + c_{2,\sigma}^\dagger), \tag{1.7}$$

thus leading to a simple form of the Hartree-Fock wave function:

$$|\Psi_{HF}\rangle = \frac{1}{2}\left(c_{1,\uparrow}^\dagger c_{1,\downarrow}^\dagger + c_{1,\uparrow}^\dagger c_{2,\downarrow}^\dagger + c_{2,\uparrow}^\dagger c_{1,\downarrow}^\dagger + c_{2,\uparrow}^\dagger c_{2,\downarrow}^\dagger\right)|0\rangle. \tag{1.8}$$

Here, the ionic configurations $c_{1,\uparrow}^\dagger c_{1,\downarrow}^\dagger |0\rangle$ and $c_{2,\uparrow}^\dagger c_{2,\downarrow}^\dagger |0\rangle$ appear with the same weight of the non-ionic configurations $c_{1,\uparrow}^\dagger c_{2,\downarrow}^\dagger |0\rangle$ and $c_{2,\uparrow}^\dagger c_{1,\downarrow}^\dagger |0\rangle$. Therefore, when the two Hydrogen atoms are pulled apart, this variational wave function does not reproduce the atomic limit. By contrast, this state becomes exact whenever the electron-electron repulsion is "switched off."

Certainly, the exact result of the H_2 molecule lies in between these two extreme cases. However, the variational wave functions of Eqs. (1.1) and (1.6) represent the prototypes of many-body states that are considered in quantum chemistry and solid-state physics. They are the simplest examples to describe localized and delocalized electrons (i.e., Mott insulators and metals in solids). The molecular orbital approach had a great success in describing many diatomic and small molecules, since most of the chemical bonds are relatively weakly correlated.

From the one hand, in quantum chemistry, these two approaches have been pursued to obtain a quantitative understanding of the chemical bond. In particular, Linus Pauling developed the concept of *resonance* among different electronic configurations to describe the benzene molecule (Pauling, 1960), generalizing the Heitler-London picture for the Hydrogen molecule. Later, this approach has been extended by Patrick Fazekas and Philip Anderson, who introduced the concept of resonating valence-bond (RVB) insulator where localized spin-$1/2$ moments couple together to form singlet pairs; here, quantum fluctuations allow tunneling (i.e., a resonance) among a large number of singlet configurations (Fazekas and Anderson, 1974). The RVB theory had a real bloom with the discovery of high-temperature superconductors, when Anderson and collaborators (Anderson, 1987; Anderson et al., 1987; Baskaran and Anderson, 1988) proposed that an insulating RVB state, with "preformed" singlet pairs, may naturally lead to superconductivity upon electron or hole doping.

On the other hand, in solid-state theory, the Hartree-Fock approximation allowed an accurate description of several metals and band insulators. Moreover, the Hartree-Fock approach has introduced the concept of self-consistent field, in which each electron experiences the average field produced by the other electrons. The methods based upon a self-consistent field, leading to an independent-electron picture, have been widely developed in the last fifty years, mainly from the local-density approximation to the density functional theory (Martin, 2004). The failure of these approaches become evident in a number of materials where unfilled d or f shells are present, like in the so-called heavy-fermion systems (Stewart, 1984). The most celebrated examples are given by the Copper-based high-temperature superconductors (Lee et al., 2006); nowadays, there is a plethora of newly synthesized oxide materials that show strong deviations from what is predicted by using an independent-electron approximation (Imada et al., 1998), thus requiring alternative methods that may include electron-electron correlations. The definition of suitable variational wave functions represents a viable and promising tool to obtain an accurate description of several materials in which the role of interactions cannot be captured by simple mean-field approaches.

1.2 The Matrix Formulation

Once approaching many-body problems, it is always convenient to consider a *finite* basis set, by truncating the original (many-body) Hilbert space in some way. Then the low-energy states, including the ground state, may be approximately described within this truncated basis set. For example, in the Heitler-London wave function of Eq. (1.1), only two states are considered, where electrons with opposite spins occupy orbitals localized around different protons; instead, in the Hartree-Fock state of Eq. (1.8) there are four possible configurations, including the ionic ones with two electrons on the same proton. In these two examples, we did not really refer to any Hamiltonian, but we just construct *Ansätze* for the ground-state wave function.

From a different perspective, once a truncated basis with $\mathcal{N}$ elements has been chosen, the Hamiltonian can be written as a $\mathcal{N} \times \mathcal{N}$ matrix. By indicating the generic state of the basis set by $|x\rangle$, the matrix element of the Hamiltonian $\mathcal{H}$ is given by:

$$\mathcal{H}_{x,x'} \equiv \langle x|\mathcal{H}|x'\rangle, \tag{1.9}$$

which, in principle, can be computed from a given Hamiltonian $\mathcal{H}$ and basis set $\{|x\rangle\}$. In the simplest approximation of the H_2 molecule, we can take only the four states with $S_z = 0$ that have been used in the Heitler-London and Hartree-Fock wave functions:

$$|1\rangle = c_{1,\uparrow}^{\dagger} c_{1,\downarrow}^{\dagger} |0\rangle, \tag{1.10}$$

$$|2\rangle = c_{1,\uparrow}^{\dagger} c_{2,\downarrow}^{\dagger} |0\rangle, \tag{1.11}$$

$$|3\rangle = c_{2,\uparrow}^{\dagger} c_{1,\downarrow}^{\dagger} |0\rangle, \tag{1.12}$$

$$|4\rangle = c_{2,\uparrow}^{\dagger} c_{2,\downarrow}^{\dagger} |0\rangle. \tag{1.13}$$

Within this truncated Hilbert space, the Hamiltonian reads as:

$$\mathcal{H} = -t \sum_{\sigma} \left(c_{1,\sigma}^{\dagger} c_{2,\sigma} + c_{2,\sigma}^{\dagger} c_{1,\sigma} \right) + U \sum_{i} n_{i,\uparrow} n_{i,\downarrow} + V \sum_{\sigma,\sigma'} n_{1,\sigma} n_{2,\sigma'}, \tag{1.14}$$

where $n_{i,\sigma} = c_{i,\sigma}^{\dagger} c_{i,\sigma}$ is the density per spin σ on the site $i = 1, 2$; t, U, and V are parameters that depend upon the overlap of orbitals $\phi(\mathbf{r} - \mathbf{R}_i)$ centered around the two atoms:

$$-t = \int d\mathbf{r} \, \phi^*(\mathbf{r} - \mathbf{R}_1) \left(-\frac{\hbar^2}{2m} \nabla^2 - \sum_{i=1,2} \frac{e^2}{|\mathbf{r} - \mathbf{R}_i|} \right) \phi(\mathbf{r} - \mathbf{R}_2), \tag{1.15}$$

$$U = \int \int d\mathbf{r} \, d\mathbf{r}' \, |\phi(\mathbf{r} - \mathbf{R}_i)|^2 \frac{e^2}{|\mathbf{r} - \mathbf{r}'|} |\phi(\mathbf{r}' - \mathbf{R}_i)|^2, \tag{1.16}$$

$$V = \int \int d\mathbf{r} \, d\mathbf{r}' \, |\phi(\mathbf{r} - \mathbf{R}_1)|^2 \frac{e^2}{|\mathbf{r} - \mathbf{r}'|} |\phi(\mathbf{r}' - \mathbf{R}_2)|^2. \tag{1.17}$$

In principle, the Hamiltonian (1.14) also contains the Coulomb interaction between the protons, however, this is a constant term, once their positions are fixed in $\mathbf{R}_1$ and $\mathbf{R}_2$.

Therefore, within the truncated basis $\{|1\rangle, |2\rangle, |3\rangle, |4\rangle\}$, the solution of the problem corresponds to the diagonalization of the 4×4 matrix:

$$\mathbf{h} = \begin{pmatrix} U & -t & -t & 0 \\ -t & V & 0 & -t \\ -t & 0 & V & -t \\ 0 & -t & -t & U \end{pmatrix}. \tag{1.18}$$

The Hartree-Fock wave function is the exact ground state for $U = V = 0$, while the Heitler-London wave function is the ground state for $t = 0$ and $U > V$. In the generic case with finite t, U, and V, the exact ground state is a superposition of these two wave functions and gives the best variational wave function in this truncated Hilbert space.

Of course, the quality of the approximation can be improved by enlarging the Hilbert space, e.g., by including more orbitals in the truncated basis. The main issue comes from the fact that the problem becomes terribly complex as the dimension of the truncated Hilbert space, and consequently the size of the matrix, increases. At present, there are no exact methods that allow the solution of a general many-body Hamiltonian with a computational effort that scales polynomially with the system size. The complexity of the many-body problem is generally exponential, and this is the main reason why strong correlation is such a difficult task.

1.3 Effective Lattice Models

In this section, we would like to report a few examples of lattice models that represent important (and in most cases still *unsolved*) cases in which correlation effects show up, producing several unconventional quantum phases. A pedagogical introduction to the field of correlated systems can be found in the book by Fazekas (1999).

The simplest example has been independently conceived by John Hubbard (1963), Martin Gutzwiller (1963), and Junjiro Kanamori (1963) and is now universally known as the Hubbard model. Here, electrons on a lattice interact among each others through a simplified "Coulomb" potential that includes only the on-site term:

$$\mathcal{H} = -t \sum_{\langle i,j \rangle, \sigma} c_{i,\sigma}^{\dagger} c_{j,\sigma} + \text{h.c.} + U \sum_{i} n_{i,\uparrow} n_{i,\downarrow}, \tag{1.19}$$

where $\langle \dots \rangle$ indicates neighboring sites on a given lattice in d spatial dimensions. For simplicity, in the following, we consider hyper-cubic (Bravais) lattices defined by primitive vectors $\mathbf{a}_\mu$, with $\mu = 1, \ldots, d$; moreover, periodic boundary conditions are taken. However, the Hubbard model can be defined on any d-dimensional lattice. Then, $c_{j,\sigma}^\dagger$ ($c_{j,\sigma}$) creates (destroys) one electron with spin σ on a Wannier orbital residing on the site j:

$$\Xi_j(\mathbf{r}) = \frac{1}{\sqrt{L}} \sum_k e^{-i\mathbf{k}\cdot\mathbf{R}_j} \Psi_k(\mathbf{r}), \tag{1.20}$$

where L is the total number of sites and $\Psi_k(\mathbf{r})$ are Bloch states constructed with the orbitals $\phi(\mathbf{r} - \mathbf{R}_i)$ centered around each site. Therefore, the operators at different sites create orthogonal states, thus satisfying the anti-commutation relation of Eqs. (1.4) and (1.5). The Hubbard model generalizes the H_2 Hamiltonian of Eq. (1.14) to a lattice of Hydrogen atoms (with $V = 0$) and is defined in the Hilbert space where each site can be empty, singly occupied (with either spin up or down), or doubly occupied. Moreover, the Hamiltonian (1.19) commutes with the total number of particles with up or down spin (i.e., $N_\uparrow$ and $N_\downarrow$, $N_e = N_\uparrow + N_\downarrow$ being the total number of electrons), thus allowing us to consider sectors with different number of particles separately.

The first term of Eq. (1.19), proportional to t, describes the electron hopping, which favors delocalized states. Indeed, for $U = 0$, the Hamiltonian can be easily diagonalized in the Fourier space (i.e., by plane waves):

$$c_{k,\sigma}^\dagger = \frac{1}{\sqrt{L}} \sum_j e^{-i\mathbf{k}\cdot\mathbf{R}_j} c_{j,\sigma}^\dagger. \tag{1.21}$$

After performing this transformation, the non-interacting Hamiltonian becomes:

$$\mathcal{H}_t = \sum_{k,\sigma} \epsilon_k c_{k,\sigma}^\dagger c_{k,\sigma}, \tag{1.22}$$

where, for the case with nearest-neighbor hopping:

$$\epsilon_k = -2t \sum_{\mu=1}^d \cos(\mathbf{k} \cdot \mathbf{a}_\mu). \tag{1.23}$$

Any state constructed from filling k-vectors with up and/or down electrons is an eigenstate of $\mathcal{H}_t$:

$$|\Phi_t\rangle = \prod_{k,\sigma} \left(c_{k,\sigma}^\dagger\right)^{\eta_{k,\sigma}} |0\rangle, \tag{1.24}$$

where $\eta_{k,\sigma} = 1$ ($\eta_{k,\sigma} = 0$) indicates that the single-particle state with momentum k and spin σ is occupied (empty); this state has an energy:

$$E = \sum_{k,\sigma} \eta_{k,\sigma} \epsilon_k. \tag{1.25}$$

For $N_\uparrow = N_\downarrow = N_e/2$, the ground state of Eq. (1.22) is obtained by filling the $N_e/2$ lowest-energy levels with both up and down electrons, e.g., $\eta_{k\sigma} = 1$ for all the momenta such that $\epsilon_k \leq \epsilon_F$ (where ϵ_F denotes the Fermi energy) and $\eta_{k\sigma} = 0$ for all the momenta such that $\epsilon_k > \epsilon_F$. On a finite lattice, the ground state is unique if there is a (finite-size) gap between the highest-energy occupied level and the lowest-energy unoccupied one. In the thermodynamic limit, the gap goes to zero and, on the hyper-cubic lattice, this state describes a metal.

By contrast, whenever the hopping term vanishes, i.e., $U/t = \infty$, the Hubbard Hamiltonian is already diagonal in real space:

$$\mathcal{H}_U = U \sum_i n_{i,\uparrow} n_{i,\downarrow}. \tag{1.26}$$

In this case, any state constructed from putting electrons on sites of the lattice is an eigenstate of $\mathcal{H}_U$:

$$|\Phi_U\rangle = \prod_{i,\sigma} \left(c_{i,\sigma}^\dagger\right)^{\xi_{i,\sigma}} |0\rangle, \tag{1.27}$$

where $\xi_{i,\sigma} = 1$ ($\xi_{i,\sigma} = 0$) indicates that the site i is (not) occupied by an electron with spin σ. Its energy is given by:

$$E = U \sum_i \xi_{i,\uparrow} \xi_{i,\downarrow} = U N_d, \tag{1.28}$$

where N_d is the number of doubly occupied sites. The ground state is highly degenerate, with $E = 0$ for $N_e \leq L$ (corresponding to all states without doubly occupied sites) and $E = U(N_e - L)$ for $L < N_e \leq 2L$ (corresponding to all states with $N_e - L$ doubly occupied sites). In particular, for $N_e = L$ (half filling) the ground state has exactly one electron per site and degeneracy equal to 2^L, corresponding to all possible configurations with up or down spin on each site. This is an insulator that is stabilized by the strong electron-electron repulsion (i.e., $U/t = \infty$), which is called Mott insulator, after the pioneering work done by Nevill Mott (1949, 1990).

Besides these trivial limits, there are very few cases in which the Hubbard model can be exactly solved on large lattices (where exact diagonalizations cannot be performed). In one spatial dimension with nearest-neighbor hopping, the so-called Bethe *Ansatz* provides us with the exact results for the energy and other thermodynamic quantities, although correlation functions are difficult to compute (Lieb and Wu, 1968): here, the ground state is an insulator for $N_e = L$ and $U > 0$, while it is metallic for $N_e \neq L$. Notably, there is an exact solution in any dimension d for $U/t = \infty$ and $N_e = L - 1$ (e.g., one hole), where the ground state is a fully polarized ferromagnet (Nagaoka, 1966). More generally, it is possible to show that the ground state is fully polarized on an arbitrary lattice whenever $t < 0$, $U/|t| = \infty$, and $N_e = L - 1$ (Tasaki, 1998). Otherwise, the exact ground-state properties of the Hubbard model with generic filling factor $n = N_e/L$ and interaction U/t are not known. Remarkably, a numerically exact Monte Carlo approach without sign problem (i.e., exact apart from statistical errors) can be performed in the half-filled case for any value of U/t in the square lattice; this approach will be discussed in Chapter 11.

From a general point of view, it is natural to expect that, in a generic d-dimensional lattice, a metal-insulator transition will appear for a given $U = U_c$ at half filling, i.e., $n = 1$ (for the exactly solvable model in one dimension, $U_c = 0$). This would be a *bona-fide* Mott transition, which is driven by the electron-electron interaction only, and not by a symmetry-breaking mechanism: in other words, electrons localize just because of the strong correlation. However, in most cases (when varying the dimensionality of the lattice and the coordination number) the metal-insulator transition comes together with the development of long-range order, most notably the formation of antiferromagnetic order. The existence of a super-exchange coupling, which favors antiferromagnetic order for large values of U/t, can be seen by performing a strong-coupling expansion with $U/t \gg 1$.

Whenever the interaction U is much larger than the hopping parameter t and $n = 1$, we can project out all the sub-spaces with one or more doubly occupied sites and obtain an effective low-energy model that describes the spin degrees of freedom. For $U/t = \infty$, the ground-state manifold is massively degenerate due to the fact that the spin of the electron can be either up or down on each site; the presence of a finite but small hopping lifts this degeneracy creating an antiferromagnetic super-exchange coupling $J = 4t^2/U$. Indeed, at the second-order of perturbation theory in t/U, two neighboring electrons with opposite spins may profit of a virtual hopping process that creates a doubly occupied site (see Fig. 1.1) while two electrons with parallel spins cannot gain any energy from that, because of the Pauli exclusion principle. As a result, the effective low-energy model that captures the spin dynamics is given by the Heisenberg Hamiltonian:

$$\mathcal{H} = J \sum_{\langle i,j \rangle} \mathbf{S}_i \cdot \mathbf{S}_j, \tag{1.29}$$

where $\mathbf{S}_j = (S_j^x, S_j^y, S_j^z)$ is the spin-$1/2$ operator of electrons:

$$S_j^x = \frac{1}{2} \left(c_{j,\uparrow}^\dagger c_{j,\downarrow} + c_{j,\downarrow}^\dagger c_{j,\uparrow} \right), \tag{1.30}$$

$$S_j^y = \frac{1}{2i} \left(c_{j,\uparrow}^\dagger c_{j,\downarrow} - c_{j,\downarrow}^\dagger c_{j,\uparrow} \right), \tag{1.31}$$

$$S_j^z = \frac{1}{2} \left(c_{j,\uparrow}^\dagger c_{j,\uparrow} - c_{j,\downarrow}^\dagger c_{j,\downarrow} \right). \tag{1.32}$$

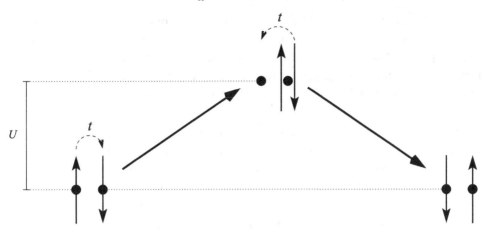

Figure 1.1 In second-order perturbation theory in t/U, if the spins of neighboring sites are antiparallel, they gain energy by a virtual process creating a double occupation.

The Heisenberg model is defined in the Hilbert space where each site is singly occupied. Again, the exact solution of the Heisenberg model can be obtained in one spatial dimension, by the Bethe *Ansatz* (Bethe, 1931): here, the ground state is not magnetically ordered and the excitation spectrum is gapless (implying power-law spin-spin correlations). The absence of a true magnetic order in the ground-state of one-dimensional systems is due to an extension of the Mermin-Wagner theorem (Pitaevskii and Stringari, 1991). In more than one spatial dimension, for bipartite lattices (i.e., where the sites can be partitioned in two sub-lattices and the super-exchange term only couples sites on different sub-lattices) there are Monte Carlo methods that provide us with numerically exact results (see Chapter 8). These stochastic approaches have been crucial to definitively show that the ground state of the Heisenberg model on the two-dimensional square lattice has long-range magnetic (Néel) order (Reger and Young, 1988). The Heisenberg model can be generalized to have an arbitrary value of the spin S (also in this case, for bipartite lattices, Monte Carlo approaches allow us to get numerically exact results).

Mobile holes can be injected in the Heisenberg model of Eq. (1.29), leading to the so-called $t-J$ model (Fazekas, 1999):

$$\mathcal{H} = -t \sum_{\langle i,j \rangle, \sigma} c_{i,\sigma}^{\dagger} c_{j,\sigma} + \text{h.c.} + J \sum_{\langle i,j \rangle} \left(\mathbf{S}_i \cdot \mathbf{S}_j - \frac{1}{4} n_i n_j \right), \qquad (1.33)$$

where all the operators act on the restricted Hilbert space without doubly occupied sites. The $t-J$ model captures the strong-coupling limit of the Hubbard model for $n < 1$ and is usually considered to give the minimal description of Cuprate superconductors (Lee et al., 2006). Moreover, the $t-J$ model can be also obtained from the strong-coupling expansion of a three-band Hubbard model, which includes both Copper and Oxygen atoms (Zhang and Rice, 1988). There are no exact solutions of the $t-J$ model for generic values of the ratio J/t and electron doping n, both in one and two spatial dimensions, except in one dimension for the super-symmetric point $J/t = 2$ (Sutherland, 1975).

We would like to mention that the Hubbard model with a *negative* interaction U can be considered to describe the case where an effective attractive electron-electron interaction is present (for example, for a non-retarded electron-phonon coupling, leading to superconductivity). Therefore, the Hubbard model with $U < 0$ can be seen as the simplest model to capture the stabilization of a superconducting phase. Also in this case, there are Monte Carlo methods that allow a numerically exact solution of the model (see Chapter 11).

Finally, the case of bosonic particles can be also considered, which is relevant for various physical systems, like for interacting ^{4}He atoms, or to represent tightly coupled pairs of fermions that may undergo a superconductor-insulator transition. The bosonic Hubbard model is defined by:

$$\mathcal{H} = -t \sum_{\langle i,j \rangle} b_i^\dagger b_j + \text{h.c.} + \frac{U}{2} \sum_i n_i(n_i - 1), \tag{1.34}$$

where $b_i^\dagger$ (b_i) creates (destroys) a boson on site i, and $n_i = b_i^\dagger b_i$ is the density of bosons on site i; these creation and annihilation operators satisfy commutation relations:

$$\left[b_i, b_j^\dagger \right] = \delta_{i,j}, \tag{1.35}$$

$$\left[b_i^\dagger, b_j^\dagger \right] = 0. \tag{1.36}$$

Again the Hamiltonian (1.34) commutes with the total number of bosons N_b, so that any sector with different number of particles can be considered separately. Notice that the Hilbert space of the bosonic Hubbard model is much larger than the fermionic one, since any number of bosons is allowed on each site. The bosonic Hubbard model is only defined for repulsive interactions, since bosons are thermodynamically unstable in presence of attractive interactions.

For $U = 0$, the non-interacting ground state is easily obtained by condensing all bosons in the lowest-energy single-particle state:

$$|\Phi_0\rangle = \frac{1}{\sqrt{N_b!}} \left(b_{k=0}^\dagger \right)^{N_b} |0\rangle. \tag{1.37}$$

By contrast, for $t = 0$, the ground state is given by placing bosons on sites and minimizing the number of multiply occupied sites (as done for the fermionic case):

for example, for $N_b \leq L$ the ground state has only empty or singly occupied sites. In contrast to the fermionic case, for $N_b = L$, the ground state, with one boson per site, is unique.

For this lattice model, there are no exact solutions (in the sense of the Bethe *Ansatz*) for generic values of the bosonic density $n = N_b/L$ and interaction U. Nevertheless, it is possible to obtain numerically exact results by using Monte Carlo techniques for any values of n and U/t; see Chapter 8.

1.4 The Variational Principle

Instead of exactly solving the truncated Hamiltonian, which is an exponentially hard problem, we can define, in analogy to Heitler-London and Hartree-Fock approaches, variational wave functions that may capture the correct low-energy properties of correlated systems. Indeed, the *variational principle* represents one important pillar when searching for reliable approximations of strongly correlated electronic or bosonic systems. Here, we discuss the basic aspects of the variational approach.

Given any approximate state $|\Psi_{\text{var}}\rangle$ for the exact ground state $|\Upsilon_0\rangle$ of a given Hamiltonian, we can define its variational energy as:

$$E_{\text{var}} = \frac{\langle \Psi_{\text{var}} | \mathcal{H} | \Psi_{\text{var}} \rangle}{\langle \Psi_{\text{var}} | \Psi_{\text{var}} \rangle}. \tag{1.38}$$

Any state in the Hilbert space can be expanded in terms of the eigenfunctions $|\Upsilon_i\rangle$ of the Hamiltonian (with energies E_i), so that the variational state can be written as:

$$|\Psi_{\text{var}}\rangle = \sum_i a_i |\Upsilon_i\rangle, \tag{1.39}$$

with $a_i = \langle \Upsilon_i | \Psi_{\text{var}} \rangle$. The normalization condition reads as:

$$\langle \Psi_{\text{var}} | \Psi_{\text{var}} \rangle = \sum_i |a_i|^2 = 1. \tag{1.40}$$

By using the expansion of Eq. (1.39), we easily obtain that:

$$\epsilon \equiv E_{\text{var}} - E_0 = \sum_{i \neq 0} |a_i|^2 (E_i - E_0) \geq 0, \tag{1.41}$$

which implies that any trial state $|\Psi_{\text{var}}\rangle$ provides an *upper bound* of the exact energy and represents the basis of controlled approximate techniques. In practice, given any approximation $|\Psi_{\text{var}}\rangle$, all computational efforts are devoted to minimizing the variational energy E_{var}.

Let us now analyze in what sense an approximate wave function, with given "distance" in energy ϵ from the exact ground state, can be considered as a good approximation of the many-body ground state $|\Upsilon_0\rangle$. A crucial role is played by the gap to the first excited state, which is always finite in a system with N particles (apart from accidental degeneracies that occur for very particular models and boundary conditions), i.e., $\Delta = E_1 - E_0 > 0$. From Eq. (1.41) and the fact that $E_i \geq E_0 + \Delta$, it follows that:

$$\epsilon \geq \Delta \sum_{i \neq 0} |a_i|^2; \tag{1.42}$$

then, by using the normalization condition (1.40), we finally have that:

$$\eta = 1 - |a_0|^2 \leq \frac{\epsilon}{\Delta}. \tag{1.43}$$

This relation tells us that, in order to have an accurate approximation of the exact ground state (i.e., $\eta \ll 1$), a sufficient condition is that the error ϵ in the variational energy has to be much smaller than the gap Δ to the first excited state.

The accuracy of correlation functions (i.e., expectation values of Hermitian operators, which do not commute with the Hamiltonian, over $|\Psi_{\mathrm{var}}\rangle$) is usually worse than the one on the ground-state energy. Let us consider a generic operator $\mathcal{O}$ and express the variational wave function as:

$$|\Psi_{\mathrm{var}}\rangle = a_0 |\Upsilon_0\rangle + \sqrt{\eta} |\Upsilon'\rangle, \tag{1.44}$$

where $|\Upsilon'\rangle$ is orthogonal to the ground state $|\Upsilon_0\rangle$ and $\eta = 1 - |a_0|^2$. Then, the difference between the expectation value calculated with the variational state and the exact one is given by:

$$|\langle \Psi_{\mathrm{var}} | \mathcal{O} | \Psi_{\mathrm{var}} \rangle - \mathcal{O}_0| = |2a_0 \sqrt{\eta} \langle \Upsilon_0 | \mathcal{O} | \Upsilon' \rangle + \eta \langle \Upsilon' | \mathcal{O} | \Upsilon' \rangle - \eta \mathcal{O}_0|, \tag{1.45}$$

where we have denoted $\mathcal{O}_0 = \langle \Upsilon_0 | \mathcal{O} | \Upsilon_0 \rangle$ and assumed, for simplicity, real wave functions. Whenever the variational state is close to the exact ground state, $\eta \ll \sqrt{\eta}$, and we can neglect all the terms that are proportional to η:

$$|\langle \Psi_{\mathrm{var}} | \mathcal{O} | \Psi_{\mathrm{var}} \rangle - \mathcal{O}_0| \approx \sqrt{\eta} |\langle \Upsilon_0 | \mathcal{O} | \Upsilon' \rangle|, \tag{1.46}$$

which shows that the accuracy in correlation functions is more problematic than the one on the ground-state energy, with a term proportional to $\sqrt{\eta}$.

1.5 Variational Wave Functions

Since no exact solutions of the fermionic and the bosonic Hubbard models are known for generic values of the coupling constants in two or higher spatial dimensions, it may be profitable to define suitable variational wave functions that may capture the main aspects of the ground-state properties. In the following, we will describe few many-body states that represent important examples in condensed-matter physics.

1.5.1 Hartree-Fock Wave Functions

For fermionic models, the simplest example is given by the Hartree-Fock approximation, where the many-body wave function is taken to be a product state of suitably optimized single-particle orbitals:

$$|\Psi_{\text{HF}}\rangle = \prod_{\alpha=1}^{N_e} \Phi_\alpha^\dagger |0\rangle; \tag{1.47}$$

here, $\Phi_\alpha^\dagger$ can be expressed in terms of the original fermionic operators as:

$$\Phi_\alpha^\dagger = \sum_i W_{\uparrow,\alpha,i}^* c_{i,\uparrow}^\dagger + \sum_i W_{\downarrow,\alpha,i}^* c_{i,\downarrow}^\dagger, \tag{1.48}$$

where $\{W_{\sigma,\alpha,i}\}$ are complex coefficients that can be optimized to get the best variational state. The condition that orbitals are normalized and orthogonal to each other implies that:

$$\sum_i \left(W_{\uparrow,\alpha,i} W_{\uparrow,\beta,i}^* + W_{\downarrow,\alpha,i} W_{\downarrow,\beta,i}^* \right) = \delta_{\alpha,\beta}. \tag{1.49}$$

In turn, the original fermionic operators can be written as:

$$c_{i,\uparrow}^\dagger = \sum_\alpha W_{\uparrow,\alpha,i} \Phi_\alpha^\dagger, \tag{1.50}$$

$$c_{i,\downarrow}^\dagger = \sum_\alpha W_{\downarrow,\alpha,i} \Phi_\alpha^\dagger. \tag{1.51}$$

The expectation value of any Hamiltonian can be easily evaluated analytically. For example, for the Hubbard model of Eq. (1.19), the variational energy is given by:

$$E_{\text{HF}} = \frac{\langle \Psi_{\text{HF}} | \mathcal{H} | \Psi_{\text{HF}} \rangle}{\langle \Psi_{\text{HF}} | \Psi_{\text{HF}} \rangle} = -t \sum_{\langle i,j \rangle, \sigma} \sum_{\alpha=1}^{N_e} W_{\sigma,\alpha,i} W_{\sigma,\alpha,j}^* + \text{h.c.}$$

$$+ U \sum_i \sum_{\alpha=1}^{N_e} \sum_{\beta=1}^{N_e} \left(|W_{\uparrow,\alpha,i}|^2 |W_{\downarrow,\beta,i}|^2 - W_{\uparrow,\alpha,i} W_{\uparrow,\beta,i}^* W_{\downarrow,\beta,i} W_{\downarrow,\alpha,i}^* \right), \tag{1.52}$$

where we have used the fact that $\langle \Psi_{\text{HF}} | \Psi_{\text{HF}} \rangle = 1$. The optimal many-body state is obtained by minimizing the variational energy with respect to all possible amplitudes $W_{\sigma,\alpha,i}^*$:

$$\frac{\partial E_{\text{HF}}}{\partial W_{\sigma,\alpha,i}^*} = -t \sum_{\langle j \rangle i} W_{\sigma,\alpha,j} + U \sum_{\beta=1}^{N_e} \left(|W_{-\sigma,\beta,i}|^2 W_{\sigma,\alpha,i} - W_{\sigma,\beta,i} W_{-\sigma,\beta,i}^* W_{-\sigma,\alpha,i} \right), \tag{1.53}$$

where $\langle j \rangle_i$ indicates all the sites j that are neighbors of the site i. Here, the electronic orbitals define self-consistent fields: the first term proportional to U

(i.e., $\sum_{\beta=1}^{N_e} |W_{-\sigma,\beta,i}|^2$) is a "direct" Coulomb term, while the second one (i.e., $\sum_{\beta=1}^{N_e} W_{\sigma,\beta,i} W^*_{-\sigma,\beta,i}$) is an "exchange" term, which is due to the anti-symmetry of the many-body wave function. The best amplitudes can be obtained by the steepest-descent procedure:

$$W_{\sigma,\alpha,i} \rightarrow W_{\sigma,\alpha,i} - \frac{\partial E_{\text{HF}}}{\partial W^*_{\sigma,\alpha,i}} \delta\tau, \tag{1.54}$$

where $\delta\tau$ is an *ad hoc* parameter that ensures to lower the energy at every iterative step. The updated orbitals are no longer orthogonal to each other; however, the variational energy does not change by taking any linear combination of the occupied orbitals and, therefore, we can obtain a new set of orthogonal states by using the Gram-Schmidt method. In this way, we can iterate the procedure until convergence.

An alternative approach is to define the Hartree-Fock wave function $|\Psi_{\text{HF}}\rangle$ as the ground state of an uncorrelated (auxiliary) Hamiltonian:

$$\mathcal{H}_0 = \sum_{i,j,\sigma,\tau} h_{i,j}^{\sigma,\tau} c^\dagger_{i,\sigma} c_{j,\tau}, \tag{1.55}$$

where $h_{i,j}^{\sigma,\tau}$ are variational parameters that must be optimized to minimize the variational energy of $|\Psi_{\text{HF}}\rangle$. The advantage of this approach with respect to the previous one is that selected symmetries can be easily imposed in the wave function, by taking a particular *Ansatz* for the $h_{i,j}^{\sigma,\tau}$'s (e.g., translational symmetry is imposed by choosing a translational invariant $\mathcal{H}_0$). By defining:

$$\Delta_{i,j}^{\sigma,\tau} = \frac{\langle \Psi_{\text{HF}} | c^\dagger_{i,\sigma} c_{j,\tau} | \Psi_{\text{HF}} \rangle}{\langle \Psi_{\text{HF}} | \Psi_{\text{HF}} \rangle}, \tag{1.56}$$

the Hartree-Fock equations for the Hubbard model of Eq. (1.19) are given by:

$$h_{i,j}^{\sigma,\tau} = -t_{i,j}\delta_{\sigma,\tau} + U\delta_{i,j}\left(\delta_{\sigma,\tau}\Delta_{i,i}^{-\sigma,-\sigma} - \delta_{\sigma,-\tau}\Delta_{i,i}^{\tau,\sigma}\right), \tag{1.57}$$

where $t_{i,j} = t$ for nearest-neighbor sites and 0 otherwise. These are also self-consistent equations, since the set of parameters $\{h_{i,j}^{\sigma,\tau}\}$ defines the many-body wave function, which in turn determines the values of the $\Delta_{i,j}^{\sigma,\tau}$'s by Eq. (1.56).

On the lattice, it is relatively simple to obtain a solution for the Hartree-Fock equations by using iterative methods; instead, it is much more difficult to reach the solution corresponding to the lowest energy, since, in the general case, there are several solutions that correspond to local minima in the variational energy. Most importantly, while the Hartree-Fock approximation may give reasonable results in the weak-coupling regime, its accuracy becomes questionable for moderate and strong interactions. For example, a Mott insulator, with no symmetry breaking, cannot be described within this approximation; moreover, it is also not possible to stabilize superconducting phases in purely repulsive Hamiltonians, thus excluding the RVB physics (Fazekas, 1999). Therefore, a step forward is needed, in order to reach a better characterization of highly correlated systems.

1.5.2 The Gutzwiller Wave Function

The simplest example of a correlated state, which goes beyond the Hartree-Fock approximation, has been conceived by Gutzwiller (1963) to describe the effect of the Hubbard U interaction in reducing the weight of configurations with multiply occupied sites. The Gutzwiller wave function is constructed by starting from the non-interacting ground state $|\Phi_0\rangle$ (either fermionic or bosonic) and then applying an operator $\mathcal{P}_G$ that suppresses the weight of configurations with multiply occupied sites:

$$|\Psi_G\rangle = \mathcal{P}_G|\Phi_0\rangle; \tag{1.58}$$

here, $\mathcal{P}_G$ is the so-called Gutzwiller factor that depends upon a single variational parameter g (e.g., $g > 0$ for the repulsive Hubbard model):

$$\mathcal{P}_G = \exp\left[-\frac{g}{2}\sum_i (n_i - n)^2\right], \tag{1.59}$$

where n is the average density. Generalizations in which $|\Phi_0\rangle$ is a generic, non-interacting (i.e., Hartree-Fock) state are often considered.

The effect of the Gutzwiller factor becomes clear once the variational state is expanded in a basis set whose elements $\{|x\rangle\}$ represent configurations with particles sitting on the lattice sites. Indeed, since the Gutzwiller factor is diagonal in this basis (it contains the density operator on each site n_i), we have that:

$$\langle x|\Psi_G\rangle = \mathcal{P}_G(x)\langle x|\Phi_0\rangle, \tag{1.60}$$

where $\mathcal{P}_G(x) \leq 1$ is a number that depends on how many doubly occupied (or multi-occupied) sites are present in the configuration $|x\rangle$. Therefore, the amplitude of the non-interacting state $\langle x|\Phi_0\rangle$ is renormalized by $\mathcal{P}_G(x)$. In the simple example of the Hydrogen molecule, we can take the Hartree-Fock wave function (1.8) as the non-interacting state $|\Phi_0\rangle$, such that, apart from a normalization constant:

$$|\Psi_G\rangle \propto \left(e^{-g}c_{1,\uparrow}^\dagger c_{1,\downarrow}^\dagger + c_{1,\uparrow}^\dagger c_{2,\downarrow}^\dagger + c_{2,\uparrow}^\dagger c_{1,\downarrow}^\dagger + e^{-g}c_{2,\uparrow}^\dagger c_{2,\downarrow}^\dagger\right)|0\rangle. \tag{1.61}$$

Therefore, within this approach, it is possible to interpolate between the Hartree-Fock approximation, which is obtained for $g = 0$, and the Heitler-London one, which is recovered in the limit of $g = \infty$.

Let us now discuss the case of an arbitrary number of lattice sites. For the Hubbard model, when the particle density is $n = 1$, we argued that a metal-insulator (for fermions) or superfluid-insulator (for bosons) transition is expected at finite values of U/t. However, a simple argument suggests that the Gutzwiller wave function can describe such a transition only when the variational parameter g tends to infinity. Indeed, for $n = 1$, on average, there is one particle per site and density excitations are represented by doublons (doubly occupied sites) and holons (empty sites). In the non-interacting state $|\Phi_0\rangle$, these objects are free to move and then responsible for the conductivity (for example, in the fermionic model, a doublon is negatively charged with respect to the average background, while the holon is positively charged). The effect of the Gutzwiller factor is to penalize the formation of such objects; however, once created, doublons and holons are no longer correlated, thus being free to move independently. Only when the energetic penalty is infinite, an insulator is obtained; here, all the density degrees of freedom are frozen and no transport is possible, implying an oversimplified description of a true insulator, where instead density fluctuations are always present. Extensive calculations have shown that the superfluid-insulator transition in bosonic systems takes place at a finite value of U/t, i.e., the optimal parameter g diverges for a finite value of U/t, (Rokhsar and Kotliar, 1991; Krauth et al., 1992); instead, for fermions, g is finite for all values of U/t and diverges only for $U/t = \infty$ (Yokoyama and Shiba, 1987a,b, 1990).

Finally, let us briefly discuss the case of $n = 1$ in the limit of $g = \infty$. In the bosonic system, only one configuration with exactly one boson per site survives after the application of the Gutzwiller factor with $g = \infty$, implying that the fully projected state is trivial. Instead, in the fermionic case, there is still an exponentially large number of states with singly occupied sites, which differ by the spin configurations. Therefore, non-trivial spin fluctuations are still allowed, possibly leading to the RVB insulator that is obtained by taking the fully projected wave function:

$$|\Psi_\infty\rangle = \mathcal{P}_\infty|\Phi_0\rangle, \tag{1.62}$$

where the full Gutzwiller projector (i.e., with $g = \infty$) is given by:

$$\mathcal{P}_\infty = \prod_i \left(n_{i,\uparrow} - n_{i,\downarrow}\right)^2. \tag{1.63}$$

Below, we will discuss few examples of RVB states that are obtained by Gutzwiller projecting particular non-interacting wave functions.

1.5.3 The Density-Density Jastrow Wave Function

As we have discussed in the previous section, the variational description of an insulator with density fluctuations is not captured by the simple Gutzwiller wave function (1.58) and requires a modification of the correlation term that is applied to the non-interacting wave function. An alternative approach is to start from a wave function that describes the localized $U/t = \infty$ limit and then includes a term that delocalizes the electrons (Eichenberger and Baeriswyl, 2007). In the following, we will pursue the former approach.

A straightforward generalization of the Gutzwiller wave function is given by the inclusion of long-range terms in the correlator:

$$|\Psi_J\rangle = \mathcal{J}|\Phi_0\rangle, \tag{1.64}$$

where $\mathcal{J}$ is the so-called Jastrow factor (Jastrow, 1955) that has been introduced in continuum models much before the Gutzwiller wave function. On the lattice, $\mathcal{J}$ takes a simple form:

$$\mathcal{J} = \exp\left[-\frac{1}{2}\sum_{i,j} v_{i,j}(n_i - n)(n_j - n)\right], \tag{1.65}$$

where $v_{i,j}$ is a pseudo-potential for density-density correlations in the variational state. For translationally invariant models, like the Hubbard Hamiltonian (1.19), $v_{i,j}$ only depends upon the relative distance of the two sites i and j, i.e., $|\mathbf{R}_i - \mathbf{R}_j|$; moreover, the on-site term $v_{i,i}$ corresponds to the Gutzwiller parameter g. The Jastrow pseudo-potential can be either parametrized in some way, in order to reduce the number of variational parameters, or optimized for all possible distances, which are $O(L)$ in translationally invariant systems; in Chapter 6, we will discuss an efficient algorithm to find the optimal Jastrow parameters $v_{i,j}$. The role of the long-range tail of the Jastrow factor is to create a bound state between holons and doublons, possibly impeding conduction, but still allowing local density fluctuations. Indeed, we have shown that such Jastrow terms may turn a non-interacting metallic state $|\Phi_0\rangle$ into an insulator in both fermionic and bosonic systems (Capello et al., 2005, 2006, 2007, 2008). In particular, by denoting the Fourier transform of the (translationally invariant) pseudo-potential $v_{i,j}$ by v_q, the gapless (i.e., metallic) phase is described by having $v_q \approx 1/|\mathbf{q}|$ for $|\mathbf{q}| \to 0$, in any spatial dimension d; by contrast, a fully gapped (i.e., insulating) phase is obtained in one-dimension with $v_q \approx 1/|\mathbf{q}|^2$ for $|\mathbf{q}| \to 0$ (Capello et al., 2005, 2008). This singular behavior of the pseudo-potential induces an exponential decay of the density-density correlations. In two and three spatial dimensions, a holon-doublon bound-state is generated by $v_q \approx \beta/|\mathbf{q}|^d$ for a sufficiently large value of β (Capello et al., 2006, 2007). However, these behaviors of the pseudo-potential, which are obtained by an energy minimization, are not sufficient to have a fully gapped phase, since a residual power-law behavior in the density-density correlations is still present.

As we mentioned above, the Jastrow wave function of Eq. (1.64) has been introduced to study continuum models (Jastrow, 1955) and has been employed to perform the first quantum Monte Carlo calculation in a many-body system

(McMillan, 1965). More precisely, a system of interacting bosons has been considered to model the ground-state properties of ^{4}He in three spatial dimensions with Lennard-Jones interactions. Here, in a first-quantization notation, the wave function with N_b bosons reads:

$$\Psi_J(\mathbf{r}_1, \ldots, \mathbf{r}_{N_b}) = \prod_{i<j} f(r_{i,j}) = \exp\left[-\sum_{i<j} u(r_{i,j})\right], \tag{1.66}$$

where $\{\mathbf{r}_i\}$ are the coordinates of the bosons and $f(r_{i,j}) = \exp[-u(r_{i,j})]$ is a function that depends upon the relative distance between two bosons $r_{i,j} = |\mathbf{r}_i - \mathbf{r}_j|$. Notice that the wave function of Eq. (1.66) is totally symmetric when exchanging two particles, thus having the correct symmetry for a bosonic wave function.

A suitable correlated wave function for N_e fermions can be obtained by applying the symmetric Jastrow factor $\prod_{i<j} f(r_{i,j})$ to a (anti-symmetric) Slater determinant, which, by using first-quantization notations, reads as:

$$\Psi_{\text{HF}}(\mathbf{r}_1, \ldots, \mathbf{r}_{N_e}) = \det\{\phi_\alpha(\mathbf{r}_j)\}, \tag{1.67}$$

where $\{\phi_\alpha(\mathbf{r}_j)\}$ is a set of one-particle orbitals. Then the Jastrow-Slater wave function is given by:

$$\Psi_{\text{JS}}(\mathbf{r}_1, \ldots, \mathbf{r}_{N_e}) = \prod_{i<j} f(r_{i,j}) \times \Psi_{\text{HF}}(\mathbf{r}_1, \ldots, \mathbf{r}_{N_e}). \tag{1.68}$$

In total, this wave function is anti-symmetric when exchanging two particles and, therefore, has the correct symmetry for a fermionic state.

1.5.4 The Backflow Wave Function

Besides the inclusion of Jastrow factors, a different way to improve the non-interacting wave function is to construct a multi-determinant (variational) wave function that is written in terms of a linear combination of different quantum states. For example, within the configuration-interaction scheme (Szabo and Ostlund, 1996), we have:

$$|\Psi_{\text{CI}}\rangle = A_0|\Psi_{\text{HF}}\rangle + \sum_{\alpha=1}^{N_c} A_\alpha|\Psi_{\text{HF},\alpha}\rangle, \tag{1.69}$$

where $|\Psi_{\text{HF}}\rangle$ is a given Hartree-Fock wave function, which is the ground-state of a suitable non-interacting Hamiltonian, and $\{|\Psi_{\text{HF},\alpha}\rangle\}$ is a set of states obtained by considering particle-hole excitations on top of $|\Psi_{\text{HF}}\rangle$; $\{A_\alpha\}$ denotes the set of variational parameters. By considering all possible virtual excitations, the approach becomes exact. Indeed, this method is equivalent to computing the best state within the basis set of the above-mentioned wave functions. In practice, this technique is suitable to describe atoms and small molecules, for which N_c can be kept sufficiently small, but not extended systems, like solids, since in the latter case an exponentially large number of states is necessary to reach an accurate approximation of correlated systems.

An alternative and more practical way to include some correlation inside the original variational state is to introduce a parametrization that allows the orbitals to depend upon the positions of the other particles, leading to the concept of *backflow* correlations. In quantum systems, a particle that moves is surrounded by a counterflow generated by all the other particles; the existence of this flow pattern pushes away the particles, thus preventing a significant overlap among them. This idea has been originally introduced by Wigner and Seitz (1934) and then developed by Feynman (1954) and Feynman and Cohen (1956) in the context of excitations in ^{4}He and the effective mass of a ^{3}He impurity in liquid ^{4}He. In the fermionic case, the Slater determinant is not constructed with the actual positions of the electrons $(\mathbf{r}_1, \ldots, \mathbf{r}_{N_e})$, see Eq. (1.67), but with new "coordinates" given by:

$$\mathbf{r}_i^b = \mathbf{r}_i + \sum_{j \neq i} \eta(|\mathbf{r}_i - \mathbf{r}_j|)(\mathbf{r}_j - \mathbf{r}_i), \tag{1.70}$$

where $\eta(|\mathbf{r}_i - \mathbf{r}_j|)$ is a suitable function that describes the effective displacement of the i-th particle due to the j-th one. The simplest wave function is built by taking plane-waves with positions given by $\{\mathbf{r}_i^b\}$. The effect of backflow correlations introduces many-body effects inside the Slater determinant, since, when the i-th electron is moved, all the new "coordinates" are modified, such that all particles respond to the movement of the single electron, adapting their positions accordingly.

Wave functions including both Jastrow factors and backflow correlations have been used to study Helium systems within the so-called hyper-netted chain approximation (Pandharipande and Itoh, 1973; Schmidt and Pandharipande, 1979). Then, they have also been used in Monte Carlo calculations to compute the properties of the homogeneous electron gas in two and three spatial dimensions (Kwon et al., 1993, 1998). The advantage of the backflow wave function is that a single Slater determinant is used, allowing us to perform calculations with a large number of particles.

More recently, the same idea of modifying the single-electron orbitals to improve variational wave functions has been extended for lattice models (Tocchio et al., 2008, 2011). Here, the transformation (1.70) cannot be applied, since electrons live on the lattice sites. Nevertheless, we can imagine that the amplitudes of the Hartree-Fock orbitals (1.48) are changed according to the many-body configuration:

$$W^b_{\sigma,i,\alpha} = \eta_0 W_{\sigma,i,\alpha} + \sum_{j \neq i} \eta_j \mathcal{O}_{i,j} W_{\sigma,j,\alpha}, \qquad (1.71)$$

where $\{\eta_j\}$ is a set of variational parameters and $\mathcal{O}_{i,j}$ is a generic many-body operator that acts on the sites i and j. For example, within the repulsive Hubbard model, the formation of holon-doublon pairs is energetically expensive for large values of U/t; therefore, these objects tend to recombine into singly occupied sites. In this case, we can consider a many-body operator $\mathcal{O}_{i,j} = D_i H_j$, where D_i (H_i) is the operator that gives 1 if the site i is doubly occupied (empty) and 0 otherwise. Then, the many-body state, which is constructed by taking the Slater determinant of these new "orbitals," will contain terms with single occupation, thus releasing the energy. More complicated expressions of the new "orbital" can be considered, as described by Tocchio et al. (2008, 2011).

1.5.5 The Haldane-Shastry Wave Function

As we have seen in the previous sections, Slater determinants describe non-interacting fermions, being constructed from single-particle orbitals, while Jastrow factors, parametrized by the pseudo-potential $f(r_{i,j})$ that depends upon the relative positions of two particles, introduce a two-body correlation in the wave function. Remarkably, there are examples in which a Slater determinant is equivalent to a Jastrow factor. This is the case when the Slater determinant is constructed from the so-called Vandermonde matrix:

$$\mathbf{V} = \begin{pmatrix} 1 & \alpha_1 & \alpha_1^2 & \cdots & \alpha_1^{N-1} \\ 1 & \alpha_2 & \alpha_2^2 & \cdots & \alpha_2^{N-1} \\ 1 & \alpha_3 & \alpha_3^2 & \cdots & \alpha_3^{N-1} \\ \vdots & \vdots & \vdots & \ddots & \vdots \\ 1 & \alpha_N & \alpha_N^2 & \cdots & \alpha_N^{N-1} \end{pmatrix}, \qquad (1.72)$$

where α_i is a generic (complex) number. More concisely $V_{i,j} = \alpha_i^{j-1}$. The determinant of this matrix is given by:

$$\det\{V_{i,j}\} = \prod_{i<j} (\alpha_j - \alpha_i). \qquad (1.73)$$

The equivalence between the determinant and a Jastrow factor is obtained once we can interpret $\{\alpha_i\}$ as the coordinates of the particles, as in Eq. (1.66). The simplest example of this correspondence is given by non-interacting spin-less fermions on a one-dimensional lattice (with L sites and periodic-boundary conditions):

$$\mathcal{H} = -t \sum_i c_i^\dagger c_{i+1} + \text{h.c.};$$ (1.74)

then, the single-particle orbitals are given by plane waves, labeled by the momentum $k = 2\pi n/L$, with $n = 0, \ldots, L - 1$:

$$c_k^\dagger = \frac{1}{\sqrt{L}} \sum_j e^{-ikX_j} c_j^\dagger.$$ (1.75)

For an odd number of electrons, the many-body ground-state is unique and is obtained by filling the N_e lowest-energy levels:

$$\Psi_{\text{SF}}(X_1, \ldots, X_{N_e}) = \det \begin{pmatrix} \alpha_1^{-L_e} & \alpha_1^{-(L_e-1)} & \cdots & \alpha_1^{L_e} \\ \alpha_2^{-L_e} & \alpha_2^{-(L_e-1)} & \cdots & \alpha_2^{L_e} \\ \vdots & \vdots & \ddots & \vdots \\ \alpha_{N_e}^{-L_e} & \alpha_{N_e}^{-(L_e-1)} & \cdots & \alpha_{N_e}^{L_e} \end{pmatrix},$$ (1.76)

where we have defined $\alpha_j \equiv \exp(-2\pi i X_j/L)$ and $L_e \equiv (N_e - 1)/2$. Then, the previous determinant can be recast in the following form:

$$\Psi_{\text{SF}}(X_1, \ldots, X_{N_e}) = \prod_i \alpha_i^{-L_e} \det \begin{pmatrix} 1 & \alpha_1 & \cdots & \alpha_1^{N_e-1} \\ 1 & \alpha_2 & \cdots & \alpha_2^{N_e-1} \\ \vdots & \vdots & \ddots & \vdots \\ 1 & \alpha_{N_e} & \cdots & \alpha_{N_e}^{N_e-1} \end{pmatrix},$$ (1.77)

which, apart from a phase factor, is a Vandermonde determinant:

$$\Psi_{\text{SF}}(X_1, \ldots, X_{N_e}) = \prod_i \alpha_i^{-L_e} \prod_{i<j} (\alpha_j - \alpha_i).$$ (1.78)

Finally, by using the definition of the α_j's, we obtain (apart from a normalization factor):

$$\Psi_{\text{SF}}(X_1, \ldots, X_{N_e}) \propto \prod_{i<j} \sin\left(\frac{\pi(X_j - X_i)}{L}\right).$$ (1.79)

A much less trivial example in which an exact ground-state wave function can be written in terms of Vandermonde determinants is given by a Heisenberg model with long-range interactions in one dimension, namely the so-called Haldane-Shastry model (Haldane, 1988; Shastry, 1988):

$$\mathcal{H} = \sum_{i,j} J_{i,j} \mathbf{S}_i \cdot \mathbf{S}_j,$$ (1.80)

where $\mathbf{S}_i$ is the spin-1/2 operator on the site i and the super-exchange coupling is $J_{i,j} = J/d_{i,j}^2$, which depends upon the chord distance between two sites (i,j):

$$d_{i,j} = \frac{L}{\pi} \left| \sin \left[\frac{\pi(i-j)}{L} \right] \right|. \qquad (1.81)$$

Here, the ground state is written as:

$$\Psi_{\mathrm{HS}}(X_1, \ldots, X_{N_e}) = \prod_i e^{\pi i X_i} \prod_{i<j} \left[\sin \left(\frac{\pi(X_j - X_i)}{L} \right) \right]^2, \qquad (1.82)$$

where $\{X_i\}$ indicate the positions of the up spins. In second-quantization notation, the Haldane-Shastry wave function can be written as a Gutzwiller projected state of spinful electrons:

$$|\Psi_{\mathrm{HS}}\rangle = \mathcal{P}_\infty \prod_{|k|<k_F} c_{k,\uparrow}^\dagger c_{k,\downarrow}^\dagger |0\rangle, \qquad (1.83)$$

where $k_F = \pi/2$, i.e., $N_e = L/2$.

The elementary excitations of the Haldane-Shastry model are *fractional $S = 1/2$* objects, called *spinons*, like in the Heisenberg model with nearest-neighbor interactions (Bethe, 1931; Faddeev and Takhtajan, 1981). However, the unique feature of the model with long-range interactions of Eq. (1.81) is that the spinon excitations are free and that the $S = 1$ spectrum is exhausted by two-spinon excitations (Haldane, 1991).

We finally mention that the Haldane-Shastry model has been motivated by a previous exact solution for a one-dimensional system with long-range interactions on the continuum (Sutherland, 1971):

$$\mathcal{H} = -\sum_i \frac{\partial^2}{\partial X_i^2} + \frac{g\pi^2}{L^2} \sum_{i<j} \left[\sin \left(\frac{\pi(X_j - X_i)}{L} \right) \right]^{-2}, \qquad (1.84)$$

where g defines the coupling strength of the potential. Also in this case, the ground-state wave function has a Jastrow-like form, which depends upon a real number λ that is fixed by the value of g, i.e., $g = 2\lambda(\lambda - 1)$:

$$\Psi_{\mathrm{Sutherland}}(X_1, \ldots, X_{N_e}) = \prod_{i<j} \left[\sin \left(\frac{\pi(X_j - X_i)}{L} \right) \right]^\lambda. \qquad (1.85)$$

1.5.6 The Laughlin Wave Function

Another example in which a Vandermonde determinant provides the exact ground-state wave function is when electrons on the continuum are confined in two spatial dimensions in the presence of an external magnetic field B. In particular, by using the symmetric gauge and a disk geometry, the single-particle orbitals in the lowest Landau level are given by (Jain, 2012):

$$\psi_k(z_j) = C_k \, z_j^k \, e^{-\frac{eB}{4\hbar c}|z_j|^2}, \qquad (1.86)$$

where k is a non-negative integer, C_k is the normalization factor, and $z_j = X_j - iY_j$ is a complex representation of the coordinates (X_j, Y_j) of the j-th particle in the plane. Therefore, the unique many-body state describing the filled lowest Landau level is given by the Vandermonde determinant with $V_{j,k} = z_j^{k-1}$ (apart from irrelevant terms):

$$\Psi_{\mathrm{LLL}}(z_1, \ldots, z_{N_e}) = \prod_{i<j}(z_j - z_i) = \det \begin{pmatrix} 1 & z_1 & \cdots & z_1^{N_e-1} \\ 1 & z_2 & \cdots & z_2^{N_e-1} \\ \vdots & \vdots & \ddots & \vdots \\ 1 & z_{N_e} & \cdots & z_{N_e}^{N_e-1} \end{pmatrix}. \qquad (1.87)$$

Robert Laughlin had a very simple and accurate description of the incompressible quantum fluid that is obtained at fractional filling factors $\nu = 1/m$ (Laughlin, 1983). He was influenced by the form of the non-interacting wave function of the lowest Landau level (1.87) and the success of the Jastrow wave function of Eq. (1.66) to describe the superfluid ^{4}He. The Laughlin wave functions, apart from irrelevant constant terms, are given by:

$$\Psi_{\mathrm{Laughlin}}(z_1, \ldots, z_{N_e}) = \prod_{i<j}(z_j - z_i)^m. \qquad (1.88)$$

Here, the correct symmetry properties when exchanging two electrons are obtained by odd integer values of m, corresponding to a filling factor $\nu = 1/m$ in the lowest Landau level. Remarkably, the Laughlin wave functions are the exact ground states of Hamiltonians with contact interactions (Haldane, 1983); moreover, they give an excellent description of the microscopic Hamiltonian with the full Coulomb interaction (Fano et al., 1986). On the one hand, the Laughlin wave function can be written as a Jastrow factor applied to the Vandermonde determinant of Eq. (1.87); indeed, by taking $m = 2p + 1$, we have that:

$$\Psi_{\mathrm{Laughlin}}(z_1, \ldots, z_{N_e}) = \prod_{i<j}(z_j - z_i)^{2p} \prod_{i<j}(z_j - z_i); \qquad (1.89)$$

here, the first term (that is symmetric under particle exchange) can be seen as a Jastrow factor, which is applied to the second term (which is anti-symmetric). On the other hand, we can use the fact that $\prod_{i<j}(z_j-z_i)^{2p} = (-1)^{pm}\prod_i\prod_{j\neq i}(z_j-z_i)^p \propto \prod_i F(z_i)$, where $m = N_e(N_e - 1)/2$, such that:

$$\Psi_{\text{Laughlin}}(z_1, \ldots, z_{N_e}) \propto \prod_i F(z_i) \prod_{i<j} (z_j - z_i) =$$

$$\det \begin{pmatrix} F(z_1) & z_1 F(z_1) & \cdots & z_1^{N_e-1} F(z_1) \\ F(z_2) & z_2 F(z_2) & \cdots & z_2^{N_e-1} F(z_2) \\ \vdots & \vdots & \ddots & \vdots \\ F(z_{N_e}) & z_{N_e} F(Z_{N_e}) & \cdots & z_{N_e}^{N_e-1} F(z_{N_e}) \end{pmatrix}. \tag{1.90}$$

In this way, the Laughlin state can be expressed as a single determinant, with "renormalized" orbitals, in the spirit of backflow corrections.

The very surprising prediction of the Laughlin wave functions is that by creating a "hole" at position z_0, an overall electric charge of e/m is lost, thus implying the existence of excitations with *fractional* quantum numbers and *fractional* (anyon) statistics (Arovas et al., 1984). Furthermore, the ground-state degeneracy, in the thermodynamic limit, depends on the topology of the space (Haldane, 1983; Haldane and Rezayi, 1985). The Laughlin state for $\nu = 1/m$ has a m^g degenerate ground state on a surface with genus g (e.g., $g = 0$ for a sphere and $g = 1$ for a torus). This degeneracy is not a consequence of a spontaneous symmetry breaking (or a symmetry of the Hamiltonian), but has a topological origin and, therefore, is robust against perturbations (Wen and Niu, 1990; Wen, 1991).

The Laughlin state (1.88) represents a suitable variational wave function also for bosonic systems, whenever m is an even integer. In particular, for $m = 2$ it represents the simplest case of an incompressible bosonic fluid. The latter state is the two-dimensional generalization of the Haldane-Shastry wave function (1.82) that has been defined in spin models.

Several other variational wave functions have been introduced in the context of the quantum Hall effect, to describe different filling factors. Among them, a particularly important one is given by the so-called Pfaffian state introduced by Moore and Read (1991), which should describe the $\nu = 5/2$ fraction (i.e., $\nu = 1/2$ in the second Landau level). The Moore-Read state has $p + ip$ pairing correlations and non-Abelian excitations with charge of $e/4$. Like the vortices in a p-wave superfluid, these quasiparticles are Majorana-fermion states at zero energy. Later, non-Abelian anyons have been further established in other quantum Hall states, e.g., the ones described by the Read-Rezayi states (Read and Rezayi, 1999).

1.5.7 The Bardeen-Cooper-Schrieffer Wave Function

One of the most celebrated variational wave functions in condensed matter physics is given by the Bardeen-Cooper-Schrieffer (BCS) approach that has been proposed to explain superconductivity originating from electron-phonon coupling (Bardeen et al., 1957; Schrieffer, 1964). In several metals, the coupling between electrons and lattice vibrations (phonons) induces an effective attraction between electrons, thus creating a bound state of particles with opposite momenta $(\mathbf{k}, -\mathbf{k})$ and opposite spins. For a translational invariant system, the BCS wave function reads as:

$$|\Psi_{BCS}\rangle = \exp\left(\sum_k f_k c_{k,\uparrow}^\dagger c_{-k,\downarrow}^\dagger\right)|0\rangle, \tag{1.91}$$

where f_k is the *pairing function*. We would like to emphasize that the BCS wave function has not a fixed number of particles, being a superposition of states with all possible (even) number of particles; therefore, it breaks the global $U(1)$ symmetry related to charge conservation.

The BCS wave function is the ground state of the non-interacting BCS Hamiltonian, which will be used often in the following as a starting point to define correlated wave functions:

$$\mathcal{H} = \sum_{k,\sigma}(\epsilon_k - \mu_0)c_{k,\sigma}^\dagger c_{k,\sigma} + \sum_k \Delta_k c_{k,\uparrow}^\dagger c_{-k,\downarrow}^\dagger + \text{h.c.}, \tag{1.92}$$

where ϵ_k gives the band structure, e.g., the one of Eq. (1.23), μ_0 is the chemical potential, and $\Delta_k = \Delta_{-k}$ is a (for simplicity, real) singlet pairing amplitude. The BCS Hamiltonian can be cast in a simple 2×2 matrix form (apart from constant terms):

$$\mathcal{H} = \sum_{k,\sigma}\begin{pmatrix} c_{k,\uparrow}^\dagger & c_{-k,\downarrow} \end{pmatrix}\begin{pmatrix} \epsilon_k - \mu_0 & \Delta_k \\ \Delta_k & -\epsilon_k + \mu_0 \end{pmatrix}\begin{pmatrix} c_{k,\uparrow} \\ c_{-k,\downarrow}^\dagger \end{pmatrix}. \tag{1.93}$$

The 2×2 matrix can be diagonalized by a unitary transformation:

$$\begin{pmatrix} \beta_k \\ \gamma_k^\dagger \end{pmatrix} = \begin{pmatrix} u_k & -v_k \\ v_k & u_k \end{pmatrix}\begin{pmatrix} c_{k,\uparrow} \\ c_{-k,\downarrow}^\dagger \end{pmatrix}, \tag{1.94}$$

where $u_k^2 + v_k^2 = 1$ preserves the anti-commutation relations. Then, denoting the BCS spectrum by:

$$E_k = \sqrt{(\epsilon_k - \mu_0)^2 + \Delta_k^2}, \tag{1.95}$$

we obtain (apart from constant terms):

$$\mathcal{H} = \sum_k E_k\left(\beta_k^\dagger \beta_k + \gamma_k^\dagger \gamma_k\right). \tag{1.96}$$

The ground state $|\Psi_{BCS}\rangle$ is obtained by imposing that:

$$\beta_k|\Psi_{BCS}\rangle = 0, \tag{1.97}$$

$$\gamma_k|\Psi_{BCS}\rangle = 0, \tag{1.98}$$

which directly lead to the form of Eq. (1.91) with $f_k = v_k/u_k$.

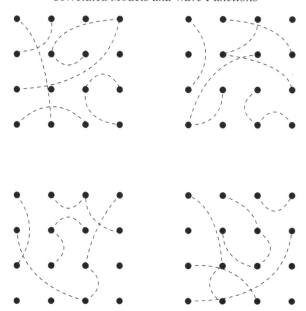

Figure 1.2 In a 4×4 cluster, we show four singlet-pair patterns of the BCS wave function. The dashed lines denote the singlet pairing among sites (i, j). The case with 8 singlets, i.e., 16 electrons, is shown. Notice that the BCS wave function allows configurations with doubly occupied sites, namely cases where two dashed lines joint in a single site.

At this point, it is useful to rewrite the BCS state in real space:

$$|\Psi_{\text{BCS}}\rangle = \exp\left(\sum_{i,j} f_{i,j} c_{i,\uparrow}^\dagger c_{j,\downarrow}^\dagger\right) |0\rangle = \sum_{N_p} \frac{1}{N_p!} \left(\sum_{i,j} f_{i,j} c_{i,\uparrow}^\dagger c_{j,\downarrow}^\dagger\right)^{N_p} |0\rangle, \quad (1.99)$$

where the pairing function in real space $f_{i,j} = f_{j,i}$ depends on $\mathbf{R}_i - \mathbf{R}_j$. Therefore, *for each sector* with $N_e = 2N_p$ electrons, the wave function is a linear combination of all possible singlet-pair patterns on the lattice (see Fig. 1.2).

Within the BCS theory, superconductivity is an instability of the metallic phase, which is described by a Fermi liquid (Nozieres, 1964); here, the direct electron-electron repulsion is efficiently screened and turns out to be treated as a small perturbation that does not drastically change the non-interacting picture. However, the presence of phonons induces an attraction among electrons, leading to an instability towards the formation of a liquid of electron pairs. In many superconducting materials, it is well established that the BCS pairing is originated by the electron-phonon coupling; in this case, the BCS theory and its further refinements give an accurate description (Schrieffer, 1964). Instead, there are several compounds in which it is less clear whether the origin of the electron attraction is mediated by phonons or by other modes (e.g., charge and/or spin fluctuations). For example, in various transition-metal oxides, there are two important aspects that have been considered crucial: the incipient presence of an insulating phase and the lack of screening effects, leading to a strong electron-electron repulsion. In a seminal paper, Anderson (1987) suggested an alternative approach, based upon a Gutzwiller projected BCS wave function, which we describe in the following section.

1.5.8 The Resonating-Valence Bond Wave Function

The RVB theory of high-temperature superconductivity is based upon the work done by Anderson and collaborators (Anderson, 1987; Anderson et al., 1987; Baskaran and Anderson, 1988), who suggested that a superconducting phase may emerge when doping a Mott insulator with "preformed" singlet pairs. In fact, the RVB state describes a liquid of spin singlets, and was proposed originally as a variational ground state of the $S = 1/2$ Heisenberg model on the triangular lattice. The key feature is that the singlets of the RVB insulator become mobile when the system is doped and thus they form real superconducting pairs. A suitable and elegant representation of the RVB state is given by Gutzwiller projecting the BCS state:

$$|\Psi_{\text{RVB}}\rangle = \mathcal{P}_\infty |\Psi_{\text{BCS}}\rangle. \tag{1.100}$$

Since the Gutzwiller projector imposes to have one electron per site, the RVB wave function is clearly insulating, with no density fluctuations. Anderson (1987) suggested that the RVB state could be very close in energy to the Néel state for undoped materials. However, for a generic (translational invariant) pairing function f_k, $|\Psi_{\text{RVB}}\rangle$ does not possess long-range magnetic order, being described by a linear combination of an exponentially large number of singlet coverings of the lattice. In fact, starting from the BCS state, which contains all possible coverings of singlets (including the ones in which two electrons with opposite spins occupy the same site – see Fig. 1.2), the Gutzwiller projector annihilates the configurations with these overlapping singlets and leaves untouched all the others (Gros, 1989), as shown in Fig. 1.3. The simplest case in which $f_{i,j} \neq 0$ only for nearest-neighbor sites is called short-range RVB state and represents an important example of a gapped insulator with *fractional* spin excitations (spinons) and topological degeneracy. Indeed, the lowest-energy spin excitation of the short-range RVB is obtained by "breaking" a singlet pair to form a triplet, which costs an energy $E = O(J)$ for the Heisenberg Hamiltonian with super-exchange coupling J. Moreover, the two spin-1/2 objects (spinons) that are created in this process are not bound together, since they can be separated without any further energy cost by rearranging the other singlets around them (see Fig. 1.4). In addition to spinons, a short-range RVB sustains also gapped excitations that do not carry spin, the so-called *visons*

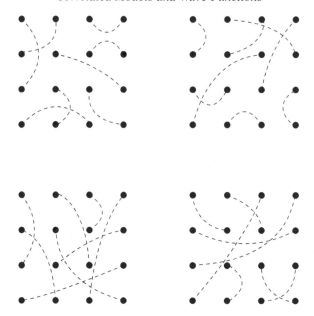

Figure 1.3 In a 4×4 cluster, we show four singlet-pair patterns of the projected BCS wave function, i.e., the RVB state. The dashed lines denote the singlet pairing among sites (i,j). The case with 8 singlets, i.e., 16 electrons, is shown. Here, in contrast to Fig. 1.2, all the configurations have singly occupied sites, namely there is always one dashed line coming out from each site.

(Kivelson et al., 1987; Read and Chakraborty, 1989; Senthil and Fisher, 2000). In the thermodynamic limit, the ground state has a degeneracy 4^g that depends upon the genus g of the surface on which the short-range RVB state is defined (Read and Chakraborty, 1989). As for the fractional quantum Hall effect, this degeneracy is not related to any spontaneous symmetry breaking, but it has a topological origin. The connection between the existence of excitations with fractional quantum numbers (and statistics) and the presence of topological order has been discussed by Oshikawa and Senthil (2006).

The generic long-range RVB wave function is obtained by allowing long-range singlets, i.e., considering $f_{i,j} \neq 0$ also for distant couples of sites (i,j). Equivalently, we can consider a generic BCS Hamiltonian and construct its ground state. The physical properties of the Gutzwiller projected state should reflect those of the BCS wave function. In particular, the BCS Hamiltonian may lead either to a gapped or to a gapless excitation spectrum, which is expected to be related to the spin excitations of the RVB state. While a gapped spectrum should imply gapped spin excitations (as for the short-range RVB), for a gapless spectrum, "breaking" very long-range singlets may have a negligible cost for a short-range Hamiltonian, thus leading to a gapless RVB state. Whether there exist two-dimensional models with

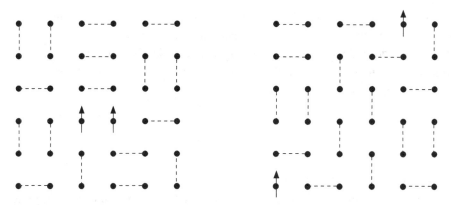

Figure 1.4 In a 6×6 cluster, a singlet pair is "broken" into a triplet, with two spinons with up spins (left). Singlets (only at nearest-neighbor sites) are shown as dashed lines. The spinons can separate themselves without any energetic cost, since the singlet background may rearrange around them (right). Therefore, they represent elementary excitations with spin $S = 1/2$ (and no charge).

an RVB ground state is still an open question; however, in various frustrated Heisenberg models, (long-range) RVB states provide very accurate variational energies (Capriotti et al., 2001; Iqbal et al., 2011).

We would like to emphasize that the definition of Eq. (1.100) is rather general and allows an efficient Monte Carlo sampling on any lattice geometry. An alternative definition of the RVB wave function is given by Liang et al. (1988); however, in this case, an efficient sampling is possible only for bipartite lattices, while for other geometries the variational Monte Carlo approach has the sign problem that usually is so serious to prevent us from obtaining reliable results on large clusters.

When mobile holes are considered, e.g., within the $t-J$ model of Eq. (1.33), the RVB wave function describes a superconducting state, as shown by using variational Monte Carlo techniques (Gros et al., 1987; Gros, 1988). Notice that, in this case, the Gutzwiller projector must allow empty sites, e.g., $\mathcal{P}_\infty \rightarrow \tilde{\mathcal{P}}_\infty = \prod_i (1 - n_{i,\uparrow} n_{i,\downarrow})$. Indeed, it has been shown that the RVB state represents an accurate variational *Ansatz* for describing the ground state of the $t-J$ model in a wide range of parameters (Sorella et al., 2002; Hu et al., 2012), even though states with charge and spin modulations also give competitive energies (White and Scalapino, 1998; Corboz et al., 2014).

1.6 Size Extensivity

In the definition of a variational wave function for a macroscopic system, i.e., a system defined by a Hamiltonian with a fixed density in the thermodynamic limit $L \rightarrow \infty$, an important concept is given by the so-called size extensivity. This term is borrowed from thermodynamics, where extensive properties are the ones that are proportional to the size of the system (Bartlett, 1981). Loosely speaking, a wave function is size extensive if the coupling between small, but macroscopic, sub-systems can be neglected; then, the wave function is written (apart from boundary corrections) as a product of independent factors, each one defined in a given sub-system:

$$|\Psi\rangle \approx \prod_{\text{sub-system}} |\Psi_{\text{sub-system}}\rangle. \tag{1.101}$$

In this way, for local Hamiltonians $\mathcal{H} \approx \sum_{\text{sub-system}} \mathcal{H}_{\text{sub-system}}$ (i.e., the coupling between different sub-systems can be neglected on a macroscopic scale), the total energy is explicitly given by the sum of the energies in each sub-system:

$$\frac{\langle \Psi | \mathcal{H} | \Psi \rangle}{\langle \Psi | \Psi \rangle} \approx \sum_{\text{sub-system}} \frac{\langle \Psi_{\text{sub-system}} | \mathcal{H}_{\text{sub-system}} | \Psi_{\text{sub-system}} \rangle}{\langle \Psi_{\text{sub-system}} | \Psi_{\text{sub-system}} \rangle}, \tag{1.102}$$

namely, an extensive quantity that is linear in the volume and the number of particles. This definition is very similar to standard concepts in classical thermodynamics, where, for example, the canonical distribution is given by the Boltzmann distribution, $\mathcal{P}(E) \propto \exp(-\beta E)$, where β is the inverse temperature. Size extensivity tells us that the entanglement between different macroscopic sub-systems is negligible in the thermodynamic limit; from this point of view, it represents a necessary condition for satisfying the so-called *area law*, a celebrated property of the low-energy spectrum of Hamiltonians with a gap (Eisert et al., 2010). With this definition, it turns out that size-extensive wave functions can be written in an exponential form (like for example Jastrow or BCS wave functions), where the exponent contains a size-extensive operator that can allow the factorization of the total wave function in the various sub-systems.

1.6.1 A Simple Model for Size Extensivity

Consider a simple model for phonons in a lattice with L sites, described by the following Hamiltonian:

$$\mathcal{H} = \omega \sum_i a_i^\dagger a_i + g \sum_i (a_i^\dagger + a_i), \tag{1.103}$$

where $a_i^\dagger$ (a_i) creates (destroys) a phonon on the site i. Here, exciting a phonon on each site costs an energy ω and g defines the stress tension that determines the equilibrium positions of the nuclei. The Hamiltonian is just the sum of non-interacting Hamiltonians acting on each site, which can be considered a sub-system in Eq. (1.101). The ground state is given by:

$$|\Upsilon_0\rangle = \prod_i e^{-\gamma a_i^\dagger}|0\rangle, \tag{1.104}$$

where $\gamma = g/\omega$. Its energy $E_0 = -Lg^2/\omega$ is correctly extensive, which is consistent with the fact that the wave function is a product of exponential factors. Therefore, the exact solution of this model is clearly size extensive, since any sub-system has no interaction with the other ones.

In order to emphasize the importance of having an exponential form of the wave function, let us consider a much simpler variational *Ansatz* given by:

$$|\Psi_{\text{lin}}\rangle = \left(1 + \alpha \sum_i a_i^\dagger\right)|0\rangle. \tag{1.105}$$

A simple calculation shows that:

$$E_{\text{lin}}(\alpha) = \frac{\langle\Psi_{\text{lin}}|\mathcal{H}|\Psi_{\text{lin}}\rangle}{\langle\Psi_{\text{lin}}|\Psi_{\text{lin}}\rangle} = \frac{\omega\alpha^2 L + 2g\alpha L}{1 + L\alpha^2}. \tag{1.106}$$

In the thermodynamic limit ($L \to \infty$), the minimization of $E_{\text{lin}}(\alpha)$ does not lead to an extensive energy:

$$E_{\text{lin}}(\alpha) \propto -\sqrt{L}, \tag{1.107}$$

which implies that the linear *Ansatz* of Eq. (1.105) is not size extensive.

1.6.2 The Classical Mapping

In this section, we discuss some general aspects of the variational wave functions. In particular, whenever we are focusing on *diagonal* operators (in the chosen basis set $\{|x\rangle\}$), the quantum expectation values over a wave function $|\Psi_{\text{var}}\rangle$ can be expressed in terms of a classical partition function at finite temperature, in the same spatial dimensions (Capello et al., 2006; Kaneko et al., 2016). This correspondence is very useful to show that quantum phase transitions are possible within the variational picture and enables us to make some general statements on the size extensivity of the wave function. To prove the mapping between quantum expectation values and classical partition functions, let us consider a basis set $\{|x\rangle\}$ in which particles have definite positions in the lattice. Then, the quantum average of any operator $\mathcal{O}$, which is diagonal in this basis, can be written in terms of the *classical* distribution:

$$\frac{\langle\Psi_{\text{var}}|\mathcal{O}|\Psi_{\text{var}}\rangle}{\langle\Psi_{\text{var}}|\Psi_{\text{var}}\rangle} = \sum_x \mathcal{P}(x)\langle x|\mathcal{O}|x\rangle, \tag{1.108}$$

where $\mathcal{P}(x)$ is given by:

$$\mathcal{P}(x) = \frac{|\langle x|\Psi_{\text{var}}\rangle|^2}{\langle\Psi_{\text{var}}|\Psi_{\text{var}}\rangle}. \tag{1.109}$$

Since $\mathcal{P}(x) \geq 0$ and $\sum_x \mathcal{P}(x) = 1$, there is a precise correspondence between the wave function and an effective classical potential $V_{cl}(x)$:

$$\mathcal{P}(x) \equiv \frac{1}{\mathcal{Z}} e^{-V_{cl}(x)}. \tag{1.110}$$

The explicit form of the potential $V_{cl}(x) = -2 \ln |\langle x|\Psi_{var}\rangle|$ depends upon the choice of the variational wave function. In particular, it is interesting to consider a Jastrow or Jastrow-Slater state $|\Psi_{var}\rangle = \mathcal{J}|\Phi_0\rangle$ of Eq. (1.64). In this case, we have that:

$$V_{cl}(x) = \sum_{i,j} v_{i,j} \left[n_i(x) - n\right] \left[n_j(x) - n\right] - 2 \ln |\Phi_0(x)|, \tag{1.111}$$

where $n_i(x)$ is the electron density at site i for the configuration $|x\rangle$, i.e., $n_i|x\rangle = n_i(x)|x\rangle$ and $\Phi_0(x) = \langle x|\Phi_0\rangle$ is the amplitude of the non-interacting state over the configuration $|x\rangle$. The first term of Eq. (1.111) is a two-body potential, which describes a classical model of charged particles (e.g., in the fermionic case, $n_i(x) = 1$ represents the "background" charge, while holons and doublons carry charge $+1$ and -1, respectively) mutually interacting through a given potential. The second term in Eq. (1.111) gives a generic potential, which includes multi-body effects. However, when density fluctuations are suppressed (for example, in presence of a large Gutzwiller factor), the quadratic term gives the most relevant contribution. In this case, we can explicitly consider the overall scale of the pseudo-potential β, i.e., $v_{i,j} = \beta \bar{v}_{i,j}$, as an inverse *classical* temperature. By using this kind of Jastrow-Slater wave functions, Kaneko et al. (2016) have shown that spontaneous symmetry breaking phenomena are possible within the variational approach when varying the strength of the Jastrow factor β. More precisely, even when both the non-interacting state $|\Phi_0\rangle$ and the Jastrow factor $\mathcal{J}$ preserve all the lattice and spin symmetries, clear signatures of order can be obtained. We would like to emphasize the fact that Jastrow wave functions are size extensive whenever the pseudo-potential $v_{i,j}$ gives rise to a stable classical model.

1.7 Projection Techniques

Variational wave functions can be improved by using the so-called *projection techniques*, which filter out the high-energy components of the "initial" (or trial) variational state. The main idea is very simple and relies on the fact that the exact ground state can be obtained by applying the so-called *power method*; let us take an initial wave function $|\Psi_0\rangle$ that may be chosen to be the best variational state $|\Psi_{var}\rangle$. Whenever the initial state $|\Psi_0\rangle$ has a finite overlap with the exact ground state $|\Upsilon_0\rangle$, we have that:

$$\lim_{n \to \infty} (\Lambda - \mathcal{H})^n |\Psi_0\rangle \propto |\Upsilon_0\rangle, \tag{1.112}$$

where Λ is a diagonal operator with $\Lambda_{x,x} = \lambda$, where λ is a real number to be specified later. In fact, Eq. (1.112) can be implemented iteratively by defining:

$$|\Psi_{n+1}\rangle = (\Lambda - \mathcal{H})|\Psi_n\rangle. \tag{1.113}$$

Then, by using the expansion of Eq. (1.39) for the initial state, we have that:

$$(\Lambda - \mathcal{H})^n|\Psi_0\rangle = \sum_i a_i(\lambda - E_i)^n|\Upsilon_i\rangle =$$

$$(\lambda - E_0)^n \left[a_0|\Upsilon_0\rangle + \sum_{i \neq 0} a_i \left(\frac{\lambda - E_i}{\lambda - E_0} \right)^n |\Upsilon_i\rangle \right]. \tag{1.114}$$

By choosing λ such that:

$$\text{Max}_i|\lambda - E_i| = |\lambda - E_0|, \tag{1.115}$$

namely for $\lambda > (E_{\text{max}} + E_0)/2$, where E_{max} is the maximum eigenvalue of $\mathcal{H}$, the ground-state component in the expansion (1.114) grows much faster than all the other ones for $n \to \infty$, leading to the ground state $|\Upsilon_0\rangle$, apart from an irrelevant normalization factor $a_0(\lambda - E_0)^n$. Notice that the convergence of $|\Psi_n\rangle$ to $|\Upsilon_0\rangle$ is obtained with an exponentially increasing accuracy in n, namely with an error that is proportional to δ^n, where $\delta = (\lambda - E_1)/(\lambda - E_0)$.

The iterative procedure of Eq. (1.113) is computationally affordable when the Hamiltonian is *sparse*, which is the case of *local* Hamiltonians that contain non-diagonal processes involving only few electrons at relatively short distances. Here, although the number of possible configurations $|x\rangle$ grows exponentially with the number of particles, whenever the Hamiltonian acts on a single configuration, it generates only a relatively small number of configurations. Therefore, the $\mathcal{N} \times \mathcal{N}$ matrix $H_{x,x'}$ has only $O(L \times \mathcal{N})$ non-zero elements over $\mathcal{N}^2$ total entries. Due to the sparseness of the matrix, each iteration (1.113) is relatively easy to perform, as only the knowledge of the non-vanishing matrix elements are required:

$$\Psi_{n+1}(x') = \sum_x (\lambda\delta_{x',x} - \mathcal{H}_{x',x})\Psi_n(x), \tag{1.116}$$

where $\Psi_n(x) = \langle x|\Psi_n\rangle$. This iterative step can be performed *exactly* whenever the whole Hilbert space can be kept in the computer memory, by using two vectors of dimension $\mathcal{N}$, i.e., $\Psi_{n+1}(x)$ and $\Psi_n(x)$. On the other hand, when the dimension of the Hilbert space is too large to be stored, the iterative procedure can be implemented stochastically, as we will describe in Chapter 8.

Finally, we mention that an alternative approach to filter out the high-energy components of the trial wave function is to consider an imaginary-time propagation:

$$\lim_{\tau \to \infty} e^{-\tau\mathcal{H}}|\Psi_0\rangle \propto |\Upsilon_0\rangle. \tag{1.117}$$

Indeed, as before, by using the expansion of Eq. (1.39) for the initial state, we have that:

$$e^{-\tau\mathcal{H}}|\Psi_0\rangle = \sum_i a_i e^{-\tau E_i}|\Upsilon_i\rangle = e^{-\tau E_0}\left[a_0|\Upsilon_0\rangle + \sum_{i\neq 0} a_i e^{-\tau(E_i-E_0)}|\Upsilon_i\rangle\right], \quad (1.118)$$

which shows that all the components with $i \neq 0$ are exponentially suppressed with respect to the ground state for $\tau \to \infty$; therefore, this procedure leads to the ground state $|\Upsilon_0\rangle$, apart from an irrelevant normalization factor $a_0 \exp(-\tau E_0)$. Notice that, whenever the energy spectrum is not bounded from above (i.e., $E_{max} = \infty$), it is necessary to consider this approach, since the condition $\lambda > (E_{max} + E_0)/2$ cannot be satisfied. The imaginary-time propagation is often implemented, with a Trotter approximation, in stochastic approaches, as we will discuss in Chapter 11.

Part II

Probability and Sampling

2

Probability Theory

2.1 Introduction

By using the laws of classical mechanics, in principle we can make *exact* predictions of events by knowing the *exact* initial conditions (i.e., all positions and velocities of the relevant degrees of freedom). However, in practice, there are several events that are unpredictable, essentially because it is impossible to have the exact knowledge of the initial conditions and a very small error in those conditions will grow exponentially in time, invalidating any attempt to follow the exact equations of motion. When tossing a coin or rolling a die, we do not know the outcome of the event, but we can give some *probability* to each event, e.g., $1/2$ for head and $1/2$ for tail when tossing a coin, or $1/6$ for each side of a die (we assume that we are playing with fair coins and dice). The probability gives the measure of the likelihood that a given event will occur. Of course, it is based upon a mathematical approach, which transforms the unpredictability into something that is somehow predictable.

This idea seems very simple, but it took several hundred years to capture it. Indeed, since the ancient times, people in Greece and in the Roman Empire (but also in Asia, for example, in India) were tenacious gamblers; nevertheless, nobody tried to understand how the random events were related to mathematical laws. Many quarrels and disputes were resolved by tossing a coin, and the result was seen as the manifestation of the "celestial will." The human superstition represented a huge obstacle to define a scientific (i.e., mathematical) approach to random events. Eventually, after several hundreds of years, superstition was overcome by an even stronger human impulse: the desire of obtaining an economical profit.

The birth of the mathematical theory of probability is due to the studies done by Girolamo Cardano, who realized that for equiprobable events (like tossing a coin or rolling a die), the probability that a single event will appear is equal to 1 over the number of all possible events (independently from any celestial will). Therefore, the probability to obtain head when tossing a coin is $1/2$ (equal to the one of obtaining tail); the probability of getting 1 when rolling a die is $1/6$, and the probability of obtaining an odd number is $3/6$. Cardano was very often short of money and kept himself solvent by being an accomplished gambler and chess player. His book *Liber de ludo aleae*, written around 1560 but not published until 1663, contains the first systematic treatment of probability.

Even though Cardano produced the first exposition of random events, the most important conceptual leap toward the modern theory of probability was given by Blaise Pascal and Pierre de Fermat in the 17th Century, when answering few questions posed by the Chevalier de Méré, who was a regular gambler in France. There, one popular game was to roll two dice several times with a bet on having at least one double 1 (ace). Chevalier de Méré wanted to know how many trials one must do to have a profitable game (i.e., that the probability of having a double ace will be larger than $1/2$). Another question was a bit more complex and opened the real basis of the theory of probability. The problem can be illustrated in this simple way: there are two players (e.g., Fermat and Pascal), who are tossing a coin: head gives one point to Fermat, while tail gives one point to Pascal. The first of the two players who achieves 3 points will win a pot of 100 francs. However, for some reason, the two gamblers must interrupt the game at the point where Fermat has 2 points and Pascal only 1; how is the 100 franc pot to be divided?

Let us briefly discuss the solutions of these questions. As far as the first one is concerned, instead of computing directly the probability P of favorable events, which requires a cumbersome calculation, it is much easier to evaluate the probability Q of unfavorable events and then take its complement, i.e., $P = 1 - Q$. In this example, the probability of obtaining a double ace (the successful event) when rolling a die is $1/36$, while the probability that this event does not appear is obviously $35/36$. The probability to obtain at least one double ace in N trials is therefore $1 - (35/36)^N$, namely it is 1 minus the probability that no successful events appear in N trials. For $N = 25$, the probability of having at least a double ace is larger than $1/2$, so it becomes profitable to bet on it. Regarding the second question, the central point is that we do not have to see what happened in the past but what could happen in the future. Indeed, the following two answers are not correct: (a) Fermat takes all because he won more than Pascal and (b) Fermat takes twice more than Pascal since the result is $2-1$. The correct one is to see what are the possibilities if they could have continued the game. Then, the possible outcomes are: head-head, head-tail, tail-head, and tail-tail; only in the latter case, Pascal would have won, while in the other three cases Fermat would have gained. Therefore, Fermat must take $3/4$ of the total pot and Pascal only $1/4$. Today, this result may look obvious; however, in the seventeenth century the idea that future events may be treated in a rigorous mathematical way has represented an incredible step forward.

There are several definitions for the probability, which are rooted in different philosophical approaches. The *classical definition* dates back to Pierre Simon Laplace, who defined the probability of a given circumstance as the ratio between the number of favorable events divided by the total number of possible events, provided all the events are equiprobable. This definition is not satisfactory since it requires the concept of equiprobability itself. Moreover, it does not give a definition when the events are not equiprobable and assumes a finite number of outcomes (and, therefore, it does not apply for continuous variables). To overcome these difficulties, the frequentist definition of probability was introduced: it defines the probability of an event as the limit of its relative frequency in a large number of trials. This interpretation needs the possibility that a given game or experiment can be repeated (with the same physical conditions) several times. In this book, we take this point of view. We assume that there exist reproducible experiments, which, under very similar initial conditions, produce different events (denoted by E_i, a Boolean variable that may be true or false, when the event i is realized or not, respectively). Within the frequentist definition, the probability assigned to any event E_i is given by the ratio between the number of events n_i in which E_i happened and the total number of trials N, when N becomes very large:

$$P(E_i \text{ is true}) \equiv P(i) = \lim_{N \to \infty} \frac{n_i}{N}; \tag{2.1}$$

clearly, the probability is a non-negative number:

$$P(i) \geq 0. \tag{2.2}$$

In the following, we give a brief overview of the theory of probability. A clear and comprehensive treatment of this subject is given in the book by Gnedenko (2014), which also includes the assiomatic approach developed by the Russian mathematician Andrey Nikolaevich Kolmogorov in 1933.

2.2 Events and Probability

Here, we describe some simple properties of events. Two events E_i and E_j are said to be *mutually exclusive* if and only if the occurrence of E_i implies that E_j does not occur and vice versa. If E_i and E_j are mutually exclusive, we have that:

$$P(E_i \text{ is true and } E_j \text{ is true}) = 0, \tag{2.3}$$

$$P(E_i \text{ is true or } E_j \text{ is true}) = P(i) + P(j). \tag{2.4}$$

The different outcomes of an experiment represent mutually exclusive events (for example, when tossing a coin, head and tail represent two mutually exclusive events). Clearly, if the number of all possible mutually exclusive events is M, then the sum of their probability over the whole space of outcomes is given by:

$$\sum_{i=1}^{M} P(i) = 1. \tag{2.5}$$

Once for a given experiment all the M mutually exclusive events are classified, each realization of the experiment is specified by a single integer i, such that E_i is verified. Therefore, we can define a *random variable* X_i, as a real-valued function associated to any possible successful event E_i. The simplest random variable is the *characteristic* random variable $X_i^{[j]}$:

$$X_i^{[j]} = \begin{cases} 1 & \text{if } i = j \ (E_j \text{ is true}) \\ 0 & \text{if } i \neq j \ (E_j \text{ is false}) \end{cases} \tag{2.6}$$

in other words, the characteristic random variable $X_i^{[j]}$ is non-zero only if the event E_j is successful: for example, if we bet that the number 36 will show up in the roulette game, the successful event, associated to a winning bet, is the appearance of 36, while all the other numbers would give rise to a loosing bet. A different example of a random variable is given by the actual outcome after rolling a die, i.e., the number that shows up:

$$X_i = i \text{ if } E_i \text{ is true,} \tag{2.7}$$

for $i = 1, \ldots, 6$.

Let us now discuss the case of composite events. For example, rolling two dice is an experiment that can be characterized by $E_i^{(1)}$ and $E_j^{(2)}$, where $E_j^{(1)}$ ($E_j^{(2)}$) refers to the possible outcomes of the first (second) die. For composite events, the probability is labeled by more than one index, in particular the *joint probability* of two events $P(i,j)$ is defined as:

$$P_{\text{joint}}(i,j) = P(E_i^{(1)} \text{ is true and } E_j^{(2)} \text{ is true}). \tag{2.8}$$

In order to obtain the probability for one variable alone (say $E_i^{(1)}$), consistently with the frequentist definition (2.1), we have to sum the joint probability over all values of the other variable (say $E_j^{(2)}$). In this way, we get the *marginal probability*:

$$P_1(i) = \sum_{j=1}^{M} P_{\text{joint}}(i,j), \tag{2.9}$$

which represents the probability of obtaining the variable $E_i^{(1)}$, without caring about the outcome of the variable $E_j^{(2)}$. Instead, we can be interested into the probability of obtaining $E_i^{(1)}$, once we know the outcome of the variable $E_j^{(2)}$; this is the *conditional probability* $\omega(i|j)$:

$$\omega(i|j) = \frac{P_{\text{joint}}(i,j)}{\sum_{i=1}^{M} P_{\text{joint}}(i,j)}, \qquad (2.10)$$

which is normalized:

$$\sum_{i=1}^{M} \omega(i|j) = 1. \qquad (2.11)$$

Since $\sum_{i=1}^{M} P_{\text{joint}}(i,j) = P_2(j)$, we have that:

$$\omega(i|j) = \frac{P_{\text{joint}}(i,j)}{P_2(j)}, \qquad (2.12)$$

and, therefore, the joint probability can be expressed as the product of the marginal probability $P_2(j)$ times the conditional probability $\omega(i|j)$:

$$P_{\text{joint}}(i,j) = \omega(i|j)P_2(j). \qquad (2.13)$$

Similarly, we can also obtain that:

$$P_{\text{joint}}(i,j) = \omega(j|i)P_1(i). \qquad (2.14)$$

By combining Eqs. (2.13) and (2.14), we obtain the Bayes formula:

$$\omega(j|i)P_1(i) = \omega(i|j)P_2(j). \qquad (2.15)$$

One random variable i is *independent* from the other one j whenever the conditional probability $\omega(i|j)$ does not depend on j; in this case, we have that $P_1(i) = \sum_{j=1}^{M} \omega(i|j)P_2(j) = \omega(i|j)$, which inserted into Eq. (2.13), gives:

$$P_{\text{joint}}(i,j) = P_1(i)P_2(j); \qquad (2.16)$$

notice that, from Eq. (2.14), we also have that $P_2(j) = \omega(j|i)$, implying that also the variable j is independent from i.

The generalization of the above definitions holds obviously also for random variables x defined on the continuum, namely variables that assume any real value within a given interval and not only discrete values. In this case, the probability to obtain a particular value x is generically vanishing, while there is a finite probability to find the random variable within a finite range (a, b). In particular, we can consider the probability that x is smaller than y, where y is a given fixed real number:

$$P(x \le y) = F(y) = \int_{-\infty}^{y} dx \, \mathcal{P}(x), \qquad (2.17)$$

which is called the *cumulative probability* of the random variable x. Clearly $F(+\infty) = 1$. The *probability density* or *distribution function* $\mathcal{P}(x)$ is then given by:

$$\mathcal{P}(x) = \left. \frac{dF(y)}{dy} \right|_{y=x}, \qquad (2.18)$$

which satisfies the following properties:

$$\mathcal{P}(x) \geq 0, \tag{2.19}$$

$$\int_{-\infty}^{+\infty} dx \, \mathcal{P}(x) = 1. \tag{2.20}$$

Notice that the above definitions can be also used in the case of discrete random variables, by taking:

$$\mathcal{P}(x) = \sum_{i=1}^{M} P(i)\delta(x - X_i). \tag{2.21}$$

Therefore, in the following, we will consider the formalism of continuous random variables, having in mind that the discrete case can be simply obtained by considering Eq. (2.21).

As before, the marginal probability of the variable x is defined as:

$$\mathcal{P}_1(x) = \int_{-\infty}^{+\infty} dy \, \mathcal{P}_{\text{joint}}(x, y), \tag{2.22}$$

and the conditional probability of obtaining x once y is known is given by:

$$\omega(x|y) = \frac{\mathcal{P}_{\text{joint}}(x, y)}{\int_{-\infty}^{+\infty} dx \, \mathcal{P}_{\text{joint}}(x, y)}, \tag{2.23}$$

such that:

$$\mathcal{P}_{\text{joint}}(x, y) = \omega(x|y)\mathcal{P}_2(y) = \omega(y|x)\mathcal{P}_1(x). \tag{2.24}$$

2.3 Moments of the Distribution: Mean Value and Variance

For any random variable x, we can define its *mean value* or *expected value*:

$$\mu \equiv \langle x \rangle = \int_{-\infty}^{+\infty} dx \, x \, \mathcal{P}(x). \tag{2.25}$$

Within the frequentist approach, the expected value of a random variable is just the average value of several repetitions of the same experiment. For example, when rolling a die, the expected value is 3.5. For the characteristic random variable $X_i^{[j]}$ of the event E_j, we simply have $\langle X_i^{[j]} \rangle = P(j)$. More generally, the n-th *moment* of the distribution is defined as the expected value of the n-th power of x:

$$\langle x^n \rangle = \int_{-\infty}^{+\infty} dx \, x^n \, \mathcal{P}(x). \tag{2.26}$$

The first moment $\langle x \rangle$ is equal to the mean value μ. The second moment allows us to define a particularly important quantity, which is the *variance*:

$$\sigma^2 \equiv \langle x^2 \rangle - \langle x \rangle^2 = \int_{-\infty}^{+\infty} dx\ (x - \langle x \rangle)^2\ \mathcal{P}(x). \tag{2.27}$$

The variance is a non-negative quantity that can be zero only when all the events having a non-vanishing probability give the same value for the variable x. In other words, whenever the variance is zero, the random character of the variable is completely lost and the experiment becomes perfectly predictable. The square root of the variance is a measure of the dispersion of the random variable and is called *standard deviation*. Notice that the existence of the variance, and of higher moments as well, is not *a priori* guaranteed in the continuous case. Indeed, the probability density $\mathcal{P}(x)$ has to decrease sufficiently fast for $x \to \pm\infty$ in order for the corresponding integrals to exist.

In the case we have two random variables, we can define their *covariance* by:

$$\sigma_{xy}^2 \equiv \langle xy \rangle - \langle x \rangle \langle y \rangle = \int_{-\infty}^{+\infty} dx \int_{-\infty}^{+\infty} dy\ (x - \langle x \rangle)\ (y - \langle y \rangle)\ \mathcal{P}_{\text{joint}}(x, y); \tag{2.28}$$

here, we must notice that the quantity σ_{xy}^2 is not guaranteed to be positive; nevertheless, in analogy to the variance, we prefer to keep the notation with the square. Obviously, for two independent variables, we have that $\sigma_{xy}^2 = 0$, as obtained from the fact that $\mathcal{P}_{\text{joint}}(x, y) = \mathcal{P}_x(x)\mathcal{P}_y(y)$.

Finally, we can also consider expected values of another random variable $y = f(x)$:

$$\langle f(x) \rangle = \int_{-\infty}^{+\infty} dx\ f(x)\ \mathcal{P}(x). \tag{2.29}$$

An important quantity related to the probability density $\mathcal{P}(x)$ is the *characteristic function* $\phi_x(t)$, which is the expected value of e^{ixt}, or equivalently the Fourier transform of the probability density:

$$\phi_x(t) = \langle e^{ixt} \rangle = \int_{-\infty}^{+\infty} dx\ e^{ixt}\ \mathcal{P}(x). \tag{2.30}$$

For small t, if the second moment $\langle x^2 \rangle$ is finite, one can expand the exponential up to second order in t, obtaining:

$$\phi_x(t) = 1 + i\langle x \rangle t - \frac{\langle x^2 \rangle}{2} t^2 + o(t^2), \tag{2.31}$$

where $o(t^2)$ are terms that are smaller than t^2: if the third moment is also finite, the next term will be proportional to t^3, while if $\langle x^3 \rangle = \infty$, this term will have an intermediate power between 2 and 3. In turn, the probability density $\mathcal{P}(x)$ is the Fourier transform of the characteristic function:

$$\mathcal{P}(x) = \frac{1}{2\pi} \int_{-\infty}^{+\infty} dt \, e^{-ixt} \, \phi_x(t). \tag{2.32}$$

Here, we would like to report few important examples of distribution functions. For a discrete random variable, the simplest possible case is given by the Bernoulli distribution, named after the Swiss scientist Jacob Bernoulli. In this case, the random variable takes the value 1 with "success" probability $\mathcal{P}$ and the value 0 with "failure" probability $\mathcal{Q} = 1 - \mathcal{P}$. It can be used to represent a coin toss where 1 and 0 would represent head and tail, respectively. In particular, fair coins have $\mathcal{P} = \mathcal{Q} = 1/2$. The mean of the Bernoulli distribution is equal to $\mathcal{P}$, while the variance is $\mathcal{P}\mathcal{Q}$. Its characteristic function is given by $\phi_{\text{bernulli}}(t) = \mathcal{Q} + \mathcal{P}e^{it}$.

Another example is given by the binomial distribution, which describes the number of successes in a sequence of N independent trials (experiments), each of which yields success with probability $\mathcal{P}$ and failure with probability $\mathcal{Q}$ (e.g., N trials of the Bernoulli variable). The probability of obtaining k successes in N trials is given by:

$$\mathcal{P}_{\text{binomial}}(k) = \frac{N!}{k! \, (N-k)!} \mathcal{P}^k \mathcal{Q}^{N-k}. \tag{2.33}$$

The mean of the binomial distribution is $N\mathcal{P}$, while the variance is $N\mathcal{P}\mathcal{Q}$. The characteristic function is $\phi_{\text{binomial}}(t) = [\phi_{\text{bernulli}}(t)]^N$. In the limit of $N \to \infty$ and $\mathcal{P} \to 0$ with $N\mathcal{P} = \lambda$, $\mathcal{P}_{\text{binomial}}(k)$ approaches the Poisson distribution:

$$\mathcal{P}_{\text{poisson}}(k) = e^{-\lambda} \frac{\lambda^k}{k!}, \tag{2.34}$$

which describes the probability of an unlike event ($\mathcal{P} \to 0$) in a large number of trials ($N \to \infty$). Both the mean and the variance of the Poisson distribution are equal to λ. Its characteristic function is $\phi_{\text{poisson}}(t) = \exp[\lambda(e^{it} - 1)]$.

For continuous random variables, the simplest possible distribution function is the uniform one, in which all the values of x in the interval $[a, b]$ are equiprobable:

$$\mathcal{P}_{\text{uniform}}(x) = \begin{cases} \frac{1}{(b-a)} & \text{if } a \leq x \leq b, \\ 0 & \text{otherwise.} \end{cases} \tag{2.35}$$

The mean of the uniform distribution is equal to $(b+a)/2$, while the variance is $(b-a)^2/12$. Its characteristic function is $\phi_{\text{uniform}}(t) = \left(e^{bt} - e^{at}\right)/[t(b-a)]$.

Finally, a pivotal role in the probability theory is given by the Gaussian (or normal) distribution, named after the German scientist Carl Friederich Gauss:

$$\mathcal{P}_{\text{gauss}}(x) = \frac{1}{\sqrt{2\pi\sigma^2}} e^{-\frac{(x-\mu)^2}{2\sigma^2}}, \tag{2.36}$$

where, μ and σ^2 denote the mean and the variance, respectively. Its characteristic function, which will be used in the following, is given by:

$$\phi_{\text{gauss}}(t) = e^{i\mu t} e^{-\frac{t^2}{2}\sigma^2}.$$ (2.37)

2.4 Changing Random Variables

In this section, we briefly show how the probability density is modified when we apply a generic transformation to a random variable. Suppose that we have a random variable x and we construct another random variable y that is a function of x:

$$y = f(x),$$ (2.38)

where, for simplicity, we consider a monotonic function $f(x)$, with positive derivative, i.e., $df(x)/dx > 0$, such that $f(x_2) = y_2 > f(x_1) = y_1$ for $x_2 > x_1$ (the case with $df(x)/dx < 0$ can be treated similarly); in the most general case of a non-monotonic function, we can take separately all the small intervals where the function is monotonic. Then, the probability to find x in a given interval $x_1 < x < x_2$ must be equal to the probability to find y in the interval $y_1 < y < y_2$:

$$P_y(y_1 < y < y_2) = P_x(x_1 < x < x_2).$$ (2.39)

Then, for a small interval $dx = (x_2 - x_1)$ for which $dy = df(x)$, we have that:

$$P_y(y)dy = P_x(x)dx.$$ (2.40)

This equation gives a punctual relation between the values of the probability densities at x and y, we can integrate both sides and get:

$$\int_{-\infty}^{y} ds\, P_y(s) = \int_{-\infty}^{x} dt\, P_x(t).$$ (2.41)

The usefulness of this relation is that it allows one to obtain the probability density for the variable y, i.e., $P_y(y)$, once we know the probability density of the variable x, i.e., $P_x(x)$, and the relation between x and y, i.e., the function $f(x)$. Notice that, whenever $df(x)/dx < 0$, the only modification comes from exchanging the limits of integration in the r.h.s. of Eq. (2.41).

The simplest application of the above relations is obtained when considering a linear transformation:

$$y = a + bx;$$ (2.42)

then, we simply get:

$$P_y(y) = \frac{1}{b} P_x\left(\frac{y-a}{b}\right).$$ (2.43)

Therefore, we immediately obtain that, if the random variable x has mean μ_x and variance σ_x^2, then the transformed random variable y has $\mu_y = a + b\mu_x$ and $\sigma_y^2 = b^2\sigma_x^2$.

2.5 The Chebyshev's Inequality

The Chebyshev's inequality gives a simple bound for the probability to obtain a result x which lies far from the mean value $\langle x \rangle$; the precise statement is that no more than $1/k^2$ of the distribution's values can be more than k standard deviations away from the mean value:

$$P(|x - \langle x \rangle| > k\sigma) \leq \frac{1}{k^2}, \tag{2.44}$$

or equivalently, by taking $\epsilon = k\sigma$:

$$P(|x - \langle x \rangle| > \epsilon) \leq \frac{\sigma^2}{\epsilon^2}. \tag{2.45}$$

The proof of the Chebyshev's inequality is very simple. Indeed, by definition:

$$P(|x - \langle x \rangle| > \epsilon) = \int_{-\infty}^{\langle x \rangle - \epsilon} dx \, \mathcal{P}(x) + \int_{\langle x \rangle + \epsilon}^{+\infty} dx \, \mathcal{P}(x); \tag{2.46}$$

in both these two intervals $[(x - \langle x \rangle)/\epsilon]^2 \geq 1$, thus we have:

$$P(|x - \langle x \rangle| > \epsilon) \leq \int_{-\infty}^{\langle x \rangle - \epsilon} dx \left(\frac{x - \langle x \rangle}{\epsilon}\right)^2 \mathcal{P}(x) + \int_{\langle x \rangle + \epsilon}^{+\infty} dx \left(\frac{x - \langle x \rangle}{\epsilon}\right)^2 \mathcal{P}(x)$$

$$\leq \int_{-\infty}^{+\infty} dx \left(\frac{x - \langle x \rangle}{\epsilon}\right)^2 \mathcal{P}(x) = \frac{\sigma^2}{\epsilon^2}. \tag{2.47}$$

The Chebyshev's bound is very general and does not assume any form of the probability distribution $\mathcal{P}(x)$; for this reason, it gives a rather weak bound on the probability to find a large fluctuation of x, i.e., the probability to find an event very far from its expected value. For specific distributions, it is possible to obtain much stronger bounds. For example, in Table 2.1, we report a comparison between the Chebyshev's bound of Eq. (2.44) and the actual values that are valid for the Gaussian distribution of Eq. (2.36).

2.6 Summing Independent Random Variables

Here, we treat the case of several independent random variables $x_1, \ldots, x_N$ for which, generalizing the arguments of section 2.2, the joint probability is the product of the marginal ones, i.e., $\mathcal{P}(x_1, \ldots, x_N) = \mathcal{P}_1(x_1) \ldots \mathcal{P}_N(x_N)$. Then, it is often

Table 2.1. *Probability that a random variable has a fluctuation away from its mean value larger than k standard deviations σ. The results of the Chebyshev's bound of Eq. (2.44) (left column) are compared with the actual values obtained from the Gaussian distribution of Eq. (2.36) (right column).*

k	Chebyshev's bound	Gaussian value
1	1.0000	0.3173
2	0.2500	0.0455
3	0.1111	0.0027
4	0.0625	6.3×10^{-5}
5	0.0400	5.8×10^{-7}
6	0.0278	2.0×10^{-9}
7	0.0204	2.5×10^{-12}
8	0.0156	1.2×10^{-15}

useful to consider a "new" random variable that is the sum of them. The main motivation of summing (independent and identically distributed) random variables is due to the fact that important and useful properties about the distribution function of the sum can be obtained in the limit of large N, i.e., for $N \to \infty$ (Gnedenko and Kolmogorov, 1954). Let us start by taking z that is the sum of two random variables:

$$z = x_1 + x_2. \tag{2.48}$$

Its probability density is simply given by:

$$P_z(z) = \int_{-\infty}^{+\infty} dx_1 \int_{-\infty}^{+\infty} dx_2 \, P_1(x_1) \, P_2(x_2) \, \delta(z - x_1 - x_2), \tag{2.49}$$

where the delta-function enforces the fact that z is the sum of x_1 and x_2. Therefore, by performing the integral over x_2, we have that $P_z(z)$ is simply given by the convolution of the probability densities:

$$P_z(z) = \int_{-\infty}^{+\infty} dx_1 \, P_1(x_1) \, P_2(z - x_1), \tag{2.50}$$

whose meaning is the following: the probability for a given value z is obtained by summing (or integrating) over all the possible outcomes of the variable x_1 the probability of finding x_1 times the probability that $x_2 = z - x_1$. From Eq. (2.50), we have that the characteristic function of the sum of two independent random variables x_1 and x_2 is the product of the two characteristic functions:

$$\phi_z(t) = \phi_1(t)\phi_2(t). \tag{2.51}$$

Let us see some examples in which we can easily obtain the probability of a sum of random variables. The simplest case is given when two random variables are uniformly distributed in $[0, 1]$, for which we get the "triangular" distribution:

$$\mathcal{P}_z(z) = \begin{cases} 0 & \text{if } z \leq 0, \\ z & \text{if } 0 \leq z \leq 1, \\ 2 - z & \text{if } 1 \leq z \leq 2, \\ 0 & \text{if } z \geq 2. \end{cases} \tag{2.52}$$

As expected, it is impossible to get a sum that is negative or exceeds 2, while the most probable value (which also coincides with the mean value) is $z = 1$. This example is the continuous version of the case where we roll two dice and consider the sum of the two outcomes: the most probable sum is 7, while it is impossible to get something that is smaller than 2 or larger than 12.

Then, the sum of three uniformly distributed random variables $z = x_1 + x_2 + x_3$ can be easily obtained by adding together two of them and get the "triangular" distribution and then adding the third one using Eq. (2.50). This procedure gives the following distribution for the sum of three uniformly distributed variables:

$$\mathcal{P}_z(z) = \begin{cases} 0 & \text{if } z \leq 0, \\ \frac{z^2}{2} & \text{if } 0 \leq z \leq 1, \\ \frac{3}{4} - \left(z - \frac{3}{2}\right)^2 & \text{if } 1 \leq z \leq 2, \\ \frac{(3-z)^2}{2} & \text{if } 2 \leq z \leq 3, \\ 0 & \text{if } z \geq 3. \end{cases} \tag{2.53}$$

Here, the most probable value is $z = 3/2$. The trend of this procedure is clear: when adding up more and more random variables, both the mean value and variance of the distribution increase. In general, the probability of the sum of random variables depends upon the number of the variables. One remarkable exception is given by Gaussian variables, for which the sum of two random numbers is still Gaussian. Indeed, by using Eq. (2.51) and the form of the characteristic function of the Gaussian distribution (2.37), we have that (indicating by μ_1 and μ_2 the means and by σ_1^2 and σ_2^2 the variances of the two variables):

$$\phi_z(t) = e^{i(\mu_1 + \mu_2)t} e^{-\frac{t^2}{2}(\sigma_1^2 + \sigma_2^2)}, \tag{2.54}$$

which is just the characteristic function of a Gaussian variable with mean $\mu_z = \mu_1 + \mu_1$ and variance $\sigma_z^2 = \sigma_1^2 + \sigma_2^2$.

If the random variables have generic distribution functions, then the distribution function of the sum of them will have a very complicated form. The remarkable fact is that, under very general conditions, it is possible to obtain an asymptotically *exact* form for $\mathcal{P}_z(z)$ when the number N of independent variables becomes very large. Before demonstrating this fundamental aspect (which goes under the name of central limit theorem), we would like to discuss some relevant issues about the mean and variance of the sum of random variables. Given the fact that both the mean and the variance are proportional to N, it is often useful to consider the average of independent and equally distributed random variables:

$$\bar{x} = \frac{1}{N} \sum_i x_i. \tag{2.55}$$

Even though, in general cases, it is not easy to work out the explicit form of the probability distribution of $\bar{x}$, both the mean and the variance of $\bar{x}$ can be easily computed. Indeed, all the N terms in the sum give an identical contribution, equal to $\langle x \rangle$, thus giving:

$$\langle \bar{x} \rangle = \langle x \rangle, \tag{2.56}$$

namely, the mean value of the average $\bar{x}$ coincides with the mean value of the single trial (experiment). We would like to emphasize that Eq. (2.56) holds also in the case where the random variables x_i are not independent. In order to compute the variance of $\bar{x}$, we simply notice that, by taking the expected value of $\bar{x}^2$, we get:

$$\langle \bar{x}^2 \rangle = \frac{1}{N^2} \sum_{i,j} \langle x_i x_j \rangle = \frac{1}{N^2} \left[\sum_i \langle x_i^2 \rangle + \sum_{i \neq j} \langle x_i \rangle \langle x_j \rangle \right]$$

$$= \frac{1}{N^2} \left[N \langle x^2 \rangle + N(N-1) \langle x \rangle^2 \right], \tag{2.57}$$

where we have used the fact that the variables are independent, leading to $\langle x_i x_j \rangle = \langle x_i \rangle \langle x_j \rangle$ for $i \neq j$; then, $\langle \bar{x}^2 \rangle$ has two contributions, the first one, coming from the terms with $i = j$, gives N terms that do not depend upon i (i.e., $\langle x_i^2 \rangle = \langle x^2 \rangle$); the second one, coming from the terms with $i \neq j$, gives $N(N-1)$ terms that are all equal (i.e., $\langle x_i \rangle \langle x_j \rangle = \langle x \rangle^2$). Therefore, the variance of $\bar{x}$ is given by:

$$\sigma_{\bar{x}}^2 = \langle \bar{x}^2 \rangle - \langle \bar{x} \rangle^2 = \frac{\sigma^2}{N}, \tag{2.58}$$

where $\sigma^2 = \langle x^2 \rangle - \langle x \rangle^2$ is the variance of the single random variable x. Therefore, for large N, the random variable $\bar{x}$, corresponding to averaging a large number of realizations of the same experiment, will have a very narrow distribution function, since $\sigma_{\bar{x}}^2 \to 0$ for $N \to \infty$, with the same mean value of the original random variables. In other words, almost all possible average measurements (each of them done by N different realizations of the same experiment) give a value for $\bar{x}$ that is closer to the true mean value than the single experiment. This important fact can be obtained in a more rigorous way by applying the Chebyshev's inequality of Eq. (2.45) to the variable $\bar{x}$:

$$P\left(\left|\frac{1}{N}\sum_i x_i - \langle x \rangle\right| > \epsilon\right) \le \frac{\sigma_{\bar{x}}^2}{\epsilon^2} = \frac{\sigma^2}{N\epsilon^2}, \tag{2.59}$$

which directly gives:

$$\lim_{N\to\infty} P\left(\left|\frac{1}{N}\sum_i x_i - \langle x \rangle\right| > \epsilon\right) = 0. \tag{2.60}$$

This result is nothing but the so-called *weak law of large numbers*, which essentially states that for any non-zero margin ϵ, for a sufficiently large N, there will be a very high probability that the average $\bar{x}$ will be close enough to the expected value, i.e., within the margin ϵ.

Eqs. (2.56) and (2.58) show that the average of many independent random variables gives an unbiased and very accurate estimation of the true expectation value. Since the exact variance of the distribution σ^2 is not generally known, it is useful to define an estimator for it, which allows us to estimate $\sigma_{\bar{x}}^2$ of Eq. (2.58). For that, we can consider:

$$s^2 = \frac{1}{N}\sum_i (x_i - \bar{x})^2 = \frac{1}{N}\sum_i x_i^2 - \left(\frac{1}{N}\sum_i x_i\right)^2, \tag{2.61}$$

which is also a random variable with:

$$\langle s^2 \rangle = \frac{1}{N}\sum_i \langle x_i^2 \rangle - \frac{1}{N^2}\sum_{i,j} \langle x_i x_j \rangle, \tag{2.62}$$

whenever the variables x_i are independent and identically distributed, we obtain:

$$\langle s^2 \rangle = \left(1 - \frac{1}{N}\right)\left(\langle x^2 \rangle - \langle x \rangle^2\right) = \left(\frac{N-1}{N}\right)\sigma^2. \tag{2.63}$$

Therefore, for $N \to \infty$, s^2 gives an asymptotically exact estimation of the variance. Notice that a slightly better estimation of the variance for any finite values of N should be given by taking $N/(N-1)s^2$ instead of s^2.

Finally, suppose that the random variable x_i is just the characteristic random variable of a given event E_j, namely $x_i \equiv X_i^{[j]}$. For this random variable, we have already noticed that the mean value is the probability of the event E_j, namely $\langle X_i^{[j]} \rangle = P(j)$. In addition, in view of the previous discussion, the mean of the random variable $\bar{x}$, obtained by averaging N independent realizations of the same experiment, gives an estimate of $P(j)$, with a standard deviation that decreases with $1/\sqrt{N}$, see Eq. (2.58). This uncertainty can be made arbitrarily small, by increasing N, so that the probability $P(j)$ of the event E_j is a well defined quantity in the limit of $N \to \infty$. This fact consistently justifies the definition of Eq. (2.1), which is the basis of the frequentist approach to probability. Notice that, within this scheme, the concept of probability is related to the *reproducibility* of the experiments.

2.7 The Central Limit Theorem

Let us finally prove a very important theorem in the theory of probability, the so-called *central limit theorem*, which provides the asymptotic probability distribution of the sum over a large number of random variables x_i, which are independent and equally distributed with probability $\mathcal{P}(x)$. The importance of this theorem relies on the fact that the asymptotic form of the distribution function is given under very general conditions, i.e., regardless of the specific form of $\mathcal{P}(x)$. The only requirement is that the variance is finite. As a first step, we define $y_i = x_i - \langle x \rangle$, which are still independent and equally distributed random variables, having $\langle y_i \rangle = 0$ by definition. Then, we construct the random variable Y:

$$Y = \frac{1}{\sqrt{N}} \sum_i y_i = \sqrt{N}(\bar{x} - \langle x \rangle), \tag{2.64}$$

which also has a vanishing expected value:

$$\langle Y \rangle = 0. \tag{2.65}$$

The characteristic function of Y is given by:

$$\phi_Y(t) = \left\langle \exp\left(\frac{it}{\sqrt{N}} \sum_i y_i \right) \right\rangle = \left\langle \prod_i \exp\left(\frac{it}{\sqrt{N}} y_i \right) \right\rangle = \left[\phi_y\left(\frac{t}{\sqrt{N}} \right) \right]^N, \tag{2.66}$$

where we have used the fact that all y_i are independent and equally distributed, so that all terms give the same factor. For small values of t, we can expand $\phi_y(\frac{t}{\sqrt{N}})$ up to second order by using Eq. (2.31) with $\mu = 0$:

$$\phi_Y(t) = \left[1 - \frac{\sigma^2 t^2}{2N} + o\left(\frac{t^2}{N} \right) \right]^N. \tag{2.67}$$

When considering the limit of a very large number of random variables, we get:

$$\lim_{N \to \infty} \phi_Y(t) = \exp\left(-\frac{\sigma^2}{2} t^2 \right), \tag{2.68}$$

which is the characteristic function of a Gaussian random variable with mean $\mu = 0$ and variance σ^2:

$$\mathcal{P}(Y) = \frac{1}{\sqrt{2\pi\sigma^2}} \exp\left(-\frac{Y^2}{2\sigma^2} \right). \tag{2.69}$$

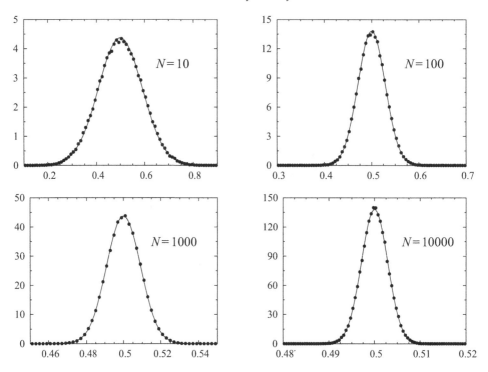

Figure 2.1 The probability distribution of $\bar{x} = 1/N \sum_i x_i$, generated over 10^5 realizations of $\bar{x}$, for different values of N. The expected Gaussian distributions, with $\mu = 0.5$ and $\sigma^2 = 1/(12N)$ are also reported.

By using the results of section 2.4, we have that also $\bar{x} = \langle x \rangle + Y/\sqrt{N}$ is a Gaussian variable with mean equal to $\langle x \rangle$ and variance equal to σ^2/N:

$$P(\bar{x}) = \sqrt{\frac{N}{2\pi\sigma^2}} \exp\left[-\frac{N(\bar{x} - \langle x \rangle)^2}{2\sigma^2}\right]. \tag{2.70}$$

In Fig. 2.1, we report few examples in which we generated 10^5 averages $\bar{x}$ of Eq. (2.55), obtained by summing N random variables that are uniformly distributed in $[0, 1)$. In particular, we divide the interval $[0, 1)$ into L small sub-intervals of width $1/L$ and compute the number of times that $\bar{x}$ falls in a given interval (normalized to the total number of trials). This quantity approaches very rapidly the Gaussian distribution (with the expected mean and variance), as predicted by the central limit theorem: already for $N = 10$, the distribution is essentially indistinguishable from a Gaussian.

The importance of the central limit theorem lies in the fact that the details (e.g., all the moments with $n > 2$) of the original distribution function $P(x)$ do not contribute to the form of the asymptotic distribution function of the average $\bar{x}$, which is then universal. Most importantly, summing many random variables gives rise to a random variable that has the same mean value of the original ones but has a much smaller variance, which tends to zero when $N \to \infty$. In other words, the random variable $\bar{x}$ becomes less and less fluctuating when increasing N, eventually becoming a deterministic number for $N \to \infty$. Therefore, the estimation of the mean is by far much more precise when summing many random variables than considering the single one. As we will discuss in the next chapter, the central limit theorem represents the heart of any Monte Carlo method to evaluate integrals.

We emphasize the fact that the Gaussian form of Eq. (2.70) should be taken with a caveat. Indeed, Eq. (2.70) holds only in the neighborhood of its maximum, i.e., for $(\bar{x} - \langle x \rangle) = O(\sigma/\sqrt{N})$, and not in the tails of the distribution, where *large deviations* usually appear. An extension of the central limit theorem is given by the so-called *large deviations theory*, which gives information not only to small deviations, i.e., $O(1/\sqrt{N})$, but also to rare events that are far away from the typical values.

We would like to finish this part by considering an application of the central limit theorem. Instead of taking the average of N random variables, let us consider the product of them:

$$z = \prod_{i=1}^{N} x_i. \tag{2.71}$$

Both the average and the variance of z can be worked out in the limit of large N. Indeed:

$$\xi = \ln z = \sum_{i=1}^{N} \ln x_i \tag{2.72}$$

is the sum of N random variables and, therefore, has a Gaussian distribution for $N \to \infty$:

$$P(\xi) = \sqrt{\frac{1}{2\pi N\sigma^2}} \exp\left[-\frac{(\xi - \langle \xi \rangle)^2}{2N\sigma^2}\right], \tag{2.73}$$

where $\langle \xi \rangle = N\langle \ln x \rangle$ and $\sigma^2 = \langle (\ln x)^2 \rangle - \langle \ln x \rangle^2$. Thus, we obtain that:

$$\langle z \rangle = \langle e^{\xi} \rangle \propto \exp\left(\langle \xi \rangle + \frac{N}{2}\sigma^2\right), \tag{2.74}$$

$$\langle z^2 \rangle = \langle e^{2\xi} \rangle \propto \exp\left(2\langle \xi \rangle + 2N\sigma^2\right), \tag{2.75}$$

which imply that for $N \to \infty$:

$$\frac{\sqrt{\langle z^2 \rangle - \langle z \rangle^2}}{\langle z \rangle} \approx \exp\left(\frac{N}{2}\sigma^2\right) \to \infty. \tag{2.76}$$

Therefore, in contrast to the sum of random variables, the product of them has a variance which diverges much faster than its expectation value.

3

Monte Carlo Sampling and Markov Chains

3.1 Introduction

Monte Carlo methods indicate a broad class of numerical algorithms that are based upon repeated random *sampling* to obtain the solution of several mathematical and physical problems. The typical issue is about the calculation of large sums or integrals and the revolutionary idea is that we do not perform an *exact* enumeration/integration but instead we generate random *samples*, which are then added together to approximate the exact result. Therefore, the concept of sampling is pivotal in any Monte Carlo approach: its meaning is just to produce random examples of configurations (e.g., N classical particles distributed in a box) that are used to give an accurate estimate of the exact quantity under examination (e.g., the internal energy at fixed temperature).

The roots of the Monte Carlo approaches date back to Enrico Fermi's attempts while studying neutron diffusion in the early thirties. He did not publish anything on the subject, but he got the credits for the idea that a time-independent Schrödinger equation can be interpreted in terms of a system of particles performing a random walk (Metropolis and Ulam, 1949). Few years after, a fundamental step forward has been done by Stanislaw Ulam, when he was working on nuclear-weapon projects at the Los Alamos National Laboratory. His first thoughts about Monte Carlo methods were suggested by a question that occurred when he was convalescing from an illness and playing solitaire. Ulam tried to calculate the likelihood of winning based on the initial layout of the cards. After exhaustive combinatorial calculations, he decided to go for a more practical approach of trying out many different layouts and observing the number of successful games. At that time, a new era of fast computers was beginning and John von Neumann understood the relevance of Ulam's suggestion and proposed a statistical approach to solve the problem of neutron diffusion in a fissionable material. Ulam and von Neumann worked together to develop algorithms including importance sampling and rejection sampling, which are at the basis of any modern Monte Carlo technique. In addition, von Neumann developed a way to obtain pseudo-random numbers by using the so-called middle-square digit method, since he realized that using a truly random number was extremely slow. Ulam and von Neumann required a code name for their secret project carried on with these stochastic methods. Then, Nicholas Metropolis, who was also working at the Los Alamos National Laboratory, suggested the name Monte Carlo, which refers to the Monte Carlo Casino in Monaco where Ulam's uncle used to gamble with cards (Metropolis and Ulam, 1949). A nice recollection of the early days of the Monte Carlo methods has been reported by Metropolis (1987).

Although the first tests of the Monte Carlo methods were done on a variety of problems in neutron transport, the real breakthrough came in 1952, when a new computer, called MANIAC, became operational at the Los Alamos laboratories (Anderson, 1986): Metropolis, together with Arianna and Marshall Rosenbluth, and Augusta and Edward Teller, studied the equation of state of the two-dimensional motion of hard spheres (Metropolis et al., 1957). They developed a strategy to enhance the computational efficiency in describing systems at thermal equilibrium, i.e., obeying the Boltzmann distribution function. According to this strategy, if a statistical move of a particle resulted in a decrease in the total energy, the new configuration was accepted. By contrast, if there was an increase in the total energy, the new configuration was accepted only if it survived a game of chance biased by a Boltzmann factor. Otherwise, the new configuration was taken equal to the old one. The so-called Metropolis algorithm, based upon random walks, is one of the most pervasive numerical algorithms used in computational approaches and was included in the top-ten list of "the greatest influence on the development and practice of science and engineering in the 20th century" (Dongarra and Sullivan, 2000). The fast improvement of computers made possible, in recent years, to reach incredible achievements in many branches of science, including mathematics, physics, chemistry, but Monte Carlo methods are frequently used also in economical sciences to predict the behavior of stock markets.

In this section, we give the justification of the Monte Carlo sampling to compute integrals, which is the simplest application of the Monte Carlo approach. Indeed, suppose that we must compute the following d-dimensional integral:

$$\mathcal{I} = \int \mathbf{dx}\, F(\mathbf{x}), \tag{3.1}$$

where $F(\mathbf{x})$ is a generic function of the vector $\mathbf{x}$ with d components. Then, without loss of generality, we can always split $F(\mathbf{x})$ into a probability density $\mathcal{P}(\mathbf{x})$ (with $\mathcal{P}(\mathbf{x}) \geq 0$ and $\int \mathbf{dx}\, \mathcal{P}(\mathbf{x}) = 1$) and a function $f(\mathbf{x}) = F(\mathbf{x})/\mathcal{P}(\mathbf{x})$:

$$\mathcal{I} = \langle f(\mathbf{x})\rangle = \int \mathbf{dx}\, f(\mathbf{x})\, \mathcal{P}(\mathbf{x}), \tag{3.2}$$

which is nothing else than the expectation value of the random variable $f(\mathbf{x})$ over the distribution function $\mathcal{P}(\mathbf{x})$, i.e. the multi-dimensional generalization of Eq. (2.29). Then, the central limit theorem implies that the *deterministic* integral $\mathcal{I}$ is equal to the *stochastic* random variable computed as the average value of $f(\mathbf{x})$ over a large number of samplings:

$$\langle f(\mathbf{x}) \rangle = \int d\mathbf{x}\, f(\mathbf{x})\, \mathcal{P}(\mathbf{x}) \approx \frac{1}{N} \sum_i f(\mathbf{x}_i), \tag{3.3}$$

where the values of $\mathbf{x}_i$ in the sum of the r.h.s. are distributed according to the probability density $\mathcal{P}(\mathbf{x})$. Indeed, for large N, the variable:

$$\bar{f} = \frac{1}{N} \sum_i f(\mathbf{x}_i) \tag{3.4}$$

is normally (Gaussian) distributed, with mean equal to $\langle f(\mathbf{x}) \rangle$ and variance σ^2/N, where $\sigma^2 = \langle f^2(\mathbf{x}) \rangle - \langle f(\mathbf{x}) \rangle^2$. Therefore, for $N \to \infty$, the random variable $\bar{f}$ tends to the deterministic number $\langle f(\mathbf{x}) \rangle$ (since fluctuations decrease to zero with $1/\sqrt{N}$). This is the meaning for Eq. (3.3), in which the l.h.s. is a deterministic number and the r.h.s. is a random variable.

In summary, the validity of the stochastic calculation (the numerical simulation) is based upon the fact that, whenever the number of samplings N is large enough, then the error due to statistical fluctuations goes to zero, implying that the errorbars of the simulations can be kept under control. In this sense, we have that:

$$\int d\mathbf{x}\, f(\mathbf{x})\, \mathcal{P}(\mathbf{x}) \approx \langle\langle f(\mathbf{x}_i) \rangle\rangle, \tag{3.5}$$

where $\langle\langle \ldots \rangle\rangle$ denotes the statistical average over many independent samples (distributed according to $\mathcal{P}(\mathbf{x})$). We emphasize that any stochastic average without errorbars is completely meaningless, since one would not have any idea of the accuracy of the simulation. While the estimation of the integral is given by $\bar{f}$, the errorbar can be obtained from the estimator of σ^2, see Eq. (2.61):

$$s^2 = \frac{1}{N} \sum_i \left[f(\mathbf{x}_i) - \bar{f} \right]^2. \tag{3.6}$$

The main issues of the stochastic calculation are (i) to generate configurations $\mathbf{x}_i$ that are distributed according to the desired probability density $\mathcal{P}(\mathbf{x})$ and then (ii) compute the function $f(\mathbf{x}_i)$ for all these configurations.

Whenever it is possible to generate configurations with the probability density $\mathcal{P}(\mathbf{x})$, we talk about *direct sampling*; in this case, all configurations are independent from each other. Unfortunately, this is only possible in a few cases for very simple probability densities that depend upon few variables $\mathbf{x}$ (e.g., the case with

$d = 1$). In the general case, we are not able to directly sample the probability density and indirect ways of obtaining such configurations must be devised. This is the case of the so-called *Markov chains*, in honor of the Russian mathematician Andrei Andreievich Markov. In the following, we will present in some detail both the direct sampling and the theory of Markov chains. In particular, while the latter approach is very general and can be applied in a large variety of cases, the former one can be implemented to sample discrete probabilities, where the total number of possible outcomes can be enumerated and stored in the computer (while it is rarely used to sample continuous variables in $d > 1$).

3.2 Reweighting Technique and Correlated Sampling

Before discussing the direct sampling, we briefly illustrate the concept of *reweighting*, which allows us to obtain the average of a function $f(\mathbf{x})$ over the probability $\mathcal{Q}(\mathbf{x})$, once a sampling over $\mathcal{P}(\mathbf{x})$ has been performed. In fact, let us suppose that we have two probabilities $\mathcal{P}(\mathbf{x})$ and $\mathcal{Q}(\mathbf{x})$ that are defined in terms of their corresponding weights $\mathcal{W}_p(\mathbf{x})$ and $\mathcal{W}_q(\mathbf{x})$, respectively:

$$\mathcal{P}(\mathbf{x}) = \frac{\mathcal{W}_p(\mathbf{x})}{\int \mathbf{dx}\, \mathcal{W}_p(\mathbf{x})}, \tag{3.7}$$

$$\mathcal{Q}(\mathbf{x}) = \frac{\mathcal{W}_q(\mathbf{x})}{\int \mathbf{dx}\, \mathcal{W}_q(\mathbf{x})}. \tag{3.8}$$

Then, from general grounds, we have that:

$$\frac{\int \mathbf{dx}\, f(\mathbf{x}) \mathcal{W}_q(\mathbf{x})}{\int \mathbf{dx}\, \mathcal{W}_q(\mathbf{x})} = \frac{\int \mathbf{dx}\, f(\mathbf{x}) \mathcal{R}(\mathbf{x}) \mathcal{W}_p(\mathbf{x})}{\int \mathbf{dx}\, \mathcal{R}(\mathbf{x}) \mathcal{W}_p(\mathbf{x})}, \tag{3.9}$$

where $\mathcal{R}(\mathbf{x}) = \mathcal{W}_q(\mathbf{x})/\mathcal{W}_p(\mathbf{x})$ is the ratio between the two weights. Therefore, the statistical sampling over the new probability $\mathcal{Q}(\mathbf{x})$ can be expressed in terms of samplings over $\mathcal{P}(\mathbf{x})$:

$$\langle\langle f(\mathbf{x})\rangle\rangle_{\mathcal{Q}} = \frac{\langle\langle f(\mathbf{x})\mathcal{R}(\mathbf{x})\rangle\rangle_{\mathcal{P}}}{\langle\langle \mathcal{R}(\mathbf{x})\rangle\rangle_{\mathcal{P}}}, \tag{3.10}$$

where $\langle\langle \ldots \rangle\rangle_{\mathcal{Q}}$ and $\langle\langle \ldots \rangle\rangle_{\mathcal{P}}$ denote the statistical samplings over $\mathcal{Q}(\mathbf{x})$ and $\mathcal{P}(\mathbf{x})$, respectively.

Then, the same sampling (i.e., set of configurations $\{\mathbf{x}_i\}$) obtained from $\mathcal{P}(\mathbf{x})$ can be used to evaluate averages over $\mathcal{Q}(\mathbf{x})$. This way of proceeding gives rise to the concept of *correlated sampling*, since two quantities are evaluated with the same set of configurations. Of course, in order to have an accurate statistics (i.e., small errorbars) on the reweighted quantity $\langle\langle f(\mathbf{x})\rangle\rangle_{\mathcal{Q}}$, the two weights must

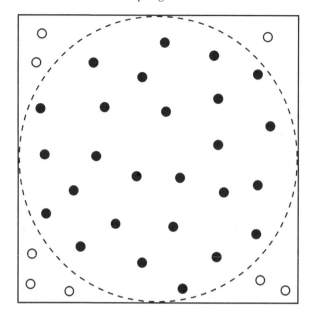

Figure 3.1 Direct sampling Monte Carlo to compute π. We shoot random bullets in the square, for a large number N of trials (i.e., $N \to \infty$) the ratio between the number of bullets inside the circle and the total number of trials converges to $\pi/4$; see text.

be quite similar, otherwise the configurations $\{\mathbf{x}_i\}$ would fall in regions where $\mathcal{Q}(\mathbf{x})$ is small and, therefore, irrelevant for the final result.

3.3 Direct Sampling

Here, in order to discuss the *direct sampling* method, we start form a simple example. Suppose that we want to compute π by a stochastic approach, then we can draw a circle with radius r and a square that exactly contains it, namely with side $L = 2r$, see Fig. 3.1. Then, we can randomly shoot bullets inside the square, counting every trial; each time a bullet falls inside the circle we increase by one the number of "hits." By keeping track of trials and hits, we can perform a direct sampling Monte Carlo calculation: the ratio between hits and trial is approaching, for a large number of trials N, the ratio of the areas of the circle and the square:

$$\frac{\pi}{4} = \frac{\int_{\text{circle}} dx\, dy}{\int_{\text{square}} dx\, dy} = \int_{\text{square}} dx\, dy\, f(x, y)\, \mathcal{P}(x, y), \qquad (3.11)$$

with

$$\mathcal{P}(x, y) = \frac{1}{\int_{\text{square}} dx\, dy}, \qquad (3.12)$$

and

$$f(x, y) = \begin{cases} 1 & \text{if } (x, y) \text{ is inside the circle,} \\ 0 & \text{otherwise.} \end{cases} \tag{3.13}$$

The function $\mathcal{P}(x, y)$ is non negative and normalized to unity, thus it may well represent a probability function. Shooting bullets in the square implies to generate a couple of random numbers (one for x and the other for y) that are uniformly distributed between 0 and L:

$$\frac{\pi}{4} \approx \frac{1}{N} \sum_i f(x_i, y_i). \tag{3.14}$$

This is the first (and simplest) example of how we can use a stochastic (i.e., Monte Carlo) approach to compute integrals. In general, any (multidimensional) integral such as Eq. (3.1) can be, in principle, computed by taking random numbers that are uniformly distributed in the whole interval of integration and then computing the value of the function in these points, i.e., $f(\mathbf{x}_i)$. However, if $f(\mathbf{x})$ is a sharply peaked function, this uniform way of sampling the interval is not efficient, since we would compute many points $\mathbf{x}_i$ for which the function $f(\mathbf{x})$ gives a negligible contribution to the integral $\mathcal{I}$, while only few points would give a finite contribution. In this case, some tricks must be employed in order to improve the sampling procedure, which we will discuss in the following.

3.4 Importance Sampling

Whenever the function $f(\mathbf{x})$ that must be evaluated has sharp peaks, a uniform sampling is not efficient, because it would lead to a considerable waste of time, spending efforts to visit regions that give a negligible contribution to the final result. A simple way to overcome this problem is to consider the so-called *importance sampling*. The idea is very general and is used in several Monte Carlo calculations. Nevertheless, it can be explained in the simple case of a one-dimensional integral. Indeed, let us suppose that we have to evaluate:

$$\mathcal{I} = \int_a^b dx \, F(x), \tag{3.15}$$

where $F(x)$ is a function that is peaked in a given point between a and b. Whenever we know the (approximated) location of the relevant regions where the function $F(x)$ is sizable, we can define probability density $\mathcal{P}(x)$ in $[a, b]$ (i.e., $\mathcal{P}(x) > 0$ and $\int_a^b dx \, \mathcal{P}(x) = 1$), which is also sizable in these regions and small everywhere else. Then, we can rewrite the original integral as:

$$\mathcal{I} = \int_a^b dx \, \frac{F(x)}{\mathcal{P}(x)} \, \mathcal{P}(x). \tag{3.16}$$

Now, if we are able to generate random numbers that are distributed according to $\mathcal{P}(x)$, we can evaluate the integral $\mathcal{I}$ as:

$$\mathcal{I} \approx \frac{1}{N} \sum_i \frac{F(x_i)}{\mathcal{P}(x_i)}; \tag{3.17}$$

the corresponding errorbar can be estimated from:

$$s^2 = \frac{1}{N} \sum_i \left[\frac{F(x_i)}{\mathcal{P}(x_i)} \right]^2 - \left[\frac{1}{N} \sum_i \frac{F(x_i)}{\mathcal{P}(x_i)} \right]^2. \tag{3.18}$$

The crucial point is that if $\mathcal{P}(x)$ is chosen to be close enough to $F(x)$, then the statistical fluctuations are highly reduced. Indeed, s^2 is a non-negative number, its minimum $s^2 = 0$ being reached for $\mathcal{P}(x) \propto F(x)$. In this case, the Monte Carlo sampling has no fluctuations. It must be emphasized that this limiting case is trivial and the Monte Carlo is useless: if we are able to extract random numbers that are distributed with $\mathcal{P}(x) \propto F(x)$ then we would have performed the integral analytically. Instead, in the realistic situation, whenever $\mathcal{P}(x)$ resembles the function $F(x)$, the statistical fluctuations are dramatically decreased, thus reducing the number of samplings that is needed to reach a given accuracy.

In general, there are sophisticated algorithms that generate very long sequences of essentially uncorrelated and uniformly distributed random numbers (or, to be more precise, *pseudo-random numbers*, see appendix A). Therefore, if we want to produce numbers that are distributed according to a different probability function, some further work is needed. In the following, we discuss some simple cases in which this task can be accomplished and the difficulties that arise in general cases.

3.5 Sampling a Discrete Distribution Probability

Here, we discuss how it is possible to sample a discrete probability distribution $P(k)$, with $k = 1, \ldots, M$ mutually exclusive events. One option is to use the so-called *acceptance-rejection* method (or more simply, just *rejection* method). This approach can be also used to sample continuous variables and is closely related to the one that we have illustrated for computing π in section 3.3. In practice, we embed the histogram of the probability $P(k)$ into a large rectangular board and then we shoot bullets in it (assuming that these are uniformly distributed in the board), see Fig. 3.2. Let us denote by $P_{\max}$ the maximum value of the $P(k)$'s; then, we generate two uniformly distributed random numbers: a first one, r_1 in $[0, 1)$, to obtain the event $k_r = \text{int}(M \times r_1) + 1$ (where $\text{int}(x)$ indicates the integer part of x),

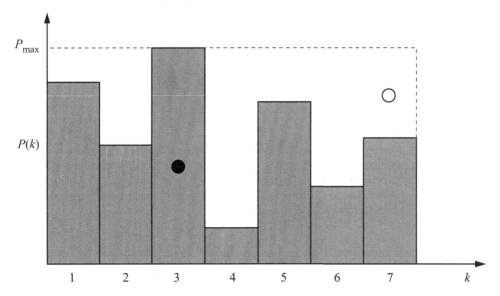

Figure 3.2 Rejection method to sample a discrete probability $P(k)$, with $k = 1, \ldots, M$ possible outcomes (here $M = 7$). Two uniformly distributed random numbers are generated: r_1 in $[0, 1)$ to obtain the event $k_r = \text{int}(M \times r_1) + 1$ and r_2 in $[0, P_{max})$. If $r_2 \leq P(k_r)$ the trial is successful and k_r is taken (the full dot selects $k_r = 3$), otherwise the trial is rejected (the empty dot indicates the rejected $k_r = 4$ event) and another couple of random numbers is generated.

and a second one, r_2 in $[0, P_{max})$. If $r_2 \leq P(k_r)$ then the trial is successful and k_r is taken as the output, otherwise the trial is rejected and another couple of random numbers is generated. It is clear that the probability to obtain a given event k is proportional to $P(k)$, since the events are generated uniformly, but accepted only if $r_2 \leq P(k)$.

This approach is not very efficient, since a given number of trials are rejected and, therefore, do not contribute to generate any output. Nevertheless, the computational cost of a rejected trial is not huge since it just requires the generation of two random numbers. From Fig. 3.2, it is clear that the rejection probability is proportional to $\sum_k (P_{max} - P_k) = MP_{max} - 1$, which is the area of the section of the rectangular board above the histogram. In the trivial (and useless) case where all events are equiprobable with $P_{max} = 1/M$ all trials are accepted. By contrast, when one event has a probability that is much larger than all the other ones, the number of rejected trials will be very large and the algorithm becomes very inefficient. In fact, a simpler approach is possible, without rejection. To visualize this approach, we must organize all probabilities in a single row, forming a sequence of boxes of length $P(k)$, see Fig. 3.3. Since $\sum_k P(k) = 1$, the total length of the row is equal to one. Then, just one uniformly distributed random number r is generated

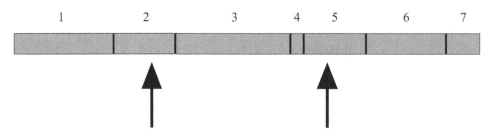

Figure 3.3 No-rejection method to sample a discrete probability $P(k)$, with $k = 1,\ldots,M$ possible outcomes (here $M = 7$). Just one uniformly distributed random number r is generated in $[0, 1)$, the event k_r is chosen, such that $\sum_{k=1,k_r-1} P(k) \leq r < \sum_{k=1,k_r} P(k)$. The two arrows indicate two attempts to select the events: the events $k_r = 2$ and 5 have been obtained.

in $[0, 1)$, which identifies a given box in the row and determines the event k_r, see Fig. 3.3. More formally, the value of k_r is the one that satisfies the following condition:

$$\sum_{k=1}^{k_r-1} P(k) \leq r < \sum_{k=1}^{k_r} P(k). \tag{3.19}$$

Clearly, also in this case, the probability to select a given k is given by $P(k)$, thus providing the correct result.

3.6 Sampling a Continuous Density Probability

Let us discuss the case of continuous random variables with probability density $\mathcal{P}(x)$. For simplicity, we consider the one-dimensional case in which there is a single random variable x. Following the discussion of section 2.4, we apply Eq. (2.41) in the case where x is uniformly distributed $[0, 1)$ (which is what random-number generators provide us) to get:

$$x = \int_{-\infty}^{y} ds \, \mathcal{P}_y(s). \tag{3.20}$$

This equation tells us that, if we want to have a random variable that is distributed according to a given probability density $\mathcal{P}_y(y)$, we must (i) perform the integral of $\mathcal{P}_y(y)$ to obtain its cumulative probability $F(y)$:

$$F(y) = \int_{-\infty}^{y} ds \, \mathcal{P}_y(s), \tag{3.21}$$

and then (ii) extract a random variable x that is uniformly distributed in $[0, 1)$ and find the value y such that $F(y) = x$; therefore, we must invert $F(y)$ and find

$y = F^{-1}(x)$. In this sense, Eq. (3.20) is nothing less than the continuous version of Eq. (3.19) that was obtained to sample a discrete distribution. For generic distributions, it is not obvious that steps (i) and (ii) can be done efficiently.

Let us consider a simple example that can be worked out analytically. Suppose that we want to generate random numbers according to the exponential distribution:

$$\mathcal{P}(y) = \begin{cases} Ae^{-Ay} & \text{for } y \geq 0, \\ 0 & \text{for } y < 0, \end{cases} \tag{3.22}$$

where A is a given constant. Then, we have the following equation:

$$x = A \int_0^y ds\, e^{-As} = 1 - e^{-Ay}, \tag{3.23}$$

which can be easily solved for y, giving:

$$y = -\frac{1}{A} \ln(1 - x). \tag{3.24}$$

In this simple example, we were able to perform both the integration and the inversion. However, this could be not achievable in a simple way for generic probabilities. Here, we show an important case that, however, can be overcome with a simple trick. Suppose that we want to generate random numbers with a Gaussian distribution (for simplicity we take $\mu = 0$ and $\sigma = 1$):

$$\mathcal{P}(y) = \frac{1}{\sqrt{2\pi}} e^{-\frac{y^2}{2}}. \tag{3.25}$$

In this case, we would solve the following equation:

$$x = \frac{1}{\sqrt{2\pi}} \int_{-\infty}^y ds\, e^{-\frac{s^2}{2}}. \tag{3.26}$$

Unfortunately, the integral does not have a simple closed form and, therefore, this approach does not give any useful outcome. Nevertheless, there is a way to overcome this problem, which goes under the name of Box-Muller trick. Let us start from noting that:

$$\left\{ \frac{1}{\sqrt{2\pi}} \int_{-\infty}^{+\infty} dx\, e^{-\frac{x^2}{2}} \right\}^2 = \frac{1}{2\pi} \int_{-\infty}^{+\infty} dx\, e^{-\frac{x^2}{2}} \int_{-\infty}^{+\infty} dy\, e^{-\frac{y^2}{2}} = 1. \tag{3.27}$$

By introducing polar coordinates:

$$x = \rho \cos\theta, \tag{3.28}$$

$$y = \rho \sin\theta, \tag{3.29}$$

we have:

$$\frac{1}{2\pi} \int_{-\infty}^{+\infty} dx\, e^{-\frac{x^2}{2}} \int_{-\infty}^{+\infty} dy\, e^{-\frac{y^2}{2}} = \int_{0}^{2\pi} \frac{d\theta}{2\pi} \int_{0}^{\infty} d\rho\, \rho\, e^{-\frac{\rho^2}{2}}. \tag{3.30}$$

Then, by considering the new variable $\xi = \rho^2/2$, we finally have:

$$\frac{1}{2\pi} \int_{-\infty}^{+\infty} dx\, e^{-\frac{x^2}{2}} \int_{-\infty}^{+\infty} dy\, e^{-\frac{y^2}{2}} = \int_{0}^{2\pi} \frac{d\theta}{2\pi} \int_{0}^{\infty} d\xi\, e^{-\xi}. \tag{3.31}$$

The meaning of this equation is that two independent variables x and y with a Gaussian distribution are equivalent to other two independent variables θ and ξ, the former one being uniformly distributed in $[0, 2\pi)$ and the latter one being exponentially distributed, see Eq. (3.24) with $A = 1$. Therefore, two Gaussian variables can be easily obtained from extracting θ and ξ:

$$\theta = 2\pi r_1, \tag{3.32}$$

$$\xi = -\ln(1 - r_2), \tag{3.33}$$

where r_1 and r_2 are two random variables uniformly distributed in $[0, 1)$. Finally, given $\rho = \sqrt{2\xi}$ we get:

$$x = \cos(2\pi r_1) \times \sqrt{-2\,\ln(1 - r_2)}, \tag{3.34}$$

$$y = \sin(2\pi r_1) \times \sqrt{-2\,\ln(1 - r_2)}, \tag{3.35}$$

In this way, it is easily possible to get Gaussian variables out of uniformly distributed numbers.

As mentioned before, the steps that must be performed in order to obtain random numbers that are distributed according to a generic $\mathcal{P}(x)$ are not always easy to be done. This is particularly true for multi-dimensional cases. Performing a direct sampling is often unfeasible both in the discrete and continuous cases. Indeed, in the former case, the total number of events cannot be very large, since they must be kept in the computer memory; instead, in the latter one, we generically face to the serious limitations that we have discussed, especially in the multi-dimensional case. Therefore, a different strategy must be considered.

3.7 Markov Chains

In the following, for simplicity of notations, we will consider the case of a single random variable x that assumes a discrete set of values, the generalization to continuous systems being straightforward. For example, $\{x\}$ may define the discrete Hilbert space of a many-body system on a finite lattice: in this case, the total number of possible configurations $\{x\}$ may easily overwhelm the computer memory, such that a direct sampling is not possible.

The idea to sample a generic probability distribution is to construct a non-deterministic, i.e., random, process for which a configuration x_n evolves as a function of a discrete iteration time n according to a stochastic dynamics:

$$x_{n+1} = F_n(x_1, \ldots, x_n, \xi_n), \tag{3.36}$$

where F_n is a function that may depend upon all the previous configurations up to n. The stochastic nature of the dynamics (3.36) is due to the fact that F_n also depends upon a random variable ξ_n that is distributed according to a probability density $\chi(\xi_n)$. At variance with the deterministic dynamics generated by the classical equations of motions (i.e., Newton's equations), in a stochastic dynamics, the concept of trajectory is not defined and the configurations x_n are random variables that are dynamically generated along the stochastic process and are distributed according to a probability distribution that evolves with n. Here, the main point is to define a suitable function F_n such that the configurations x_n will be distributed (for large enough time n) according to the probability that we want to sample. In this way, we can overcome the fact that we are not able to perform a direct sampling of the probability distribution.

A particularly simple case is given by the so-called Markov chains, where the configuration at time $n + 1$ just depends upon the one at time n:

$$x_{n+1} = F(x_n, \xi_n), \tag{3.37}$$

where the function F is taken to be time independent. Although ξ_n and ξ_{n+1} are independent random variables, $x_n \equiv x$ and $x_{n+1} \equiv x'$ are not independent; the joint probability distribution of these variables can be decomposed into the product of the marginal and the conditional probability, see section 2.2:

$$\mathcal{P}_{\text{joint},n}(x', x) = \omega(x'|x)\,\mathcal{P}_n(x). \tag{3.38}$$

Here, the conditional probability is such that $\omega(x'|x) \geq 0$ for all x and x' and satisfies the following normalization:

$$\sum_{x'} \omega(x'|x) = 1. \tag{3.39}$$

It represents the probability that, having the configuration x at the iteration n, x' appears at $n + 1$; its actual form depends upon the function $F(x, \xi)$ and the probability distribution $\chi(\xi)$.

We are now in the position of deriving the so-called *Master equation* associated to the Markov chain. Indeed, the marginal probability of the variable x' is given by:

$$\mathcal{P}_{n+1}(x') = \sum_{x} \mathcal{P}_{\text{joint},n}(x', x), \tag{3.40}$$

so that, by using Eq. (3.38), we get:

$$\mathcal{P}_{n+1}(x') = \sum_x \omega(x'|x)\,\mathcal{P}_n(x). \tag{3.41}$$

This equation allows us to calculate the evolution of the marginal probability $\mathcal{P}_n(x)$ as a function of n, since the conditional probability $\omega(x'|x)$ is determined by the stochastic dynamics in Eq. (3.37) and does not depend upon n. More precisely, although the actual value of the random variable x is not known deterministically, the probability distribution of x is instead known at each iteration n, once an initial condition is given, i.e., $\mathcal{P}_0(x)$. The solution for $\mathcal{P}_n(x)$ is obtained iteratively by solving the Master equation, starting from the given initial condition up to the desired value of n. It is simple to simulate a Markov chain on a computer, by generating random numbers for ξ at each iteration n, and this is the reason why Markov chains are particularly important for Monte Carlo calculations.

Before discussing in detail the properties of Markov chains and how they can be used to sample a given distribution function, we would like to show how this approach can be applied to compute π, similarly to what we have shown for the direct sampling before. In this case, instead of randomly shooting bullets in the square and counting the "hits" inside the circle (see Fig. 3.1), we imagine to perform a random walk in the square and lay down a bullet on each position that we visit, see Fig. 3.4. Starting from a random place inside the square, we move on at discrete times n: for example, the new position at time $n + 1$ can be chosen randomly in a small square of side δ, centered around the position at time n (see Fig. 3.4):

$$x^{\text{new}} = x^{\text{old}} + \xi_x, \tag{3.42}$$
$$y^{\text{new}} = y^{\text{old}} + \xi_y, \tag{3.43}$$

where ξ_x and ξ_y are two independent random numbers that are uniformly distributed in $[-\delta/2, \delta/2]$. At the end of a long random walk, the ratio between the number of bullets inside the circle and the total number of them will approach $\pi/4$. In this approach, it is clear that the new position $(x^{\text{new}}, y^{\text{new}})$ is correlated to the old one $(x^{\text{old}}, y^{\text{old}})$, since the new coordinates cannot be farer than $\delta/2$ from the old ones. In particular, the correlation increases when δ becomes smaller and smaller. By contrast, in the direct sampling of Fig. 3.1 the position of each bullet was independent from the other ones. Nevertheless, also here, for a long enough random walk, we visit uniformly all the area of the square and the number of bullets inside the circle over the total number of bullets must be proportional to the ratio between the area of the circle and the one of the square, i.e., $\pi/4$.

There is a small caveat in this approach: what should we do if our step brings us outside the square? Surprisingly, the correct answer, as it will be clear when we will discuss the Metropolis algorithm, is that the trial must be rejected

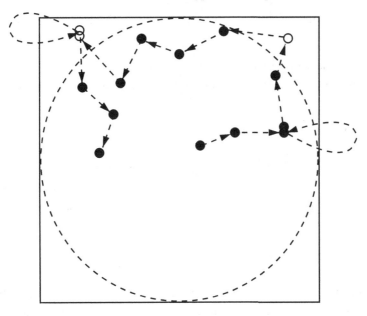

Figure 3.4 Markov chain sampling to compute π. We perform a random walk and drop off bullets behind us. The new position at time $n + 1$ is chosen randomly in a small square of side δ centered around the position at time n. For a large number N of steps the ratio between the number of bullets inside the circle and the total number of trials converges to $\pi/4$; see text. Notice that in two cases (third and eleventh steps) would have brought the bullet outside the big square; in this case, the steps is not done and a second bullet is put on top of the previous one.

(we do not move) and a bullet must be put on top of the previous one, indicating that we occupied the same position for two consecutive times. Then we try another move, until we succeed in moving inside the square.

3.8 Detailed Balance and Approach to Equilibrium

At this point, the natural and important question about the Markov process is to understand under which conditions the sequence of distributions $\mathcal{P}_n(x)$ converges to some limiting (i.e., equilibrium) distribution $\mathcal{P}_{\text{eq}}(x)$ or not. In the following, we will assume that $\mathcal{P}_{\text{eq}}(x) > 0$ for all the configurations x. Indeed, the configurations for which $\mathcal{P}_{\text{eq}}(x) = 0$ do not contribute to the final result and can be effectively discarded. For example, in classical statistical mechanics, $\mathcal{P}_{\text{eq}}(x) \propto \exp[-\beta V(x)]$ (where β is the inverse of the temperature and $V(x)$ is the potential energy) and $\mathcal{P}_{\text{eq}}(x) \neq 0$ for all configurations with $V(x) < \infty$; those configurations with $V(x) = \infty$ do not contribute to the partition function. The general theory of Markov chains and their approach to equilibrium is a vast field of research

(Norris, 1997), which we do not want to discuss here. The questions that we want to address now are:

1. Does a stationary distribution $\mathcal{P}_{eq}(x)$ exist?
2. Is the convergence to $\mathcal{P}_{eq}(x)$ guaranteed when starting from a given *arbitrary* $\mathcal{P}_0(x)$?

The first question requires that:

$$\mathcal{P}_{eq}(x') = \sum_x \omega(x'|x)\, \mathcal{P}_{eq}(x). \tag{3.44}$$

In order to satisfy this stationarity requirement, it is sufficient (but not necessary) to satisfy the so-called *detailed balance* condition:

$$\omega(x'|x)\, \mathcal{P}_{eq}(x) = \omega(x|x')\, \mathcal{P}_{eq}(x'). \tag{3.45}$$

This relationship indicates that the number of processes undergoing a transition $x \rightarrow x'$ has to be exactly compensated, to maintain a stable stationary condition, by the same amount of reverse processes $x' \rightarrow x$. It is very simple to show that the detailed balance condition allows a stationary solution of the Master equation. Indeed, if for some n we have that $\mathcal{P}_n(x) = \mathcal{P}_{eq}(x)$, then:

$$\mathcal{P}_{n+1}(x') = \sum_x \omega(x'|x)\, \mathcal{P}_{eq}(x) = \mathcal{P}_{eq}(x') \sum_x \omega(x|x') = \mathcal{P}_{eq}(x'), \tag{3.46}$$

where we used the detailed balance condition of Eq. (3.45) and the normalization condition for the conditional probability (3.39). Therefore, we have shown that the Master equation (3.41) admits stationary solutions, namely solutions that do not depend upon the (discrete) time n.

Now, we would like to understand under which conditions the equilibrium solution is unique and when a generic initial condition $\mathcal{P}_0(x)$ converges to it. First of all, we would like to define the concept of *periodicity* in the Markov chains. A state x has a period k if any return to it occurs in multiples of k steps; if $k = 1$ the state is said to be *aperiodic*: in this case returns to x occurs at irregular time steps. A Markov chain is aperiodic if all states are aperiodic. In Fig. 3.5, we show two examples of periodic and aperiodic Markov chains. The second concept is *reducibility*. A state x is *accessible* from another one x' if there is a non-zero probability to visit x starting the Markov chain from x'. Notice that it is not required to have a finite transition probability that directly couples the two states but only that they are connected by a sequence of elementary steps. A Markov chain is *irreducible* if any state is accessible from any other one (in other words, whenever it is possible to reach any state starting from any other state). Finally, a state is said

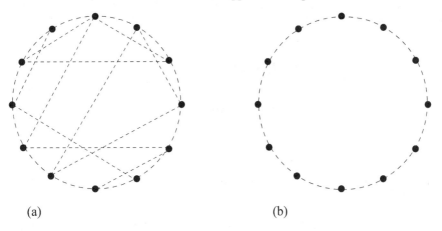

(a)   (b)

Figure 3.5 Aperiodic (a) and periodic with $k = 2$ (b) examples of Markov chains. Dashed lines indicate allowed transitions governed by the conditional probability $\omega(x'|x)$. In (b), only nearest-neighbor transitions exist; therefore, starting from any site, it is possible to come back to it only after an even number of steps (i.e., $k = 2$).

to be *positive recurrent* if the return time is finite (in other words, if it is possible to come back to it in a finite number of steps). A Markov chain is *ergodic* if it is aperiodic and all states are positive recurrent.

Let us now address the second question. First of all, the transition probability $\omega(x'|x)$ can be seen as a *non-symmetric* matrix. However, in presence of the detailed balance condition of Eq. (3.45), $\omega(x'|x)$ can be rewritten in terms of a *symmetric* matrix $\mathbf{H}$, through a similarity transformation:

$$H_{x,x'} = H_{x',x} = \omega\left(x'|x\right) \frac{\Upsilon_0(x)}{\Upsilon_0(x')}, \tag{3.47}$$

where $H_{x,x'} \geq 0$ and $\Upsilon_0(x) = \sqrt{\mathcal{P}_{eq}(x)}$ defines a vector, whose components are strictly positive for all configurations x (see at the beginning of section 3.8).

The first observation is that $\Upsilon_0(x)$ is an eigenstate of $\mathbf{H}$ with eigenvalue $\lambda_0 = 1$. Indeed, from the previous definition of Eq. (3.47), we have that:

$$\sum_{x'} H_{x,x'} \Upsilon_0(x') = \Upsilon_0(x) \sum_{x'} \omega(x'|x) = \Upsilon_0(x), \tag{3.48}$$

where we used the normalization of the conditional probability.

Now, we will prove that there are no other eigenvalues of $\mathbf{H}$ that, in absolute value, are larger than λ_0, namely $|\lambda_\alpha| \leq 1$. In order to prove this statement, we consider the square of $\mathbf{H}$ that obviously has positive eigenvalues λ_α^2. This is necessary in order to exclude the existence of an eigenvalue equal to -1, as in the case reported in Fig. 3.5(b) that is periodic with period $k = 2$. Then, we take the eigenvector $\Psi(x)$ with the largest eigenvalue λ_Ψ^2:

$$\sum_{x'} (H^2)_{x,x'} \Psi(x') = \lambda_\Psi^2 \Psi(x), \tag{3.49}$$

which implies (assuming that $\Psi(x)$ is normalized, i.e., $\sum_x \Psi^2(x) = 1$):

$$\sum_{x,x'} \Psi(x)(H^2)_{x,x'} \Psi(x') = \lambda_\Psi^2. \tag{3.50}$$

By taking the absolute values of both sides, we obtain:

$$\lambda_\Psi^2 = \left| \sum_{x,x'} \Psi(x)(H^2)_{x,x'} \Psi(x') \right| \leq \sum_{x,x'} |\Psi(x)|(H^2)_{x,x'} |\Psi(x')| . \tag{3.51}$$

For the Min-Max property of a Hermitian matrix, it follows that also $|\Psi(x)|$ is an eigenstate of $\mathbf{H}^2$ with eigenvalue equal to λ_Ψ^2. However, the matrix $\mathbf{H}^2$ is symmetric and, therefore, eigenvectors corresponding to different eigenvalues must be orthogonal. Since from Eq. (3.48) we have that $\Upsilon_0(x) = \sqrt{\mathcal{P}_{eq}(x)}$ is an eigenvector of $\mathbf{H}^2$, with eigenvalue $\lambda_0^2 = 1$, then $|\Psi(x)|$ cannot be orthogonal to it. Therefore, we arrive to the conclusion that $\lambda_\Psi^2 = \lambda_0^2 = 1$.

Nevertheless, it is possible that the eigenvalue $\lambda_0^2 = 1$ is not unique, since for degenerate eigenvalues the eigenvectors are not forced to be orthogonal. However, the possibility to have a degenerate eigenvalue is ruled out by imposing the requirement that $\mathbf{H}^2$ is irreducible. Indeed, let us suppose that another eigenstate $\Upsilon_0'(x)$ of $\mathbf{H}^2$ has eigenvalue $\lambda_0^2 = 1$. Then, for any constant α, $\Upsilon_0(x) + \alpha \Upsilon_0'(x)$ is also an eigenstate with the same eigenvalue. Moreover, from the previous discussion, also $\Phi(x) = |\Upsilon_0(x) + \alpha \Upsilon_0'(x)|$ is an eigenstate, and the constant α can be chosen to have $\Phi(\bar{x}) = 0$ for a particular configuration $x \equiv \bar{x}$. Then, since $\Phi(x)$ is an eigenstate of $\mathbf{H}^2$, we have that:

$$\sum_{x} (H^2)_{\bar{x},x} \Phi(x) = \lambda_0^2 \Phi(\bar{x}) = 0, \tag{3.52}$$

which directly implies that $\Phi(x) = 0$ for all configurations connected to $\bar{x}$ by $(H^2)_{\bar{x},x}$, since $\Phi(x)$ is non-negative and $(H^2)_{\bar{x},x}$ is strictly positive. By applying iteratively the previous condition to the new configurations connected with $\bar{x}$, we can show that all configurations that are generated have $\Phi(x) = 0$. Irreducibility implies that *all* configurations will be reached by this procedure and $\Phi(x) = 0$ is verified in the whole space of configurations. Therefore, $\Upsilon_0'(x)$ is just proportional to $\Upsilon_0(x)$ and is not a different eigenvector. This fact implies that the maximum eigenvalue $\lambda_0^2 = 1$ is non-degenerate and, in particular, $\lambda_0 = -1$ does not exist and $\lambda_0 = 1$ is unique.

We would like to mention that this result is related to the Perron-Frobenius theorem (Meyer, 2000), which applies to non-negative and ergodic matrices that, however, are not necessarily symmetric. In this case, it is possible to show that the maximum eigenvalue λ_0 is real, positive, and unique (i.e., all the other eigenvalues λ_i, which can be complex, are such that $|\lambda_i| < \lambda_0$). Moreover, all the components of the left eigenvector corresponding to λ_0 are positive.

Going back to the symmetric case, the previous results imply that any initial $P_0(x)$ will converge toward the stationary distribution $P_{eq}(x) = \Upsilon_0^2(x)$. Indeed, by using Eq. (3.47), the Master equation (3.41) is written as:

$$P_n(x) = \sum_{x'} H_{x,x'} \frac{\Upsilon_0(x)}{\Upsilon_0(x')} P_{n-1}(x'); \tag{3.53}$$

then, by iterating this procedure, i.e., by expressing the marginal probability at every step in terms of the previous one, we obtain a relation between $P_n(x)$ and the initial probability at step $n = 0$:

$$P_n(x) = \sum_{x'} (H^n)_{x,x'} \frac{\Upsilon_0(x)}{\Upsilon_0(x')} P_0(x'), \tag{3.54}$$

here the n-th power of the matrix $\mathbf{H}$ can be expanded in terms of its eigenvectors:

$$(H^n)_{x,x'} = \sum_{\alpha} \lambda_\alpha^n \Upsilon_\alpha(x') \Upsilon_\alpha(x). \tag{3.55}$$

By replacing this expansion in Eq. (3.54) we obtain:

$$P_n(x) = \Upsilon_0(x) \sum_{\alpha} \lambda_\alpha^n \Upsilon_\alpha(x) \left[\sum_{x'} \frac{\Upsilon_\alpha(x')}{\Upsilon_0(x')} P_0(x') \right]$$

$$= \Upsilon_0^2(x) \left[\sum_{x'} P_0(x') \right] + \Upsilon_0(x) \sum_{\alpha>0} \lambda_\alpha^n \Upsilon_\alpha(x) \left[\sum_{x'} \frac{\Upsilon_\alpha(x')}{\Upsilon_0(x')} P_0(x') \right]. \tag{3.56}$$

For large n all terms with $\alpha > 0$ decay exponentially, as $|\lambda_\alpha| < 1$ for $\alpha > 0$ and, therefore, only the first one survives in the above summation. Given the fact that the initial probability distribution is normalized, i.e., $\sum_{x'} P_0(x') = 1$, we finally get that the probability distribution converges to the desired one:

$$\lim_{n \to \infty} P_n(x) = \Upsilon_0^2(x) = P_{eq}(x). \tag{3.57}$$

In practical implementations, we can assume that, after a *thermalization time* n_{therm}, the probability distribution $P_n(x)$ is essentially converged to the equilibrium one $P_{eq}(x)$, so that the configurations x_n (with $n > n_{\text{therm}}$) can be used to evaluate the quantity of our interest. In most of the cases, however, subsequent configurations are not independent from each other and a finite number of steps is needed to reduce the degree of correlation among them. The *correlation time* is the time that is necessary to have (essentially) independent configurations. Correlation and thermalization times coincide, being directly related to the spectrum of the transition probability (3.47). Indeed, from Eq. (3.56), which gives the approach to equilibrium of the probability distribution, it is evident that the largest eigenvalue (smaller than 1) determines the number of Markov steps that are necessary to loose memory of a given state and obtain an (essentially) independent one.

3.9 Metropolis Algorithm

In this section, we present a practical way of constructing a conditional probability $\omega(x'|x)$ that satisfies the detailed balance condition (3.45), such that, for large values of n, the configurations x_n are distributed according to a given probability distribution $\mathcal{P}_{eq}(x)$. Metropolis and collaborators (Metropolis et al., 1957) introduced a very simple scheme, which is also very general and can be applied to many different cases. Later, the so-called Metropolis algorithm has been extended to more general cases by W. Keith Hastings (1970) (very often, the name of "Metropolis-Hastings algorithm" is also used), As a first step, we split the transition probability $\omega(x'|x)$ into two pieces:

$$\omega(x'|x) = T(x'|x)A(x'|x), \tag{3.58}$$

where $T(x'|x)$ defines a *trial probability* that proposes the new configuration x' from the present one x and $A(x'|x)$ is the *acceptance probability*. In the original work by Metropolis and co-workers, the trial probability has been taken symmetric, i.e., $T(x'|x) = T(x|x')$. However, in the generalized version of the algorithm $T(x'|x)$ can be chosen with large freedom, as long as ergodicity is ensured. Then, in order to define a Markov process that satisfies the detailed balance condition, the proposed configuration x' is accepted with a probability:

$$A(x'|x) = \text{Min} \left\{ 1, \frac{\mathcal{P}_{eq}(x')T(x|x')}{\mathcal{P}_{eq}(x)T(x'|x)} \right\}. \tag{3.59}$$

Without loss of generality, we can always choose $T(x|x) = 0$, namely we never propose to remain with the same configuration. Nevertheless, $\omega(x|x)$ can be finite, since the proposed move can be rejected. The actual value of $\omega(x|x)$ is fixed by the normalization condition $\sum_{x'} \omega(x'|x) = 1$.

In most cases (as in the original work by Metropolis and collaborators), it is useful to consider symmetric trial probabilities $T(x'|x) = T(x|x')$. In this case, the acceptance probability simplifies into:

$$A(x'|x) = \text{Min} \left\{ 1, \frac{\mathcal{P}_{eq}(x')}{\mathcal{P}_{eq}(x)} \right\}. \tag{3.60}$$

The proof that detailed balance is satisfied by considering the acceptance probability of Eq. (3.59) is very simple. Indeed, let us consider the case in which x and $x' \neq x$ are such that $\mathcal{P}_{eq}(x')T(x|x')/[\mathcal{P}_{eq}(x)T(x'|x)] > 1$. In this case, we have that:

$$A(x'|x) = 1, \tag{3.61}$$

$$A(x|x') = \frac{\mathcal{P}_{eq}(x)T(x'|x)}{\mathcal{P}_{eq}(x')T(x|x')}; \tag{3.62}$$

then, we can directly verify that the detailed balance is satisfied:

$$T(x'|x)A(x'|x)\mathcal{P}_{eq}(x) = T(x|x')A(x|x')\mathcal{P}_{eq}(x'). \tag{3.63}$$

A similar proof can be obtained in the opposite case where x and x' are such that $\mathcal{P}_{eq}(x')T(x|x')/[\mathcal{P}_{eq}(x)T(x'|x)] < 1$.

Summarizing, if x_n is the configuration at time n, the Markov chain iteration is defined in two steps:

1. Propose a move by generating a configuration x' according to the transition probability $T(x'|x_n)$;
2. Accept or reject the trial move. The move is *accepted* and the new configuration x_{n+1} is taken to be equal to x', if a random number η, uniformly distributed in $[0, 1)$, is such that $\eta < A(x'|x_n)$; otherwise the move is *rejected* and one keeps $x_{n+1} = x_n$.

The important simplifications introduced by the Metropolis algorithm are:

- It is enough to know the equilibrium probability distribution $\mathcal{P}_{eq}(x)$ up to a normalization constant: indeed, only the ratio $\mathcal{P}_{eq}(x')/\mathcal{P}_{eq}(x)$ is needed in calculating the acceptance rate $A(x'|x)$ in Eq. (3.59). This allows us to avoid to evaluate a computationally prohibitive normalization.
- The transition probability $T(x'|x)$ can be chosen to be very simple. For example, in a one-dimensional problem on the continuum, a new coordinate of a particle x' can be taken with the rule $x' = x + \xi$, where ξ is a random number uniformly distributed in $[-a, a]$, yielding $T(x'|x) = 1/(2a)$ for $x - a < x' < x + a$. Notice that in this case $T(x'|x) = T(x|x')$.
- Whenever the new configuration x' is very close to the old one x (e.g., for the example described in the previous point for a small enough) all the moves have a high probability to be accepted, since $\mathcal{P}_{eq}(x')/\mathcal{P}_{eq}(x) \approx 1$, and the rejection mechanism is ineffective. However, in this case the configurations that are generated along the Markov chain are highly correlated among themselves. By contrast, proposing a new configuration that is very far from the old one can be dangerous, since $\mathcal{P}_{eq}(x')/\mathcal{P}_{eq}(x)$ could be very small; nevertheless, once accepted, the new

configuration will be very weakly correlated to the previous one. A good rule of thumb to decrease the correlation time is to tune the trial probability $T(x'|x)$ (for example, by increasing a in the above example), in order to have an average acceptance rate of about 0.5, which corresponds to accepting, on average, only half of the total proposed moves. Although there is no reason that this represents the optimal choice, it usually provides a very good option.

Finally, we would like to come back to the example given in Fig. 3.4, where we have mentioned that, whenever we do a step that brings us outside the big square, we have to reject the move and put a second bullet on top of the previous one. Within the Metropolis algorithm this fact becomes clear: the trial probability $T(x'|x)$ is uniform within the small square of side δ centered around our present position; since $\mathcal{P}_{eq}(x)$ is constant inside the big square and zero outside, the trial move is always accepted if falls inside the big square and, instead, it is rejected if it goes outside. In this case, the new configuration is the same as the previous one, i.e., $x_{n+1} \equiv x_n$ so that it will enter twice in computing the average. Notice that an alternative procedure (that is equally correct) would be to consider a trial probability that does not consider moves outside the big square: in this case, when the position is close to the border, instead of having a square of side δ we must consider a small rectangle. However, in this case, we must pay attention because, in general, the trial probability will not be always symmetric and, therefore, the full acceptance probability of Eq. (3.59) must be considered.

3.10 How to Estimate Errorbars

Here, we would like to discuss in some detail how to determine *non-linear* functions of averages and estimate their errorbars in Monte Carlo simulations. In section 2.6, we have seen how to estimate a simple average, i.e., $\mu_x = \langle x \rangle$, and its errorbar, from a set of measurements $\{x_i\}$, with $i = 1, \ldots, N$, see Eqs. (2.56) and (2.58). The same method applies to any *linear* combinations of different averages; however, in some cases, we need *non-linear* functions of the averages, such as the fluctuation of a given quantity $\langle x^2 \rangle - \langle x \rangle^2$ or a combination of different moments such as $1 - \langle x^4 \rangle / 3 \langle x^2 \rangle^2$ (i.e., the so-called Binder cumulant, which is used to locate a phase transition in classical statistical mechanics, being equal to zero in a symmetric phase and non-zero in a symmetry-broken phase).

Here, we consider non-linear functions of averages of one or more variables, $f(\mu_x, \mu_y, \ldots)$. For the first example that we mentioned above, we have:

$$f\left(\mu_x, \mu_y\right) = \mu_y - \mu_x^2, \tag{3.64}$$

where $y = x^2$ and, therefore, $\mu_y = \langle x^2 \rangle$. Instead, for the second example:

$$f(\mu_y, \mu_z) = 1 - \frac{\mu_z}{3\mu_y^2}, \qquad (3.65)$$

where $y = x^2$ and $z = x^4$, leading to $\mu_y = \langle x^2 \rangle$ and $\mu_z = \langle x^4 \rangle$, respectively.

The most natural way to estimate $f(\mu_x, \mu_y, \dots)$ from a given set of data point is to take $f(\bar{x}, \bar{y}, \dots)$, where $\bar{x} = 1/N \sum_i x_i$ and $\bar{y} = 1/N \sum_i y_i$. Indeed, this is the correct procedure; however, for non-linear functions, this kind of estimate has a bias, which is order $1/N$. Certainly, for large values of N, the bias is much smaller than the statistical error, which is order $1/\sqrt{N}$, and can be neglected in the calculation. Most importantly, the calculation of the errorbar on $f(\bar{x}, \bar{y}, \dots)$ requires some specification that we are going to discuss in the following. The traditional way to evaluate the incertitude is to consider the error propagation, while much more straightforward (and efficient) ways are based upon the bootstrap or jackknife procedures (Young, 2012). Here, we start by explaining the error propagation (which is very rarely used in practice) and then discuss the bootstrap and jackknife approaches which are commonly used in most cases.

3.10.1 Error Propagation

In the following, for simplicity, we consider a function that only depends upon two expectation values, i.e., $f(\mu_x, \mu_y)$. The generalization to the most general case is straightforward. In order to find the bias and the errorbar of $f(\bar{x}, \bar{y})$, we can expand this quantity around $f(\mu_x, \mu_y)$:

$$f(\bar{x}, \bar{y}) \approx f(\mu_x, \mu_y) + (\partial_{\mu_x} f) \Delta_x + (\partial_{\mu_y} f) \Delta_y$$
$$+ \frac{1}{2} (\partial^2_{\mu_x, \mu_x} f) \Delta_x^2 + (\partial^2_{\mu_x, \mu_y} f) \Delta_x \Delta_y + \frac{1}{2} (\partial^2_{\mu_y, \mu_y} f) \Delta_y^2, \quad (3.66)$$

where $\Delta_x = (\bar{x} - \mu_x)$ and $\Delta_y = (\bar{y} - \mu_y)$. The leading contribution to the bias comes from the second-order terms, since the first-order terms in Δ_x and Δ_y average to zero when the procedure is repeated many times, i.e., $\langle \Delta_x \rangle = \langle \Delta_y \rangle = 0$. Instead, the second-order terms have, in general, finite expectation values:

$$\langle \Delta_x^2 \rangle = \langle \bar{x}^2 \rangle - \langle \bar{x} \rangle^2 = \sigma_{\bar{x}}^2 = \frac{\sigma_x^2}{N}, \qquad (3.67)$$

$$\langle \Delta_y^2 \rangle = \langle \bar{y}^2 \rangle - \langle \bar{y} \rangle^2 = \sigma_{\bar{y}}^2 = \frac{\sigma_y^2}{N}, \qquad (3.68)$$

$$\langle \Delta_x \Delta_y \rangle = \langle \overline{xy} \rangle - \langle \bar{x} \rangle \langle \bar{y} \rangle = \sigma_{\overline{xy}}^2 = \frac{\sigma_{xy}^2}{N}, \qquad (3.69)$$

where we used Eq. (2.58) and $\sigma_x^2 = \langle x^2 \rangle - \langle x \rangle^2$, $\sigma_y^2 = \langle y^2 \rangle - \langle y \rangle^2$, and $\sigma_{xy}^2 = \langle xy \rangle - \langle x \rangle \langle y \rangle$. Therefore, we have that:

$$\langle f(\bar{x}, \bar{y}) \rangle - f(\mu_x, \mu_y) \approx \frac{1}{2N} \left[\left(\partial^2_{\mu_x, \mu_x} f \right) \sigma_x^2 + 2 \left(\partial^2_{\mu_x, \mu_y} f \right) \sigma_{xy}^2 + \left(\partial^2_{\mu_y, \mu_y} f \right) \sigma_y^2 \right],$$

$$(3.70)$$

which demonstrates that the difference between the exact value $f(\mu_x, \mu_y)$ and its estimation given by $f(\bar{x}, \bar{y})$ is $O(1/N)$. Notice that, whenever the function is linear in the expectation values, the second derivatives vanish and, therefore, there is no bias. Usually, we do not care about this bias, since it is much smaller than the statistical error, which is proportional to $1/\sqrt{N}$. In fact, the leading contribution to the errorbar associated to $f(\bar{x}, \bar{y})$ can be easily computed by considering the expansion (3.66). Then, the variance is given by:

$$\sigma_f^2 = \langle f^2(\bar{x}, \bar{y}) \rangle - \langle f(\bar{x}, \bar{y}) \rangle^2$$
$$= (\partial_{\mu_x} f)^2 \langle \Delta_x^2 \rangle + 2(\partial_{\mu_x} f)(\partial_{\mu_y} f) \langle \Delta_x \Delta_y \rangle + \left(\partial_{\mu_y} f \right)^2 \langle \Delta_y^2 \rangle$$
$$= \frac{1}{N} \left[\left(\partial_{\mu_x} f \right)^2 \sigma_x^2 + 2(\partial_{\mu_x} f) \left(\partial_{\mu_y} f \right) \sigma_{xy}^2 + \left(\partial_{\mu_y} f \right)^2 \sigma_y^2 \right], \qquad (3.71)$$

where all second-order terms in Eq. (3.66) cancel when considering the difference between $\langle f^2(\bar{x}, \bar{y}) \rangle$ and $\langle f(\bar{x}, \bar{y}) \rangle^2$. As usual, σ_f^2 can be computed by substituting σ_x^2, σ_{xy}^2, and σ_y^2 with their estimations obtained from s_x, s_{xy}, and s_y, see Eq. (2.63). The main drawback of this approach is that it requires the exact calculation of all partial derivatives, besides keeping track of all the variances and covariances. In the following, we present two simple techniques that do not require any analytical calculation.

3.10.2 The Bootstrap Method

Within the bootstrap approach, we generate N_{boot} data sets each containing *exactly* N points just by selecting randomly the points from the original data set, which is denoted by $\{x_i\}$, with $i = 1, \ldots, N$. Along this procedure, the probability that a data point is selected is $1/N$ and, therefore, on average it will appear once in each data set. However, in the new sample, each point of the original data set can appear more than once (or it may not appear at all).

Let us denote by $n_{i,\alpha}$ the number of times that x_i appears in the bootstrap α, with $\alpha = 1, \ldots, N_{\text{boot}}$. Since each bootstrap data set contains N data points, we have the following constraint:

$$\sum_{i=1}^{N} n_{i,\alpha} = N. \qquad (3.72)$$

Whenever the number of data sets N_{boot} is large enough, it reproduces the correct averages; in particular, we denote:

$$[n_i]_{\text{boot}} = \frac{1}{N_{\text{boot}}} \sum_{\alpha=1}^{N_{\text{boot}}} n_{i,\alpha}, \tag{3.73}$$

$$[n_i^2]_{\text{boot}} = \frac{1}{N_{\text{boot}}} \sum_{\alpha=1}^{N_{\text{boot}}} n_{i,\alpha}^2. \tag{3.74}$$

Since the probability that x_i occurs $n_{i,\alpha}$ times in the bootstrap α is given by the binomial probability:

$$P(n_{i,\alpha}) = \frac{N!}{n_{i,\alpha}! \, (N - n_{i,\alpha})!} p^{n_{i,\alpha}} (1 - p)^{N - n_{i,\alpha}}, \tag{3.75}$$

where $p = 1/N$, the mean and variance are given by:

$$[n_i]_{\text{boot}} = Np = 1, \tag{3.76}$$

$$[n_i^2]_{\text{boot}} - [n_i]_{\text{boot}}^2 = Np(1 - p) = 1 - \frac{1}{N}. \tag{3.77}$$

Notice that, because of the presence of the constraint of Eq. (3.72), the values of $n_{i,\alpha}$ and $n_{j,\alpha}$ for $i \neq j$ in the same bootstrap data set are not independent, but they have a (small) correlation (that goes to zero as $N \to \infty$). Indeed, by squaring Eq. (3.72) and averaging over N_{boot}, we obtain:

$$\frac{1}{N^2} \sum_{i,j} [n_i n_j]_{\text{boot}} = 1, \tag{3.78}$$

which can be rewritten by using the fact that $[n_i]_{\text{boot}} = 1$ as:

$$\frac{1}{N^2} \sum_{i,j} \left([n_i n_j]_{\text{boot}} - [n_i]_{\text{boot}} [n_j]_{\text{boot}} \right) = 0. \tag{3.79}$$

By splitting the terms with $i = j$ from the others (that give all equal contributions) and using Eq. (3.77), we obtain:

$$\frac{1}{N} \left(1 - \frac{1}{N} \right) + \left(\frac{N-1}{N} \right) \left([n_i n_j]_{\text{boot}} - [n_i]_{\text{boot}} [n_j]_{\text{boot}} \right) = 0, \tag{3.80}$$

which finally leads to:

$$[n_i n_j]_{\text{boot}} - [n_i]_{\text{boot}} [n_j]_{\text{boot}} = -\frac{1}{N}. \tag{3.81}$$

We are now in the position to be able to compute the averages of different quantities. In particular, let us start with the simple case of $\mu = \langle x \rangle$. The average for a given bootstrap data set is given by:

$$x_\alpha^B = \frac{1}{N} \sum_{i=1}^{N} n_{i,\alpha} x_i. \tag{3.82}$$

The final bootstrap estimate of μ is then given by:

$$\overline{x^B} = \frac{1}{N_{\text{boot}}} \sum_{\alpha=1}^{N_{\text{boot}}} x_\alpha^B = \frac{1}{N} \sum_{i=1}^{N} [n_i]_{\text{boot}} \, x_i = \frac{1}{N} \sum_{i=1}^{N} x_i = \bar{x}, \tag{3.83}$$

where we have used Eq. (3.76). Therefore, the average over the bootstrap data sets gives exactly the average of the original data set $\{x_i\}$. Then, to compute the variance, we notice that:

$$\overline{(x^B)^2} = \frac{1}{N_{\text{boot}}} \sum_{\alpha=1}^{N_{\text{boot}}} \left(x_\alpha^B \right)^2 = \frac{1}{N^2} \sum_{i,j} [n_i n_j]_{\text{boot}} \, x_i x_j. \tag{3.84}$$

By using Eqs. (3.77) and (3.81), we get :

$$s_{x^B}^2 = \overline{(x^B)^2} - \left(\overline{x^B} \right)^2 = \frac{1}{N^2} \left(1 - \frac{1}{N} \right) \sum_{i=1}^{N} x_i^2 - \frac{1}{N^3} \sum_{i \neq j} x_i x_j = \frac{s^2}{N}, \tag{3.85}$$

where s^2 is the variance defined in Eq. (2.61). Therefore, the expectation values are:

$$\langle \overline{x^B} \rangle = \langle \bar{x} \rangle = \mu, \tag{3.86}$$

$$\langle s_{x^B}^2 \rangle = \left(\frac{N-1}{N^2} \right) \sigma^2 = \left(\frac{N-1}{N} \right) \sigma_{\bar{x}}^2, \tag{3.87}$$

where we used Eqs. (2.58) and (2.63). In summary, the bootstrap estimate of the variance $\sigma_{\bar{x}}^2$ is given by $N/(N-1)s_{x^B}^2$. This procedure does not require the calculation of the partial derivatives of Eq. (3.71); therefore, it is much easier to implement than the straightforward error propagation.

Similarly, we can easily compute the bootstrap estimate of $f(\mu_x, \mu_y)$:

$$f_\alpha^B = f \left(x_\alpha^B, y_\alpha^B \right). \tag{3.88}$$

The final estimate is given by averaging the bootstrap data set:

$$\overline{f^B} = \frac{1}{N_{\text{boot}}} \sum_{\alpha=1}^{N_{\text{boot}}} f_\alpha^B, \tag{3.89}$$

while the errorbar is obtained from:

$$s_{f^B}^2 = \overline{(f^B)^2} - \left(\overline{f^B}\right)^2. \tag{3.90}$$

Indeed, it can be shown that by expanding f_α^B around $f(\mu_x, \mu_y)$, similarly to what has been done in Eq. (3.66):

$$f_\alpha^B = f\left(x_\alpha^B, y_\alpha^B\right) \approx f(\mu_x, \mu_y) + (\partial_{\mu_x} f)\Delta_{\alpha,x}^B + (\partial_{\mu_y} f)\Delta_{\alpha,y}^B, \tag{3.91}$$

where $\Delta_{\alpha,x}^B = (x_\alpha^B - \mu_x)$ and $\Delta_{\alpha,y}^B = (y_\alpha^B - \mu_y)$, and then, following the steps of Eqs. (3.83) and (3.85), we obtain that the bootstrap estimate of σ_f^2 is given by $N/(N-1)s_{f^B}^2$.

The drawback of the bootstrap method is that about 37% of the points are not selected in a single bootstrap. Indeed, the probability that a data point is not taken in a given bootstrap α is $(1-1/N)^N$, which for large N approaches $e^{-1} \approx 0.37$. Therefore, much of the information of the original data set is not used. In the following, we describe an alternative approach that does not suffer from this problem.

3.10.3 The Jackknife Method

Within the jackknife approach, we define the i-th estimate to be the average over all the data in the original data set *except* the point i. As before, we start the discussion for the simple case of $\mu = \langle x \rangle$. Here, we have:

$$x_i^J = \frac{1}{N-1} \sum_{\substack{j=1 \ (j\neq i)}}^N x_j = \frac{N}{N-1}\bar{x} - \frac{1}{N-1}x_i. \tag{3.92}$$

The final jackknife estimate of μ is given by:

$$\overline{x^J} = \frac{1}{N}\sum_{i=1}^N x_i^J = \frac{N}{N-1}\bar{x} - \frac{1}{N-1}\bar{x} = \bar{x}. \tag{3.93}$$

In order to compute the errorbar associated to it, we notice that:

$$\overline{(x^J)^2} = \frac{1}{N}\sum_{i=1}^N \left(x_i^J\right)^2 = \bar{x}^2 + \frac{1}{(N-1)^2}\left(\overline{x^2} - \bar{x}^2\right). \tag{3.94}$$

The jackknife estimate of the variance is obtained from:

$$s_{x^J}^2 = \overline{(x^J)^2} - \left(\overline{x^J}\right)^2 = \frac{1}{(N-1)^2}\left(\overline{x^2} - \bar{x}^2\right) = \frac{s^2}{(N-1)^2}. \tag{3.95}$$

The expectation value of Eq. (3.93) is obviously the mean μ, while the expectation value of Eq. (3.95) is the variance of the mean divided by $(N-1)$:

$$\langle \overline{x^J} \rangle = \langle \overline{x} \rangle = \mu, \tag{3.96}$$

$$\langle s_{x^J}^2 \rangle = \frac{\sigma^2}{N(N-1)} = \frac{\sigma_{\overline{x}}^2}{N-1}, \tag{3.97}$$

where we used again Eqs. (2.58) and (2.63). In summary, the jackknife estimate of the variance $\sigma_{\overline{x}}^2$ is given by $(N-1)s_{x^J}^2$. Notice that in the jackknife approach there is a $(N-1)$ factor that multiplies $s_{x^J}^2$, this is due to the fact that the new samples are very correlated, since they would be all equal except that each one neglects just one point.

Similarly to what has been discussed for the bootstrap approach, the jackknife technique can be used to estimate any function of expectation values. We have just to define:

$$f_i^J = f\left(x_i^J, y_i^J\right). \tag{3.98}$$

The final estimate is given by averaging the jackknife data sets:

$$\overline{f^J} = \frac{1}{N}\sum_{i=1}^{N} f_i^J, \tag{3.99}$$

while the errorbar is obtained from:

$$s_{f^J}^2 = \overline{(f^J)^2} - \left(\overline{f^J}\right)^2. \tag{3.100}$$

Again, by expanding f_i^J around $f(\mu_x, \mu_y)$:

$$f_i^J = f\left(x_i^J, y_i^J\right) \approx f(\mu_x, \mu_y) + (\partial_{\mu_x} f)\Delta_{i,x}^J + (\partial_{\mu_y} f)\Delta_{i,y}^J, \tag{3.101}$$

where $\Delta_{i,x}^J = (x_i^J - \mu_x)$ and $\Delta_{i,y}^J = (y_i^J - \mu_y)$, and then following the procedure of Eqs. (3.93) and (3.95), we obtain that the jackknife estimate of the variance σ_f^2 is $(N-1)s_{f^J}^2$.

3.11 Errorbars in Correlated Samplings

Let us now discuss the case where the original data set consists of *correlated* points, i.e., the set of $\{x_i\}$ with $i = 1, \ldots, N$ are not independent. This situation appears whenever the data set is not generated by a direct sampling but instead by a Markov process. Indeed, in this case, the subsequent points will possess some degree of correlation, since, in general, it is very hard to accept a new configuration which is completely decorrelated from the previous one. Nevertheless, also for a correlated data set, the average gives an unbiased estimation of the exact mean:

$$\langle \bar{x} \rangle = \frac{1}{N} \sum_{i=1}^{N} \langle x_i \rangle = \langle x \rangle, \tag{3.102}$$

since the average is a linear function of the data set $\{x_i\}$. By contrast, the quantity s^2 defined in Eq. (2.61) is no longer an unbiased estimator of the exact variance:

$$\langle s^2 \rangle = \frac{1}{N} \sum_{i} \langle x_i^2 \rangle - \frac{1}{N^2} \sum_{i,j} \langle x_i x_j \rangle \neq \left(\frac{N-1}{N} \right) \sigma_x^2, \tag{3.103}$$

which is due to the fact that $\langle x_i x_j \rangle \neq \langle x_i \rangle \langle x_j \rangle$. In general, the estimation of the variance using s^2 leads to underestimating the errorbars.

To overcome this problem, we can perform the so-called *binning technique* (also called *block analysis*). We divide up the data set $\{x_i\}$ derived from a long Markov chain into several (N_{bin}) segments (i.e., bins), each of length $L_{\mathrm{bin}} = N/N_{\mathrm{bin}}$. On each bin j, with $j = 1, \ldots, N_{\mathrm{bin}}$, we define the partial average:

$$x^j = \frac{1}{L_{\mathrm{bin}}} \sum_{i=(j-1)L_{\mathrm{bin}}+1}^{jL_{\mathrm{bin}}} x_i. \tag{3.104}$$

Clearly, the average over the bins is equal to the original average:

$$\overline{x^j} = \frac{1}{N_{\mathrm{bin}}} \sum_{j=1}^{N_{\mathrm{bin}}} x^j = \bar{x}. \tag{3.105}$$

However, the probability distribution of the "new" (binned) variables x^j is different from the one of the x_i's. Given the definition of Eq. (3.104), the variance of the x^j's is generally smaller than the one of the x_i's. This fact can be easily understood in the case where the original variables are already independent and N_{bin} is large: in this case, the central limit theorem, discussed in section 2.7, holds and implies that the variance of the binned variables is $1/L_{\mathrm{bin}}$ smaller than the one of the original variables. In the general case, by increasing the bin length L_{bin}, the new variables x^j will be more and more uncorrelated among each other, eventually becoming independent random variables. Indeed, after the equilibration part of the Markov process, that we assume already performed at the step $i = 1$, the average correlation function:

$$C(n-m) = \langle x_n x_m \rangle - \langle x_n \rangle \langle x_m \rangle, \tag{3.106}$$

depends only on the discrete time difference $n - m$ (since stationarity implies time-homogeneity) and approaches zero exponentially as $C(n-m) \propto e^{-|n-m|/\tau}$, where τ (in units of the discrete time-step) is the correlation time in the Markov chain.

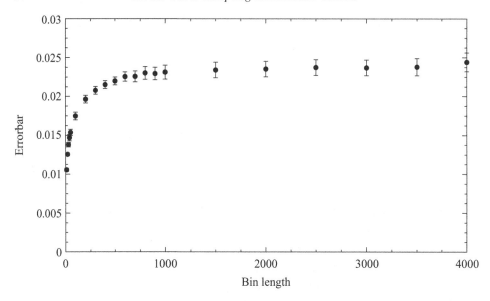

Figure 3.6 The errorbar $s_{\bar{x}}$ estimated with the binning technique, as a function of the bin length L_{bin} for a typical simulation based upon a Markov chain process.

Therefore, if we take L_{bin} to be sufficiently larger than τ, then the different bin averages x^j can be reasonably considered to be independent random variables and the variance can be easily estimated:

$$s_{\text{bin}}^2 = \frac{1}{N_{\text{bin}}} \sum_{j=1}^{N_{\text{bin}}} \left(x^j - \bar{x} \right)^2 . \tag{3.107}$$

Then, the variance of the average value (3.105) is given by:

$$s_{\bar{x}}^2 = \frac{s_{\text{bin}}^2}{N_{\text{bin}}} . \tag{3.108}$$

The errorbar on the average value is given by the square root of $s_{\bar{x}}^2$. Notice that, in the case where the original variables are already uncorrelated, the reduced (i.e., $1/L_{\text{bin}}$) variance of the binned variables is compensated with the smaller number (i.e., N_{bin}) of them, leading to the same variance of the mean values before and after the binning procedure. In Fig. 3.6, we report a typical case of correlated measures, where the errorbar $s_{\bar{x}}$ depends upon the length of the bin and shows a plateaux for sufficiently large values of L_{bin}.

4

Langevin Molecular Dynamics

4.1 Introduction

In this chapter, we describe one of the simplest, but nevertheless robust, method for simulating a given number N_p of classical particles in a finite volume Ω for a temperature T, namely for computing physical quantities in the so-called *NVT* ensemble (Allen and Tildesley, 1987; Tuckerman, 2010). This is obtained by means of the *first-order* Langevin dynamics, which is defined by the solution of a set of stochastic differential equations:

$$\frac{d\mathbf{R}(t)}{dt} = \mathbf{f}[\mathbf{R}(t)] + \boldsymbol{\eta}(t), \tag{4.1}$$

where $\mathbf{R}(t)$ is a (time dependent) D dimensional vector with components R_α ($\alpha = 1, \ldots, D$, with $D = N_p \times d$, d being the spatial dimensionality). In other words, this D-dimensional vector is a concise notation for the positions of all the particles interacting via a classical potential $V(\mathbf{R})$, which defines the deterministic "force" with components $f_\alpha(\mathbf{R})$:

$$f_\alpha(\mathbf{R}) = -\frac{\partial V(\mathbf{R})}{\partial R_\alpha}; \tag{4.2}$$

finally, $\boldsymbol{\eta}(t)$ is a random vector with components $\eta_\alpha(t)$ ($\alpha = 1, \ldots, D$) that represents a random noise with a vanishing mean value and no correlations between components $\alpha \neq \beta$ and times $t \neq t'$, i.e., it is a so-called *white noise*:

$$\langle \eta_\alpha(t) \rangle = 0, \tag{4.3}$$

$$\langle \eta_\alpha(t)\eta_\beta(t') \rangle = 2T\delta_{\alpha,\beta}\delta(t - t'), \tag{4.4}$$

where $\langle \ldots \rangle$ indicates the expectation value. We would like to emphasize that the first-order Langevin equations (4.1) are substantially different from the Newton equation of motions, which are *second-order* (deterministic) differential equations connecting the actual force to the acceleration of the particles. The dynamics that is generated by the Newton equations may be used to simulate a set of classical particles in a finite volume Ω at fixed energy E, namely within the so-called *NVE* ensemble (Allen and Tildesley, 1987; Tuckerman, 2010). From one side, the first-order Langevin equations can be seen as an *ad hoc* approach to generate configurations that are equilibrated according to the Boltzmann distribution (see below). On the other side, they can be obtained from a coarse-grained description of the Brownian motion; here, a small grain with mass m in a fluid experiences, besides an external force $\mathbf{f(R)}$, a friction force with coefficient γ and a random force, which is due to random density fluctuations in the fluid. This approach leads to the *second-order* Langevin dynamics:

$$m\frac{d\mathbf{V}(t)}{dt} = \mathbf{f}[\mathbf{R}(t)] - \gamma\mathbf{V}(t) + \boldsymbol{\eta}(t), \qquad (4.5)$$

$$\frac{d\mathbf{R}(t)}{dt} = \mathbf{V}(t). \qquad (4.6)$$

Then, the first-order Langevin equations are obtained in the limit of $m \to 0$ (in this limit, γ just sets the time scale). Usually, this limit is called *over-damped* regime of the second-order Langevin dynamics.

Contrary to ordinary differential equations, the stochastic differential equations do not provide a unique solution starting from a given initial condition, but the possible solutions acquire a statistical meaning: starting at $t = t_0$ from $\mathbf{R}_0 \equiv \mathbf{R}(t_0)$, several stochastic trajectories are possible depending on the particular realization of the noise. Thus, a meaningful quantity that is defined is the probability $\mathcal{P}(\mathbf{R}, t)$ to find a given configuration $\mathbf{R}$ at time t, with the initial condition:

$$\mathcal{P}(\mathbf{R}, t_0) = \delta(\mathbf{R} - \mathbf{R}_0). \qquad (4.7)$$

In the following, we will show that, after an equilibration time, $\mathcal{P}(\mathbf{R}, t)$ converges to an equilibrium distribution that is independent from the initial condition $\mathbf{R}_0$. In particular, the equilibrium probability is given by the Boltzmann distribution:

$$\mathcal{P}_{\mathrm{eq}}(\mathbf{R}) = \frac{1}{\mathcal{Z}} \exp\left[-\frac{V(\mathbf{R})}{T}\right], \qquad (4.8)$$

where $\mathcal{Z}$ is the partition function, needed for the normalization condition of the probability:

$$\mathcal{Z} = \int d\mathbf{R} \exp\left[-\frac{V(\mathbf{R})}{T}\right]. \qquad (4.9)$$

Then, the solution of the differential equations (4.1) can be used to sample a large number N of configurations $\mathbf{R}_n$ at discrete times t_n to represent the canonical distribution $\mathcal{P}_{\mathrm{eq}}(\mathbf{R})$, which allows us to compute any correlation function $\mathcal{O}(\mathbf{R})$:

$$\int \mathbf{dR} \; \mathcal{O}(\mathbf{R}) \; \mathcal{P}_{eq}(\mathbf{R}) \approx \frac{1}{N} \sum_n \mathcal{O}(\mathbf{R}_n) \tag{4.10}$$

similarly to the ordinary Monte Carlo method discussed in section 3.1.

For a given system, there is no general rule telling whether it is more convenient to work with molecular dynamics or Monte Carlo approaches. However, in some cases, molecular dynamics represents the best choice. For example, in several *ab-initio* methods, the classical potential $V(\mathbf{R})$ is not given by a simple form, but it is the result of complicated algorithms (e.g., when it represents the ion-ion interaction that comes from a density-functional theory in the Born-Oppenheimer approximation). Here, a similar amount of computer time is taken to compute all components of the forces or to employ a single calculation of the energy necessary for a Metropolis acceptance step. In such a situation molecular dynamics is typically more convenient than Monte Carlo, because with the same computational effort all the positions of the atoms are changed at once. Moreover, it is not necessary to employ local moves limited to a single or a few atoms to remain with a good acceptance rate. By contrast, the advantage of the Monte Carlo method is that there is no other bias than the statistical one, whereas in molecular dynamics it is always necessary to introduce a time discretization, which implies a systematic, but controllable, error (as we will discuss in the following).

4.2 Discrete-Time Langevin Dynamics

In a computer simulation, a time discretization is necessary to obtain the solution of differential equations. Therefore, in order to define an appropriate algorithm for the simulation of classical particles at finite temperature, we integrate both sides of Eq. (4.1) over a finite interval (t_n, t_{n+1}), where $t_n = t_0 + \Delta n$ are discretized times (t_0 is the initial time and Δ represents the discrete time step). In this way, we obtain (at the lowest-order approximation in Δ):

$$\mathbf{R}_{n+1} - \mathbf{R}_n = \int_{t_n}^{t_{n+1}} dt \; \{\mathbf{f}[\mathbf{R}(t)] + \boldsymbol{\eta}(t)\} = \Delta \mathbf{f}_n + \int_{t_n}^{t_{n+1}} dt \; \boldsymbol{\eta}(t), \tag{4.11}$$

where $\mathbf{R}_n = \mathbf{R}(t_n)$ and $\mathbf{f}_n = \mathbf{f}[\mathbf{R}(t_n)]$; here, we have approximated the integral of the force in this interval with the lowest-order approximation $\Delta \mathbf{f}_n$, because the force is approximately constant within the small time interval, see Eqs. (4.15) and (4.16). Notice that this is the only approximation done in the above integrations. The time-integral in the r.h.s. of Eq. (4.11) can be estimated by noticing that a sum of many random variables (and, therefore, also an integral) is a Gaussian random number (see section 2.7). In particular, by introducing a random vector $\mathbf{z}_n$, which is normally (Gaussian) distributed with zero mean and unit variance, we have:

$$\int_{t_n}^{t_{n+1}} dt \, \eta(t) = \sqrt{2T\Delta} \, \mathbf{z}_n, \tag{4.12}$$

where the coefficient in the r.h.s. gives the correct variance of the integral:

$$\langle z_{\alpha,n} z_{\beta,n} \rangle = \frac{1}{2T\Delta} \int_{t_n}^{t_{n+1}} dt \int_{t_n}^{t_{n+1}} dt' \, \langle \eta_\alpha(t) \eta_\beta(t') \rangle = \frac{\delta_{\alpha,\beta}}{\Delta} \int_{t_n}^{t_{n+1}} dt = \delta_{\alpha,\beta}, \tag{4.13}$$

where we have used Eq. (4.4). Moreover, we have that $\langle z_{\alpha,n} z_{\beta,m} \rangle = 0$ for $n \neq m$, since the integrand is always zero in such cases.

By collecting all these results, we can write down the final expression for the discretized-time Langevin equation:

$$\mathbf{R}_{n+1} = \mathbf{R}_n + \Delta \mathbf{f}_n + \sqrt{2T\Delta} \, \mathbf{z}_n, \tag{4.14}$$

which defines an iteration scheme that gives the new coordinates $\mathbf{R}_{n+1}$, at time t_{n+1}, in terms of the old ones $\mathbf{R}_n$, at time t_n, the discretized force $\mathbf{f}_n$, and a set of Gaussian variables $\mathbf{z}_n$. This iteration represents just a Markov process, which can be implemented by a simple iterative algorithm, since it is only required to have an algorithm that evaluates the force for any given positions $\mathbf{R}_n$ of the N classical particles. It is important to emphasize that, for $\Delta \to 0$, the noise (which is proportional to $\sqrt{\Delta}$) dominates over the deterministic force (which is linear in Δ). Notice that the presence of the noisy term $\eta(t)$ makes the solution of the Langevin equation non-continuous and non-differentiable, since:

$$\frac{\mathbf{R}_{n+1} - \mathbf{R}_n}{\Delta} = O\left(\frac{1}{\sqrt{\Delta}}\right), \tag{4.15}$$

which justifies the approximation in the integral of the deterministic force in Eq. (4.11):

$$\int_{t_n}^{t_{n+1}} dt \, \mathbf{f}[\mathbf{R}(t)] = \Delta \mathbf{f}_n + O(\Delta^{3/2}). \tag{4.16}$$

As a consequence of Eq. (4.15), the actual trajectory for $\Delta \to 0$ is not defined. Nevertheless, we will show that the time discretization of Eq. (4.14) allows us to determine the evolution of the probability distribution $\mathcal{P}(\mathbf{R}, t)$ with no uncertainty in the limit $\Delta \to 0$. We finally mention that, in the limit of zero temperature $T = 0$, the noisy term disappears from the equations of motion and, therefore, the algorithm becomes deterministic. In particular, it turns into the steepest descent method (Press et al., 2007), yielding the minimum (or, more generally, a local minimum) of the potential $V(\mathbf{R})$ in a deterministic way for $n \to \infty$ and Δ small enough.

4.3 From the Langevin to the Fokker-Planck Equation

Here, we derive the Master equation for the discretized Langevin dynamics of Eq. (4.14) that defines the evolution of the probability $\mathcal{P}_n(\mathbf{R})$ of the classical variables $\mathbf{R}$ for a finite time step Δ. Most importantly, the limit $\Delta \to 0$ can be performed, leading to the so-called Fokker-Planck equation that represents the Master equation of the Langevin equation (4.1).

Since the discretized Langevin dynamics defines a Markov process, the probability $\mathcal{P}_n(\mathbf{R})$ is fully determined in terms of the conditional probability $K(\mathbf{R}'|\mathbf{R})$ associated to the Markov step:

$$\mathcal{P}_{n+1}(\mathbf{R}') = \int d\mathbf{R}\, K(\mathbf{R}'|\mathbf{R})\mathcal{P}_n(\mathbf{R}). \tag{4.17}$$

The conditional probability can be determined by noticing that, in Eq. (4.14), only the variable $\mathbf{z}_n$ is stochastic, while the force is fully deterministic. Therefore, given $\mathbf{R}$, the new variable $\mathbf{R}'$ is given by the deterministic part $\mathbf{R} + \Delta \mathbf{f}(\mathbf{R})$ plus a random noise that is normally distributed with zero mean and variance equal to $2T\Delta$. Then:

$$K(\mathbf{R}'|\mathbf{R}) = \prod_\alpha \int \frac{dz_\alpha}{\sqrt{2\pi}} e^{-z_\alpha^2/2} \delta\left(R'_\alpha - R_\alpha - \Delta f_\alpha - \sqrt{2T\Delta} z_\alpha\right), \tag{4.18}$$

which is clearly normalized with $\int d\mathbf{R}' K(\mathbf{R}'|\mathbf{R}) = 1$. By replacing this form of the conditional probability in the Master equation, we obtain:

$$\mathcal{P}_{n+1}(\mathbf{R}') = \prod_\alpha \int \frac{dz_\alpha}{\sqrt{2\pi}} e^{-z_\alpha^2/2} \int dR_\alpha \delta\left(R'_\alpha - R_\alpha - \Delta f_\alpha - \sqrt{2T\Delta} z_\alpha\right) \mathcal{P}_n(\mathbf{R}). \tag{4.19}$$

Now, we carry out the integral over the R_α's. Thus, we are led to find the zeros of the argument of the δ-function in Eq. (4.19), for fixed $\mathbf{R}'$ and $\mathbf{z}$:

$$\mathbf{R}' - \mathbf{R} - \Delta \mathbf{f}(\mathbf{R}) - \sqrt{2T\Delta}\, \mathbf{z} = 0, \tag{4.20}$$

which represents a set of D non-linear equations in D unknown (the R_α's). This problem is apparently difficult to solve; however, the solution can be found by performing a systematic expansion for small Δ:

$$\mathbf{R} = \mathbf{R}' - \sqrt{2T\Delta}\, \mathbf{z} - \Delta \mathbf{f}(\mathbf{R}') + O\left(\Delta^{3/2}\right). \tag{4.21}$$

In this way, we can carry out the integration over the R_α's in the Master equation (4.19) and obtain:

$$\mathcal{P}_{n+1}(\mathbf{R}') = \prod_\alpha \int \frac{dz_\alpha}{\sqrt{2\pi}} \frac{e^{-z_\alpha^2/2}}{|1 + \Delta f'_\alpha(\mathbf{R}')|} \mathcal{P}_n(\mathbf{R}' - \Delta \mathbf{f}(\mathbf{R}') - \sqrt{2T\Delta}\, \mathbf{z}), \tag{4.22}$$

where $f'_\alpha(\mathbf{R}') = \partial f_\alpha / \partial R'_\alpha$, computed in $\mathbf{R}'$; this relation is valid up to order $O(\Delta^{3/2})$. By further expanding it to leading order in Δ, we get:

$$
\mathcal{P}_{n+1}(\mathbf{R}') = \prod_\alpha \int \frac{dz_\alpha}{\sqrt{2\pi}} e^{-z_\alpha^2/2} \left[1 - \Delta \sum_\beta \frac{\partial f_\beta(\mathbf{R}')}{\partial R'_\beta} \right]
$$
$$
\times \, \mathcal{P}_n(\mathbf{R}' - \Delta \mathbf{f}(\mathbf{R}') - \sqrt{2T\Delta}\,\mathbf{z});
\tag{4.23}
$$

finally, we can expand the probability distribution:

$$
\mathcal{P}_n(\mathbf{R}' - \Delta \mathbf{f}(\mathbf{R}') - \sqrt{2T\Delta}\mathbf{z}) \approx \mathcal{P}_n(\mathbf{R}')
$$
$$
- \sum_\beta \left[\Delta f_\beta(\mathbf{R}') + \sqrt{2T\Delta}z_\beta \right] \frac{\partial \mathcal{P}_n(\mathbf{R}')}{\partial R'_\beta} + T\Delta \sum_{\alpha,\beta} z_\alpha z_\beta \frac{\partial^2 \mathcal{P}_n(\mathbf{R}')}{\partial R'_\alpha \partial R'_\beta}.
\tag{4.24}
$$

In principle, the validity of the Taylor expansion is not justified for large values of z_α, however, this is not a problem because all the integrals in $\{z_\alpha\}$ are dominated in the region where $|z_\alpha| < 1$. By substituting the above expansion in Eq. (4.23) and carrying out the Gaussian integrations over $\{z_\alpha\}$, we get (changing $\mathbf{R}' \to \mathbf{R}$):

$$
\mathcal{P}_{n+1}(\mathbf{R}) = \mathcal{P}_n(\mathbf{R}) + \Delta \sum_\alpha \left[T\frac{\partial^2 \mathcal{P}_n(\mathbf{R})}{\partial R_\alpha^2} - \frac{\partial f_\alpha(\mathbf{R})}{\partial R_\alpha} \mathcal{P}_n(\mathbf{R}) - f_\alpha(\mathbf{R}) \frac{\partial \mathcal{P}_n(\mathbf{R})}{\partial R_\alpha} \right].
\tag{4.25}
$$

Now, the limit $\Delta \to 0$ can be obtained. Indeed, for small values of Δ:

$$
\mathcal{P}_{n+1}(\mathbf{R}) - \mathcal{P}_n(\mathbf{R}) \approx \Delta \frac{\partial \mathcal{P}(\mathbf{R},t)}{\partial t},
\tag{4.26}
$$

which brings us to the Fokker-Planck equation for the probability density $\mathcal{P}(\mathbf{R},t)$:

$$
\frac{\partial \mathcal{P}(\mathbf{R},t)}{\partial t} = T \sum_\alpha \frac{\partial^2 \mathcal{P}(\mathbf{R},t)}{\partial R_\alpha^2} - \sum_\alpha \frac{\partial}{\partial R_\alpha} \left[\mathcal{P}(\mathbf{R},t) f_\alpha(\mathbf{R}) \right].
\tag{4.27}
$$

It is easy to show that the Boltzmann distribution of Eq. (4.8) is a stationary solution of the Fokker-Planck equation; indeed, we can rewrite Eq. (4.27) as:

$$
\frac{\partial \mathcal{P}(\mathbf{R},t)}{\partial t} = \sum_\alpha \frac{\partial}{\partial R_\alpha} \left[T\frac{\partial \mathcal{P}(\mathbf{R},t)}{\partial R_\alpha} - \mathcal{P}(\mathbf{R},t) f_\alpha(\mathbf{R}) \right],
\tag{4.28}
$$

whose r.h.s. is vanishing for the Boltzmann distribution. Another important property of the Fokker-Planck equation is that the r.h.s. is a total divergence. This is just the consequence that the normalization of the probability $\int d\mathbf{R} \, \mathcal{P}(\mathbf{R}, t) = N(t) = 1$, which represents a constant of motion of the equation. Indeed, by integrating both sides of the equation over a given volume and applying the Gauss theorem for the r.h.s., we get:

$$\frac{\partial N(t)}{\partial t} = \int d\mathbf{S} \, \mathbf{A}(\mathbf{R}, t) \cdot \mathbf{n}(\mathbf{R}), \tag{4.29}$$

where

$$A_\alpha(\mathbf{R}, t) = T\frac{\partial \mathcal{P}(\mathbf{R}, t)}{\partial R_\alpha} - \mathcal{P}(\mathbf{R}, t) f_\alpha(\mathbf{R}), \tag{4.30}$$

and $\mathbf{n}(\mathbf{R})$ is the unit vector perpendicular to the surface $\mathbf{S}$. Then, $\mathbf{A}(\mathbf{R}, t)$ vanishes at infinity (since the probability vanishes at infinity), implying that $N(t)$ is independent from t.

4.4 Fokker-Planck Equation and Quantum Mechanics

Here, we show how the Fokker-Planck equation allows us to compute the probability $\mathcal{P}(\mathbf{R}, t)$ at any given time, once the initial condition $\mathcal{P}(\mathbf{R}, t_0)$ is given. Moreover, we also show that $\mathcal{P}(\mathbf{R}, t)$ approaches the equilibrium distribution for large time t in an exponential way, starting from any initial probability.

There is a deep relationship between the Fokker-Planck equation and the Schrödinger equation in imaginary time (Nelson, 1966; Parisi and Wu, 1981). This is obtained by writing the solution of Eq. (4.27) in the following way:

$$\mathcal{P}(\mathbf{R}, t) = \Upsilon_0(\mathbf{R})\Phi(\mathbf{R}, t) \tag{4.31}$$

where $\Upsilon_0(\mathbf{R}) = \sqrt{\mathcal{P}_{eq}(\mathbf{R})}$ represents a normalized quantum state:

$$\int d\mathbf{R} \, \Upsilon_0^2(\mathbf{R}) = 1. \tag{4.32}$$

By substituting the above definition of $\mathcal{P}(\mathbf{R}, t)$ into the Fokker-Planck equation (4.27), we obtain that $\Phi(\mathbf{R}, t)$ satisfies the Schrödinger equation in imaginary time:

$$-\frac{\partial \Phi(\mathbf{R}, t)}{\partial t} = \mathcal{H}_{eff}\Phi(\mathbf{R}, t), \tag{4.33}$$

where $\mathcal{H}_{eff}$ is an effective Hamiltonian given by:

$$\mathcal{H}_{eff} = -T\sum_\alpha \frac{\partial^2}{\partial R_\alpha^2} + V_{eff}(\mathbf{R}); \tag{4.34}$$

here, the inverse (classical) temperature plays the role of the mass of the particles. Moreover, $V_{eff}(\mathbf{R})$ is an effective potential that depends upon the classical potential $V(\mathbf{R})$:

$$V_{eff}(\mathbf{R}) = \frac{1}{2} \sum_\alpha \left\{ \frac{1}{2T} \left[\frac{\partial V(\mathbf{R})}{\partial R_\alpha} \right]^2 - \frac{\partial^2 V(\mathbf{R})}{\partial R_\alpha^2} \right\}. \tag{4.35}$$

Notice that, in the limit of zero temperature, the minima of the original potential $V(\mathbf{R})$ are also minima of $V_{eff}(\mathbf{R})$; indeed, the first term of the above equation (which dominates for $T \to 0$) vanishes at the minima of the original potential $V(\mathbf{R})$. The effective potential can be rewritten as:

$$V_{eff}(\mathbf{R}) = \frac{T}{\Upsilon_0(\mathbf{R})} \sum_\alpha \frac{\partial^2 \Upsilon_0(\mathbf{R})}{\partial R_\alpha^2}. \tag{4.36}$$

Now, $\Upsilon_0(\mathbf{R})$ is an eigenstate of $\mathcal{H}_{eff}$, as verified by a direct inspection of Eqs. (4.34) and (4.36), with energy $E_0 = 0$. Since it has no nodes (e.g., $\mathcal{P}_{eq}(\mathbf{R})$ is positive for all the configurations $\mathbf{R}$), it is also the actual ground state of the effective Hamiltonian. Then, the solution of the Schrödinger equation, and the corresponding Fokker-Planck equation, can be formally given in closed form by expanding the initial condition in terms of the eigenstates $\Upsilon_n(\mathbf{R})$ of $\mathcal{H}_{eff}$:

$$\mathcal{P}(\mathbf{R}, t_0) = \Upsilon_0(\mathbf{R}) \sum_n a_n \Upsilon_n(\mathbf{R}), \tag{4.37}$$

where

$$a_n = \int d\mathbf{R} \, \frac{\Upsilon_n(\mathbf{R})}{\Upsilon_0(\mathbf{R})} \mathcal{P}(\mathbf{R}, t_0), \tag{4.38}$$

which implies that $a_0 = 1$ from the normalization condition on $\mathcal{P}(\mathbf{R}, t_0)$. We thus obtain the full evolution of the probability $\mathcal{P}(\mathbf{R}, t)$ as:

$$\mathcal{P}(\mathbf{R}, t) = \Upsilon_0(\mathbf{R}) \sum_n a_n e^{-E_n t} \Upsilon_n(\mathbf{R}). \tag{4.39}$$

Therefore, for large times t, $\mathcal{P}(\mathbf{R}, t)$ converges *exponentially* to the stationary equilibrium distribution $\Upsilon_0^2(\mathbf{R}) = \mathcal{P}_{eq}(\mathbf{R})$. The characteristic time τ for equilibration is given by the inverse gap to the first excitation, i.e., $\tau = 1/E_1$.

Let us finish this section by showing that the evolution generated by the Fokker-Planck equation satisfies the detailed balance condition. The formal solution of the Schrödinger equation is given by:

$$\Phi(\mathbf{R}', t) = \int d\mathbf{R} \, \langle \mathbf{R}' | e^{-\mathcal{H}_{eff} t} | \mathbf{R} \rangle \Phi(\mathbf{R}, 0). \tag{4.40}$$

Therefore, by using Eq. (4.31), the evolution of the probability density can be written as:

$$\mathcal{P}(\mathbf{R}', t) = \int d\mathbf{R} \, K_t(\mathbf{R}'|\mathbf{R}) \mathcal{P}(\mathbf{R}, 0)$$

$$= \int d\mathbf{R} \, \langle \mathbf{R}' \left| e^{-\mathcal{H}_{\text{eff}} t} \right| \mathbf{R} \rangle \sqrt{\frac{\mathcal{P}_{\text{eq}}(\mathbf{R}')}{\mathcal{P}_{\text{eq}}(\mathbf{R})}} \mathcal{P}(\mathbf{R}, 0), \qquad (4.41)$$

which implies that the conditional probability is given by:

$$K_t(\mathbf{R}'|\mathbf{R}) = \langle \mathbf{R}' \left| e^{-\mathcal{H}_{\text{eff}} t} \right| \mathbf{R} \rangle \sqrt{\frac{\mathcal{P}_{\text{eq}}(\mathbf{R}')}{\mathcal{P}_{\text{eq}}(\mathbf{R})}}. \qquad (4.42)$$

Since $\mathcal{H}_{\text{eff}}$ is symmetric, i.e., $\langle \mathbf{R}' | e^{-\mathcal{H}_{\text{eff}} t} | \mathbf{R} \rangle = \langle \mathbf{R} | e^{-\mathcal{H}_{\text{eff}} t} | \mathbf{R}' \rangle$, we have that:

$$\frac{K_t(\mathbf{R}'|\mathbf{R})}{K_t(\mathbf{R}|\mathbf{R}')} = \frac{\mathcal{P}_{\text{eq}}(\mathbf{R}')}{\mathcal{P}_{\text{eq}}(\mathbf{R})}, \qquad (4.43)$$

which shows that the detailed balance condition is indeed satisfied. We would like to emphasize that the time discretization introduces an error that spoils the detailed balance condition. In this sense, for any finite values of the discrete time step Δ, the equilibrium distribution is not given by the Boltzmann one (4.8), but reduces to it when $\Delta \to 0$.

4.4.1 Exact Solution for the Harmonic Case

To have some insight on how time discretization affects the dynamics, let us consider a generic quadratic potential of the type:

$$V(\mathbf{R}) = \frac{1}{2} \sum_{\alpha, \beta} K_{\alpha, \beta} (R_\alpha - R_{\text{eq}, \alpha})(R_\beta - R_{\text{eq}, \beta}), \qquad (4.44)$$

where $\mathbf{R}_{\text{eq}}$ are the equilibrium positions. The force-constant matrix $\mathbf{K}$ is symmetric and, therefore, can be diagonalized by a unitary matrix $\mathbf{U}$:

$$\bar{\mathbf{K}} = \mathbf{U} \mathbf{K} \mathbf{U}^\dagger, \qquad (4.45)$$

and its eigenvalues will be denoted by $\bar{K}_\alpha$, with $\bar{K}_1 \leq \bar{K}_2 \leq \cdots \leq \bar{K}_D$, i.e., the diagonal elements of the diagonal matrix $\bar{\mathbf{K}}$. The effective potential considered in Eq. (4.35) can be explicitly given in this case and the effective Hamiltonian $\mathcal{H}_{\text{eff}}$ remains harmonic with:

$$V_{\text{eff}}(\mathbf{R}) = \frac{1}{2} \sum_{\alpha, \beta} K_{\alpha, \beta}^{\text{eff}} (R_\alpha - R_{\text{eq}, \alpha})(R_\beta - R_{\text{eq}, \beta}) - \frac{1}{2} \sum_\alpha K_{\alpha, \alpha}, \qquad (4.46)$$

where the effective harmonic coupling is:

$$K_{\alpha,\beta}^{\text{eff}} = \frac{1}{2T} \sum_\gamma K_{\alpha,\gamma} K_{\gamma,\beta} = \frac{1}{2T} \left[K^2\right]_{\alpha,\beta}. \tag{4.47}$$

Since the effective Hamiltonian is quadratic, we obtain a harmonic problem with the above force-constant matrix and the mass $m = 1/2T$ for all degrees of freedom. Thus, according to standard calculations, the eigenvalues of the dynamical matrix $\mathbf{K}$ provide the eigenmodes of the harmonic problem (up to an energy shift):

$$E(\{n_\alpha\}) = \sum_\alpha \bar{K}_\alpha \left(n_\alpha + \frac{1}{2}\right) - \frac{1}{2} \sum_\alpha \bar{K}_\alpha = \sum_\alpha \bar{K}_\alpha n_\alpha, \tag{4.48}$$

where n_α are non-negative integers (the ground state corresponding to $n_\alpha = 0$ for all values of α). Here, we have used that the constant term in Eq. (4.46) is the trace of the matrix $\mathbf{K}$, which is the sum of its eigenvalues. The correlation time in this case can be explicitly given by:

$$\tau = \frac{1}{\bar{K}_1}, \tag{4.49}$$

where $\bar{K}_1$ is the minimum eigenvalue of the harmonic potential. Remarkably the correlation time τ does not depend on the temperature T.

The discretized version of the Langevin equation (4.14) can be solved explicitly in the harmonic case, as it will be explained in the following. In fact, by multiplying both sides of Eq. (4.14) by $\mathbf{U}$ and introducing normal-mode coordinates $\mathbf{Q} = \mathbf{U}(\mathbf{R} - \mathbf{R}_{\text{eq}})$, the discretized Langevin dynamics becomes:

$$Q_{\alpha,n+1} = \left(1 - \Delta \bar{K}_\alpha\right) Q_{\alpha,n} + \sqrt{2T\Delta} y_{\alpha,n}, \tag{4.50}$$

where $\mathbf{y}_n = \mathbf{U}\mathbf{z}_n$ are normal distributed random variables satisfying:

$$\langle y_{\alpha,n} y_{\beta,n} \rangle = \delta_{\alpha,\beta}. \tag{4.51}$$

Since normal modes do not couple, we can deal with each mode separately. The Master equation corresponding to Eq. (4.50) depends on the conditional probability $K(Q'_\alpha | Q_\alpha)$ that, in this case, can be explicitly derived:

$$K(Q'_\alpha | Q_\alpha) = \frac{1}{\sqrt{4\pi T\Delta}} \exp\left\{-\frac{1}{4T\Delta}\left[Q'_\alpha - (1 - \Delta\bar{K}_\alpha)Q_\alpha\right]^2\right\}. \tag{4.52}$$

Then, we have that the probability distribution:

$$\mathcal{P}_{\text{eq},\alpha}(Q_\alpha) = \frac{1}{\mathcal{Z}} \exp\left[-\frac{K'_\alpha Q_\alpha^2}{2T}\right], \tag{4.53}$$

with

$$K'_\alpha = \left(1 - \frac{\Delta \bar{K}_\alpha}{2}\right) \bar{K}_\alpha, \qquad (4.54)$$

satisfies the detailed balance condition:

$$\frac{K(Q'_\alpha | Q_\alpha)}{K(Q_\alpha | Q'_\alpha)} = \frac{\mathcal{P}_{\text{eq},\alpha}(Q'_\alpha)}{\mathcal{P}_{\text{eq},\alpha}(Q_\alpha)}. \qquad (4.55)$$

Indeed, by using Eq. (4.52), we find:

$$\frac{K(Q'_\alpha | Q_\alpha)}{K(Q_\alpha | Q'_\alpha)} = \exp\left\{\frac{1}{4T\Delta}\left[(\Delta \bar{K}_\alpha)^2 - 2\Delta \bar{K}_\alpha\right]\left[(Q'_\alpha)^2 - Q^2_\alpha\right]\right\}, \qquad (4.56)$$

which is consistent with the distribution (4.53) with K'_α given by Eq. (4.54). Thus, collecting all modes together, the equilibrium distribution is given by:

$$\mathcal{P}_{\text{eq}}(Q_\alpha) = \frac{1}{\mathcal{Z}} \exp\left[-\frac{1}{2T}\sum_\alpha \left(1 - \frac{\Delta \bar{K}_\alpha}{2}\right) \bar{K}_\alpha Q^2_\alpha\right]. \qquad (4.57)$$

An important remark about this result is that $\mathcal{P}_{\text{eq}}(Q_\alpha)$ is only defined when:

$$1 - \frac{\Delta \bar{K}_\alpha}{2} \geq 0, \qquad (4.58)$$

for all α, otherwise the Langevin iteration will produce unbounded values of the coordinates Q_α. Therefore, we arrive to the condition that the time step Δ must satisfy the following condition:

$$\Delta < \frac{2}{\bar{K}_D}, \qquad (4.59)$$

where $\bar{K}_D$ is the maximum eigenvalue of the force-constant matrix **K**. Indeed, this represents a very general result for the Langevin molecular dynamics, which is stable only for a small enough time step.

In summary, it is important to make the following remarks:

- The error in the discretization of the Langevin equation scales correctly to zero for $\Delta \to 0$ since the exact distribution is obtained in this limit as $K'_\alpha \to \bar{K}_\alpha$ for $\Delta \to 0$. Notice that the relative error in the determination of the equilibrium distribution, namely the error in the spring constant $\bar{K}_\alpha$, does not depend on the temperature T; therefore, the time step Δ can be kept independent from the temperature for given target accuracy.

- At finite values of Δ, the error in the discretization determines a (slightly) different equilibrium distribution that, however, remains of the same Gaussian form. Only for a single mode (i.e., for $D = 1$), this change can be interpreted as a renormalization of the effective temperature, $T \to T/(1 - \frac{\bar{K}_1 \Delta}{2})$.

- The number of iterations n_{corr} that are needed to generate an independent configuration $\mathbf{R}'$ from a given one $\mathbf{R}$ is then given by:

$$n_{corr} = \frac{\tau}{\Delta} \geq \frac{\bar{K}_D}{\bar{K}_1} \equiv \bar{K}_{cond}, \qquad (4.60)$$

where the correlation time τ is determined by Eq. (4.49) and $\bar{K}_{cond}$ is the *condition number* of the matrix $\mathbf{K}$, namely the ratio between its largest and smallest (non-zero) eigenvalues. Notice that, for the harmonic potential, n_{corr} does not depend on the temperature.

4.5 Accelerated Langevin Dynamics

When the condition number $\bar{K}_{cond}$ is large, the matrix is ill conditioned, implying that the number of iterations to generate a new independent configuration is extremely large. Therefore, the method is inefficient and some trick is necessary to speed up the algorithm (Parisi, 1984). One simple idea is to introduce in the Langevin equation (4.1) an acceleration matrix $\mathbf{S}$ such that:

$$\frac{d\mathbf{R}(t)}{dt} = \mathbf{S}^{-1}\mathbf{f}[\mathbf{R}(t)] + \boldsymbol{\eta}(t), \qquad (4.61)$$

where the noise satisfies the following conditions:

$$\langle \eta_\alpha(t) \rangle = 0, \qquad (4.62)$$

$$\langle \eta_\alpha(t)\eta_\beta(t') \rangle = 2TS_{\alpha,\beta}^{-1}\delta(t - t'). \qquad (4.63)$$

The modified algorithm gives a substantial improvement with respect to the original one of Eqs. (4.1), (4.3), and (4.4) when $\mathbf{S}$ is chosen as close as possible to the Hessian matrix:

$$H_{\alpha,\beta}(\mathbf{R}) = \frac{1}{2}\frac{\partial^2 V(\mathbf{R})}{\partial R_\alpha \partial R_\beta}. \qquad (4.64)$$

In the following, we consider the case in which the matrix $\mathbf{S}$ does not depend upon the coordinates $\{\mathbf{R}\}$, although a generalization in this sense is possible (Mazzola and Sorella, 2017). The advantage of considering the scheme of Eq. (4.61) can be understood by exploiting the limit of small temperatures. Indeed, for $T \to 0$, the standard Langevin equation (4.1) reduces to the steepest descent method that, starting from a given initial point $\mathbf{R}$, gives a practical way to find the closest minimum of the potential $V(\mathbf{R})$ (Press et al., 2007). The number of steps that are necessary to convergence is related to the condition number. Instead, by using the Newton-Raphson method, we can reach the target with one step in the harmonic case, no matter how large is the condition number. Therefore, by considering $\mathbf{S}$ that closely approximates the Hessian matrix, the dynamics generated by Eq. (4.61) represents the generalization at finite temperatures of the Newton-Raphson approach, which minimizes the thermalization/correlation time.

Following the same steps as in section 4.2, we can arrive at the discretized version of the Langevin equation:

$$\mathbf{R}_{n+1} = \mathbf{R}_n + \Delta \mathbf{S}^{-1}\mathbf{f}_n + \sqrt{2T\Delta}\,\mathbf{z}_n, \tag{4.65}$$

with

$$\langle z_{\alpha,n} z_{\beta,n} \rangle = S_{\alpha,\beta}^{-1}. \tag{4.66}$$

The previous relations imply that the following modified Fokker-Planck equation holds in the limit $\Delta \to 0$:

$$\frac{\partial \mathcal{P}(\mathbf{R},t)}{\partial t} = \sum_\alpha \frac{\partial}{\partial R_\alpha} \left\{ \sum_\beta S_{\alpha,\beta}^{-1} \left[T\frac{\partial \mathcal{P}(\mathbf{R},t)}{\partial R_\beta} - \mathcal{P}(\mathbf{R},t)f_\beta(\mathbf{R}) \right] \right\}, \tag{4.67}$$

which has the same equilibrium distribution as before, i.e., the Boltzmann one given in Eq. (4.8).

The efficiency of this acceleration scheme can be appreciated by considering the harmonic case, which is a good approximation for low enough temperatures, where the thermal fluctuations are limited to configurations close to the minimum of the potential. In this case:

$$\mathbf{f}[\mathbf{R}(t)] = -\mathbf{K}[\mathbf{R}(t) - \mathbf{R}_{eq}]. \tag{4.68}$$

Hence, by choosing $\mathbf{S} = \mathbf{K}$, we get:

$$\mathbf{S}^{-1}\mathbf{f}_n = -(\mathbf{R}_n - \mathbf{R}_{eq}); \tag{4.69}$$

then, by taking $\Delta = 1$ (as in the Newton-Raphson method), Eq. (4.65) drastically simplifies into:

$$\mathbf{R}_{n+1} = \mathbf{R}_{eq} + \sqrt{2T}\,\mathbf{z}_n, \tag{4.70}$$

which represents a very efficient method for generating new configurations with very short correlation time, i.e., $n_{corr} = 1$; indeed, the above equation sets a new independent configuration according to:

$$\mathcal{P}_{eq}(\mathbf{R}) = \frac{1}{\mathcal{Z}} \exp\left[-\frac{1}{4T} \sum_{\alpha,\beta} K_{\alpha,\beta}(R_\alpha - R_{eq,\alpha})(R_\beta - R_{eq,\beta}) \right]; \tag{4.71}$$

notice that this equilibrium distribution corresponds to the correct (Gaussian) results for the harmonic potential of Eq. (4.44), apart from the presence of an extra

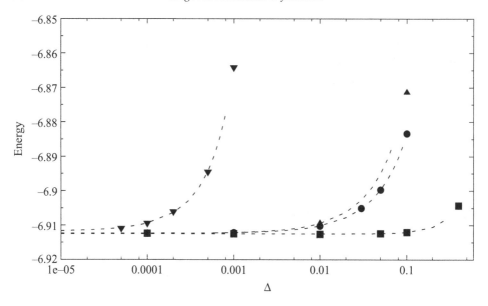

Figure 4.1 The convergence of the total energy as a function of the discretization step Δ that is used to integrate the Langevin equations is shown for a classical system with 64 particles interacting through a Lennard-Jones potential. The side of the cubic box is $L = 7.8$ and the temperature is $T = 0.4$. The standard scheme of Eq. (4.14) (downward triangles), the accelerated dynamics of Eq. (4.65) (upward triangles), and the one of Eq. (4.76) with $A = 1$ (full circles) and $A = 4$ (full squares) are reported.

factor 2 multiplying the temperature, which is due to the $\sqrt{2T}$ term in Eq. (4.70). This fact is a drawback of the approximation introduced by the time discretization. Nevertheless, for general cases, when the potential is not harmonic, this method is able to reduce substantially the correlation time and represents a very efficient and simple method that could be taken in mind before starting a molecular dynamics simulation with the first-order Langevin dynamics. An example of the accelerated Langevin dynamics for a classical systems of 64 particles, interacting through a Lennard-Jones potential, is reported in Fig. 4.1, for a sufficiently low temperature inside the solid phase.

An improvement in the discretization of the accelerated Langevin equation of Eq. (4.61) can be performed by assuming that the chosen acceleration matrix $\mathbf{S}$ gives a good approximation of the Hessian at equilibrium $\mathbf{K} \approx A\mathbf{S}$, apart from an overall constant A that can be empirically tuned to achieve the smallest possible error in the time discretization. Thus, we can consider that, even when the potential is not harmonic:

$$\mathbf{S}^{-1}\mathbf{f}[\mathbf{R}(t)] = -A\mathbf{R}(t) + \mathbf{C}[\mathbf{R}(t)], \qquad (4.72)$$

where $C[R(t)]$ is weekly dependent on $R(t)$, once we neglect non-harmonic contributions at equilibrium; then, a good approximation is given by taking $C[R(t)] = AR_{eq}$. Therefore, we are led to consider:

$$\frac{dR(t)}{dt} = -A[R(t) - R_{eq}] + \eta(t). \tag{4.73}$$

Since this differential equation is linear, it can be integrated *exactly* (i.e., without any error in Δ) by considering the new variable $R'(t)$, such that $R(t) = X(t)R'(t)$, where $X(t)$ is a suitable time-dependent matrix:

$$\frac{dX(t)}{dt} = -AX(t), \tag{4.74}$$

$$X(t)\frac{dR'(t)}{dt} = AR_{eq} + \eta(t). \tag{4.75}$$

Then, the result in the interval (t_n, t_{n+1}) is given by:

$$R_{n+1} = R_n - A\Delta_A(R_n - R_{eq}) + \sqrt{2\Delta_{2A}T}\, z_n, \tag{4.76}$$

where

$$\Delta_A = \frac{1 - e^{-A\Delta}}{A}, \tag{4.77}$$

$$\langle z_{\alpha,n} z_{\beta,n} \rangle = S_{\alpha,\beta}^{-1}. \tag{4.78}$$

This scheme represents a more accurate integration of the accelerated dynamics, which for large values of Δ gives the exact dynamics of the harmonic potential of Eq. (4.44). Indeed, for $\Delta \to \infty$, we have:

$$R_{n+1} = R_{eq} + \sqrt{\frac{T}{A}}\, z_n, \tag{4.79}$$

which leads to the correct Boltzmann distribution at temperature T (since $\langle z_{\alpha,n} z_{\beta,n} \rangle = AK_{\alpha,\beta}^{-1}$):

$$\mathcal{P}_{eq}(R) = \frac{1}{\mathcal{Z}} \exp\left[-\frac{1}{2T} \sum_{\alpha,\beta} K_{\alpha,\beta}(R_\alpha - R_{eq,\alpha})(R_\beta - R_{eq,\beta}) \right]. \tag{4.80}$$

In the general non-harmonic case, we can replace the second term in the r.h.s. of Eq. (4.76) by $\Delta_A S^{-1} f_n$ and obtain the iterative integration scheme:

$$R_{n+1} = R_n + \Delta_A S^{-1} f_n + \sqrt{2\Delta_{2A}T}\, z_n, \tag{4.81}$$

which becomes exact if non-harmonic terms can be neglected and K is exactly given by AS. Notice that Eq. (4.81) is equivalent to change $\Delta \to \Delta_A$ and $T \to T\Delta_{2A}/\Delta_A$ in Eq. (4.65). In practice, given an approximation of the Hessian matrix S, we can empirically select the value of A to minimize the time step error and work with a large value of Δ. Although the error for a generic potential $V(\mathbf{R})$ remains linear in Δ, it is usually quite small as it depends only on the non-harmonic terms for an accurate approximation of the Hessian.

The results obtained with this approach for a classical systems of 64 particles interacting through a Lennard-Jones potential are reported in Fig. 4.1. We notice that the optimal choice for A is not given by $A = 1$, which corresponds to the exact Hessian. This is because the non-harmonic terms are quite strong in this case and a rescaled Hessian is more accurate to describe the overall shape of the potential.

Part III

Variational Monte Carlo

5

Variational Monte Carlo

5.1 Quantum Averages and Statistical Samplings

In this chapter, we discuss the general framework in which the variational Monte Carlo methods are defined and few important implementations for interacting systems of bosons and fermions on the lattice. The main advantage of considering this approach relies on the variational principle that has been shown in section 1.4: the energy of a given quantum state is always bounded from below by the exact ground-state one, giving us the route to obtain the best possible solution to the problem. In most cases, a suitable parametrization of the variational wave function allows us to consider a wide range of different quantum phases (e.g., metals, superconductors, and insulators). By performing an optimization of the parameters, we can reach the lowest-energy state, which is expected to capture the correct ground-state behavior. Therefore, variational wave functions represent a flexible and valuable approach to get important insights into the low-energy properties of models that cannot be solved by exact methods. By contrast, the main limitation of this approach is the fact that it is based on a given *Ansatz*, which may contain a relevant bias that cannot be removed within the chosen parametrization.

We also mention that variational wave functions can be easily defined and treated for a wide class of models, irrespective of the range of interactions and the dimension of the local (i.e., single-site) Hilbert space, which can be even infinite and does not need the use of an uncontrolled cutoff to work with a finite-dimensional space. In this respect, variational Monte Carlo is better than other methods, like density-matrix renormalization group or tensor-network methods in which the complexity dramatically increases with both the range of the interaction and the dimension of the local Hilbert space (White, 1992; Schollwöck, 2005, 2011).

Let us start by describing the general framework in which variational Monte Carlo methods are defined. First of all, we fix a complete basis set $\{|x\rangle\}$ in the

Hilbert space, in which (for simplicity) the states are taken to be *orthogonal* and *normalized* such that:

$$\sum_x |x\rangle\langle x| = \mathbb{I}. \tag{5.1}$$

Then, any quantum state $|\Psi\rangle$ can be written as:

$$|\Psi\rangle = \sum_x |x\rangle\langle x|\Psi\rangle = \sum_x \Psi(x)|x\rangle. \tag{5.2}$$

In turn, the expectation value of an operator $\mathcal{O}$ over a given variational wave function $|\Psi\rangle$ takes the following form:

$$\langle\mathcal{O}\rangle = \frac{\langle\Psi|\mathcal{O}|\Psi\rangle}{\langle\Psi|\Psi\rangle} = \frac{\sum_x\langle\Psi|x\rangle\langle x|\mathcal{O}|\Psi\rangle}{\sum_x\langle\Psi|x\rangle\langle x|\Psi\rangle}. \tag{5.3}$$

The main problem in evaluating the expectation value is that the number of configurations in the sum is exponentially large with the number of particles. Although the dimension of the Hilbert space can be slightly reduced by employing few conservation laws (e.g., the conservation of the total number of particles and/or the total spin component along the quantization axis), it still remains exponentially large with the number of particles. Therefore, for large systems, it is impossible to perform an exact enumeration of the configurations to compute $\langle\mathcal{O}\rangle$ exactly. Nevertheless, Eq. (5.3) can be recast into a form that can be easily treated by standard Monte Carlo methods. Indeed, we have that:

$$\langle\mathcal{O}\rangle = \frac{\sum_x |\langle\Psi|x\rangle|^2 \frac{\langle x|\mathcal{O}|\Psi\rangle}{\langle x|\Psi\rangle}}{\sum_x |\langle\Psi|x\rangle|^2} = \frac{\sum_x |\Psi(x)|^2 \mathcal{O}_L(x)}{\sum_x |\Psi(x)|^2}, \tag{5.4}$$

where we have defined the *local estimator* of the operator $\mathcal{O}$:

$$\mathcal{O}_L(x) = \frac{\langle x|\mathcal{O}|\Psi\rangle}{\langle x|\Psi\rangle}. \tag{5.5}$$

The important point is that

$$\mathcal{P}(x) = \frac{|\Psi(x)|^2}{\sum_x |\Psi(x)|^2} \tag{5.6}$$

can be interpreted as a probability, since it is a non-negative quantity for all configurations $|x\rangle$ and is normalized, i.e., $\sum_x \mathcal{P}(x) = 1$. Therefore, the problem of computing a quantum average of the operator $\mathcal{O}$ can be rephrased into the calculation of the average of the random variable $\mathcal{O}_L(x)$ of Eq. (5.5) over the distribution probability $\mathcal{P}(x)$ given by Eq. (5.6). In particular, if we consider the expectation value of the Hamiltonian, the local estimator corresponds to the so-called *local energy*, which is defined by:

$$e_L(x) = \frac{\langle x|\mathcal{H}|\Psi\rangle}{\langle x|\Psi\rangle}. \tag{5.7}$$

In summary, it is possible to define a stochastic algorithm (e.g., a Markov process) in which a sequence of configurations $\{|x_n\rangle\}$ is generated (for example by using the Metropolis algorithm described in section 3.9). Then, after an equilibration time, they are distributed according to the desired probability $\mathcal{P}(x)$. Then, the quantum expectation value $\langle\mathcal{O}\rangle$ is evaluated from the mean value of the random variable $\mathcal{O}_L(x)$ over the visited configurations:

$$\langle\mathcal{O}\rangle \approx \frac{1}{N}\sum_{n=1}^{N}\mathcal{O}_L(x_n). \tag{5.8}$$

Finally, errorbars can be computed as described in section 3.11.

5.2 The Zero-Variance Property

An important feature of the variational Monte Carlo approach is the *zero-variance property*. Let us suppose that the variational state $|\Psi\rangle$ coincides with an exact eigenstate of $\mathcal{H}$ (not necessarily the ground state), namely $\mathcal{H}|\Psi\rangle = E|\Psi\rangle$. Then, it follows that the local energy $e_L(x)$ is constant:

$$e_L(x) = \frac{\langle x|\mathcal{H}|\Psi\rangle}{\langle x|\Psi\rangle} = E\frac{\langle x|\Psi\rangle}{\langle x|\Psi\rangle} = E. \tag{5.9}$$

Therefore, the random variable $e_L(x)$ does not depend on $|x\rangle$, which immediately implies that its variance is zero, while its mean value E coincides with the exact eigenvalue (in other words, $e_L(x)$ is not a random variable). Clearly, this is an extreme case that is very rare for generic correlated models. However, in general, the variance of $e_L(x)$ will decrease its value whenever the variational state $|\Psi\rangle$ will approach an exact eigenstate. This fact is very important to reduce the statistical fluctuations and improve the numerical efficiency. The zero-variance property is a feature that exists only for quantum expectation values, while it is absent in classical calculations, where observables have thermal fluctuations.

Finally, we would like to notice that the average of the square of the local energy corresponds to the quantum average of the Hamiltonian squared:

$$\frac{\langle\Psi|\mathcal{H}^2|\Psi\rangle}{\langle\Psi|\Psi\rangle} = \frac{\sum_x\langle\Psi|\mathcal{H}|x\rangle\langle x|\mathcal{H}|\Psi\rangle}{\sum_x\langle\Psi|x\rangle\langle x|\Psi\rangle} = \frac{\sum_x|\Psi(x)|^2|e_L(x)|^2}{\sum_x|\Psi(x)|^2}. \tag{5.10}$$

Thus, the variance of the random variable $e_L(x)$ is exactly equal to the quantum variance of the Hamiltonian over the variational state $|\Psi\rangle$:

$$\sigma^2_{e_L} = \frac{\langle\Psi|(\mathcal{H}-E)^2|\Psi\rangle}{\langle\Psi|\Psi\rangle}. \tag{5.11}$$

5.3 Jastrow and Jastrow-Slater Wave Functions

In the Monte Carlo evaluation of quantum averages, see Eq. (5.4), we must compute:

- The ratio of probabilities with different configurations, which implies the ratio of overlaps between the given variational state and two configurations of the basis set:

$$\frac{\mathcal{P}(x')}{\mathcal{P}(x)} = \left|\frac{\langle x'|\Psi\rangle}{\langle x|\Psi\rangle}\right|^2, \tag{5.12}$$

as required in the Metropolis algorithm.
- The local estimator $\mathcal{O}_L(x)$, which, in turn, implies ratios of overlaps and matrix elements of the observable between states of the basis set. For example, when considering the energy, we have:

$$e_L(x) = \frac{\langle x|\mathcal{H}|\Psi\rangle}{\langle x|\Psi\rangle} = \sum_{x'}\langle x|\mathcal{H}|x'\rangle\frac{\langle x'|\Psi\rangle}{\langle x|\Psi\rangle}. \tag{5.13}$$

Naively, the computation of the local estimator looks a tremendously hard task, since it requires a summation over all the states of the many-body Hilbert space; however, thanks to the locality of the Hamiltonian (or, similarly, any other local operator or correlation function), only few terms actually contribute to the sum. Indeed, given the configuration $|x\rangle$, the matrix element $\langle x|\mathcal{H}|x'\rangle$ is non-zero only for $O(L)$ configurations $|x'\rangle$. As an example, let us consider the fermionic Hubbard model: by using the local basis, $|x\rangle$ is connected only to few other configurations that differ for the hopping of one electron from a given site to one of its neighbors; then, the maximum number of such processes is equal to the number of sites L times the number of bonds times 2 (due to the spin). Therefore, the computation of the local estimator only requires a small number of operations, usually proportional to the number of sites/particles.

As we have seen, the building block of the variational Monte Carlo approach is the computation of $\langle x|\Psi\rangle$, which is the amplitude of the variational state over a generic element of the basis set. More precisely, along all the Markov process, only ratios of these overlaps must be computed. This calculation must be done for each configuration that is visited along the Markov process and, therefore, it must be done as fast as possible. This fact imposes some constraint on the form of the variational wave function. Usually, fermionic states require the calculation of determinants or Pfaffians, while bosonic wave functions require permanents (the definition of the permanent of a matrix $\mathbf{M}$ differs from that of the determinant in the fact that the signatures of the permutations are not taken into account). Fortunately, there are fast (i.e., polynomial) algorithms to evaluate determinants and Pfaffians, thus allowing us to consider these states as variational *Ansätze* for electron systems; by contrast, the calculation of permanents remains an exponentially-hard problem, which strongly limits the number of bosons that can be handled in a reasonable computation time. Nevertheless, the wave functions in which all bosons are condensed in a single state do not need such a calculation and can be considered efficiently in variational Monte Carlo approaches.

Within the variational technique that has been described before, it is possible to treat a large variety of fermionic and bosonic states, which may interpolate between weak and strong correlation regimes. In the following, we would like to give some detailed description on how to construct the variational states and how to compute the building blocks that are necessary along the Markov process. We will consider the Jastrow (bosonic) or Jastrow-Slater (fermionic) states that have been introduced and discussed in Chapter 1:

$$|\Psi_J\rangle = \mathcal{J}|\Phi_0\rangle, \tag{5.14}$$

where $|\Phi_0\rangle$ is a generic uncorrelated (bosonic or fermionic) state and $\mathcal{J}$ is the Jastrow factor, which takes into account the electron correlation. A particularly simple but important case is given by the case where $\mathcal{J}$ contains density-density correlations:

$$\mathcal{J} = \exp\left(-\frac{1}{2}\sum_{i,j} v_{i,j} n_i n_j\right), \tag{5.15}$$

where the $v_{i,j}$'s can be taken as variational parameters, whose total number is L^2. Here, the pseudo-potential $v_{i,j}$ couples densities at different sites i and j and not density fluctuations, as in Eq. (1.65); for systems with conserved number of particles, the two definitions of the Jastrow factors coincide, apart for an irrelevant multiplicative factor. For a translational invariant model, $v_{i,j}$ only depends upon the distance between i and j, thus the number of parameters can be reduced to $O(L)$. We would like to remark that, within the Monte Carlo approach, it is possible to treat exactly (but still having statistical errors) the limit of singular Jastrow factors with $v_{i,i} = \infty$, e.g., the Gutzwiller projector that eliminates all multiply occupied sites (this case being relevant for an infinite Hubbard-U interaction). Indeed, this case can be easily incorporated by building a Markov chain where only configurations $|x\rangle$ that satisfy this constraint are visited.

The advantage of considering the wave function of Eq. (5.14) in the variational Monte Carlo technique is that the calculations can be extremely efficient and fast. Indeed, whenever the Jastrow factor is diagonal in the chosen basis, we have that:

$$\langle x|\Psi_J\rangle = \mathcal{J}(x)\langle x|\Phi_0\rangle, \tag{5.16}$$

where $\mathcal{J}(x)$ is the value of the Jastrow operator computed for the configuration $|x\rangle$, i.e., $\mathcal{J}|x\rangle = \mathcal{J}(x)|x\rangle$. Therefore, given the electronic configuration, $\mathcal{J}(x)$ is a number that can be evaluated in $O(L^2)$ operations for the Jastrow term of Eq. (5.15). We emphasize that, in order to have a polynomial algorithm, the Jastrow factor must only contain operators that are diagonal in the basis $|x\rangle$, otherwise $\mathcal{J}|x\rangle$ would generate an exponentially large number of states, ruling out any calculation on large systems. In addition, $\langle x|\Phi_0\rangle$ can be also easily evaluated in $O(L^3)$ operations (for fermions) or $O(1)$ operations (for condensed bosons), see sections 5.5, 5.6, and 5.7.

5.4 The Choice of the Basis Sets

In most cases (but this is not mandatory), the basis set is chosen to be a product state in the real space. On a lattice with L sites, we must first give an ordering to them, $i = 1,\ldots,L$ and then take a basis set on each lattice site $\{|\xi\rangle_i\}$, whose dimension depends upon the model under consideration. For example, in the single-band Hubbard model $|\xi\rangle_i$ can be any of the following four states:

$$|0\rangle_i, \quad c_{i,\uparrow}^\dagger|0\rangle_i, \quad c_{i,\downarrow}^\dagger|0\rangle_i, \quad c_{i,\uparrow}^\dagger c_{i,\downarrow}^\dagger|0\rangle_i, \tag{5.17}$$

namely, the empty state on the site i, the two singly occupied states (with spin up or down along the quantization axis z), or the doubly occupied state.

For the bosonic Hubbard model, the local Hilbert space is infinite with an arbitrary number of bosons on the same site:

$$|0\rangle_i, \quad b_i^\dagger|0\rangle_i, \quad \frac{1}{\sqrt{2!}}(b_i^\dagger)^2|0\rangle_i, \quad \frac{1}{\sqrt{3!}}(b_i^\dagger)^3|0\rangle_i, \quad \ldots \tag{5.18}$$

Finally, for the spin-S Heisenberg model, the local basis can be taken as the $2S+1$ states with $S_i^z = -S,\ldots,S$:

$$|-S\rangle_i, \quad |-S+1\rangle_i, \quad \ldots, \quad |S-1\rangle_i, \quad |S\rangle_i. \tag{5.19}$$

Once the local Hilbert space has been fixed, a generic element of the basis set of the whole lattice $\{|x\rangle\}$ is given by the product state:

$$|x\rangle = |\xi\rangle_1 \ldots |\xi\rangle_L. \tag{5.20}$$

For the subsequent use, we also define the vacuum of the whole lattice as:

$$|0\rangle = |0\rangle_1 \ldots |0\rangle_L. \tag{5.21}$$

The advantage of taking a local basis, defined on each site of the lattice, comes from the fact that the Hamiltonian is usually local in space and its matrix elements are easily computed: the interaction terms are often diagonal, like for example the Hubbard-U interaction, or local, like the Hund's coupling in multi-band models, while the hopping terms are usually short-range, resulting in a small number (i.e., proportional to the number of particles) of non-vanishing matrix elements for each many-body configuration $|x\rangle$.

Within the Monte Carlo sampling (i.e., along the Markov process), it is convenient to generate the new (proposed) configuration among the ones that are obtained by applying the Hamiltonian. In this way, the generic configuration $|x_n\rangle$ at step n may differ from the one of the basis set by an overall sign (for fermionic systems). A similar issue is present when computing local estimators $\mathcal{O}_L(x)$, such as the local energy (5.7). This fact does not represent a problem, provided we bookkeep the sign change that may appear. Let us consider N_e spinless electrons on L sites, the configuration

$$|x_n\rangle = c^\dagger_{R_1} \ldots c^\dagger_{R_l} \ldots c^\dagger_{R_{N_e}} |0\rangle \tag{5.22}$$

can be described by a vector κ of length L, whose i-th element, if non-zero, gives the position (from left to right) of the corresponding creation operator in the above string defining the configuration. For example, by taking $N_e = 3$ and $L = 6$, the configuration $c^\dagger_3 c^\dagger_1 c^\dagger_6 |0\rangle$ is associated to the vector $(2, 0, 1, 0, 0, 3)$. Then, the application of the hopping term $c^\dagger_{R'_l} c_{R_l}$ gives:

$$|x_{n+1}\rangle = c^\dagger_{R_1} \ldots c^\dagger_{R'_l} \ldots c^\dagger_{R_{N_e}} |0\rangle; \tag{5.23}$$

showing that the only changes in the string appear at the position R_l and R'_l. In the example above, by applying $c^\dagger_2 c_1$, we get $c^\dagger_3 c^\dagger_2 c^\dagger_6 |0\rangle$ and the vector κ becomes $(0, 2, 1, 0, 0, 3)$. Along the Markov chain, it is useful to store and bookkeep κ, for the correct definition of the many-body state, see sections 5.6 and 5.7.

5.5 Bosonic Systems

Here, we describe in detail how to evaluate efficiently the ratio of wave functions on different configurations to apply the Metropolis algorithm, see Eq. (5.12). First of all, we consider the computation of the non-interacting part $\langle x|\Phi_0\rangle$, then we discuss the contribution coming from $\mathcal{J}(x)$.

5.5.1 Definition of the Non-Interacting State

The natural choice for a bosonic non-interacting wave function is to condense all the particles in a single state. The simplest example is to consider the single-particle "orbital" state defined by:

$$\phi_\alpha^\dagger |0\rangle = \left(\sum_i V_{i,\alpha} b_i^\dagger \right) |0\rangle, \tag{5.24}$$

where $\{V_{i,\alpha}\}$ are generic amplitudes for having the boson on site i; α is an index that specifies the orbital: for example, the case of a boson in a zero-momentum state is given by $V_{i,\alpha} = 1/\sqrt{L}$. Then, the many-body state with N_b bosons occupying the same orbital is:

$$|\Phi_0\rangle = \frac{1}{\sqrt{N_b!}} \left(\sum_i V_{i,\alpha} b_i^\dagger \right)^{N_b} |0\rangle. \tag{5.25}$$

By expanding the summation, we can rewrite the many-body state over the basis set of Eq. (5.18):

$$|\Phi_0\rangle = \frac{1}{\sqrt{N_b!}} \sum_{m_1,\ldots,m_L} \frac{N_b!}{m_1! \ldots m_L!} V_{1,\alpha}^{m_1} \ldots V_{L,\alpha}^{m_L} (b_1^\dagger)^{m_1} \ldots (b_L^\dagger)^{m_L} |0\rangle, \tag{5.26}$$

which allows us to have a very simple expression for the overlap $\langle x|\Phi_0\rangle$, where the configuration $|x\rangle \equiv |n_1,\ldots,n_L\rangle$ contains n_i bosons on site i:

$$|n_1,\ldots,n_L\rangle = \frac{1}{\sqrt{n_1! \ldots n_L!}} (b_1^\dagger)^{n_1} \ldots (b_L^\dagger)^{n_L} |0\rangle, \tag{5.27}$$

leading to:

$$\langle n_1,\ldots,n_L|\Phi_0\rangle = \sqrt{\frac{N_b!}{n_1! \ldots n_L!}} V_{1,\alpha}^{n_1} \ldots V_{L,\alpha}^{n_L}. \tag{5.28}$$

5.5.2 Fast Computation of the Non-Interacting State

Given the form of the non-interacting state, for which $\langle n_1,\ldots,n_L|\Phi_0\rangle$ is given by Eq. (5.28), the ratio of two overlaps with different configurations can be obtained by a simple calculation. For example, if we consider the ratio between configurations that only differ by a boson hopping between sites l and k, namely $|x\rangle \equiv |n_1,\ldots,n_k,\ldots,n_l,\ldots,n_L\rangle$ and $|x'\rangle \equiv |n_1,\ldots,n_k+1,\ldots,n_l-1,\ldots,n_L\rangle$, we have:

$$\frac{\langle n_1,\ldots,n_k+1,\ldots,n_l-1,\ldots,n_L|\Phi_0\rangle}{\langle n_1,\ldots,n_k,\ldots,n_l,\ldots,n_L|\Phi_0\rangle} = \sqrt{\frac{n_l}{n_k+1}} \left(\frac{V_{k,\alpha}}{V_{l,\alpha}} \right). \tag{5.29}$$

Therefore, whenever single-boson hopping processes are considered along the Markov chain, the contribution of the non-interacting part of the wave function to the Metropolis ratio (5.12) is very simple, since it requires $O(1)$ operations. The case in which more than one boson is moved can be obtained by generalizing the previous analysis in a straightforward way.

We would like to emphasize that the calculation of $\langle x|\Phi_0\rangle$ becomes dramatically complicated if the bosons are not condensed but occupy different orbitals:

$$|\Phi_0\rangle = \prod_{\alpha=1}^{N_b}\left(\sum_i V_{i,\alpha}b_i^\dagger\right)|0\rangle. \tag{5.30}$$

Indeed, in this case, we obtain that the overlap is given by a permanent of an $N_b \times N_b$ matrix:

$$\langle n_1,\ldots,n_L|\Phi_0\rangle = \mathrm{per}(V_{R_j,\alpha}), \tag{5.31}$$

where the R_j's (with $j = 1,\ldots,N_b$) take the values of the occupied sites. Notice that the values of R_j may appear more than once, according to the occupation of each site, e.g., $R_j = k$ appears n_k times. The calculation of the permanent (or the ratio of two permanents) is exponentially hard, thus preventing us to perform calculations with large values of N_b.

5.5.3 Fast Computation of the Jastrow Factor

Let us now consider the contribution coming from the Jastrow factor. A straightforward calculation of $\mathcal{J}(x)$ requires $O(L^2)$ calculations for the Jastrow factor of Eq. (5.15), thus leading to the same complexity when computing the ratio $\mathcal{J}(x')/\mathcal{J}(x)$ that appears in the Metropolis ratio (5.12). However, whenever the two configurations differ only by few boson hoppings, it is possible to apply a fast computation of the ratio, which involves $O(L)$ operations. Indeed, let us consider: $|x\rangle \equiv |n_1,\ldots,n_k,\ldots,n_l,\ldots,n_L\rangle$ and $|x'\rangle \equiv |n_1,\ldots,n_k+1,\ldots,n_l-1,\ldots,n_L\rangle$:

$$\frac{\mathcal{J}(n_1,\ldots,n_k+1,\ldots,n_l-1,\ldots,n_L)}{\mathcal{J}(n_1,\ldots,n_k,\ldots,n_l,\ldots,n_L)} = \frac{\exp\left(-\sum_i v_{i,k}n_i\right)}{\exp\left(-\sum_i v_{i,l}n_i\right)}e^{v_{k,l}-v_{k,k}}, \tag{5.32}$$

where we used the fact that the pseudo-potential $v_{i,j}$ is symmetric, i.e., $v_{i,j} = v_{j,i}$ and translational invariant, i.e., $v_{k,k} = v_{l,l}$. The second term in the r.h.s. of Eq. (5.32) does not depend upon the bosonic configuration and can be computed at the beginning of the simulation, once and for all. Instead, the first term depends upon the bosonic configuration, which is sampled along the Markov chain. The computation of the ratio can be done in $O(1)$ operations once we compute and store a vector of dimension L that depends upon $|x\rangle = |n_1,\ldots,n_k,\ldots,n_l,\ldots,n_L\rangle$:

$$T_{\mathrm{Jastrow}}(j) = \sum_i v_{i,j}n_i. \tag{5.33}$$

Then, once the new proposed $|x'\rangle = |n_1, \ldots, n_k + 1, \ldots, n_l - 1, \ldots, n_L\rangle$ is accepted, the vector $T_{\text{Jastrow}}(j)$ must be updated, which requires an $O(L)$ operations. Indeed, since $n'_i = n_i + \delta_{i,k} - \delta_{i,l}$, we have that:

$$T'_{\text{Jastrow}}(j) = T_{\text{Jastrow}}(j) + v_{k,j} - v_{l,j}. \tag{5.34}$$

Therefore, we can compute from scratch $T_{\text{Jastrow}}(j)$ at the beginning of the Markov chain for all the sites $j = 1, \ldots, L$ and then update it by using Eq. (5.34) every time a new configuration is accepted along the Markov process. As a safe habit, from time to time, it is recommended to recompute $T_{\text{Jastrow}}(j)$ from scratch, since the rounding error of the fast update can accumulate and give rise to numerical errors.

In summary, by using this scheme for the Jastrow factor, the single-move algorithm scales with $O(L)$; however, by performing these local updates, we need to perform $O(L)$ moves to obtain an almost independent configuration, for systems with a finite correlation time. Therefore, the bosonic code has a quadratic scaling with the number of sites/bosons.

5.6 Fermionic Systems with Determinants

Let us now move to fermionic systems, described by Jastrow-Slater wave functions. Since the part on the calculation of the Jastrow term is similar to the previous bosonic case, we do not repeat it here. The same fast updating can be used also for fermions. Instead, the part involving the calculation of the non-interacting state $\langle x|\Phi_0\rangle$ is totally different from the bosonic case. In the following, we describe how it is possible to devise an efficient algorithm to deal with such an object.

5.6.1 Definition of the Non-Interacting State

Any non-interacting wave function can be obtained as the ground state of a suitable quadratic Hamiltonian $\mathcal{H}_0$. First of all, we contract the spin index σ and the lattice site i into a single index I running from 1 to $2L$:

$$c_{i,\uparrow} \equiv d_i, \tag{5.35}$$

$$c_{i,\downarrow} \equiv d_{i+L}. \tag{5.36}$$

Then, we start from the simple case in which the non-interacting Hamiltonian is written as:

$$\mathcal{H}_0 = \sum_{I,J} t_{I,J} d_I^\dagger d_J, \tag{5.37}$$

which contains hopping terms only, also including processes in which the spin along z is not conserved, i.e., the terms with $I \leq L$ and $J > L$ and vice-versa. In a compact form, the non-interacting Hamiltonian of Eq. (5.37) can be written as:

$$\mathcal{H}_0 = \mathbf{d}^\dagger \mathbf{T} \mathbf{d} \ , \tag{5.38}$$

where

$$\mathbf{d}^\dagger = \begin{pmatrix} d_1^\dagger & \cdots & d_{2L}^\dagger \end{pmatrix}, \tag{5.39}$$

and

$$\mathbf{T} = \begin{pmatrix} t_{1,1} & \cdots & t_{1,2L} \\ \vdots & \ddots & \vdots \\ t_{2L,1} & \cdots & t_{2L,2L} \end{pmatrix}. \tag{5.40}$$

Since $\mathcal{H}_0$ commutes with the total number of electrons $N_e = \sum_I d_I^\dagger d_I$, the eigenstates are single-particle orbitals. In practice, the $2L \times 2L$ matrix $\mathbf{T}$ can be easily diagonalized by using standard libraries (e.g., LAPACK routines):

$$\mathcal{H}_0 = \mathbf{d}^\dagger \mathbf{U} \, \mathbf{U}^\dagger \mathbf{T} \mathbf{U} \, \mathbf{U}^\dagger \mathbf{d} \ = \mathbf{\Phi}^\dagger \mathbf{E} \mathbf{\Phi} \ = \sum_\alpha \varepsilon_\alpha \phi_\alpha^\dagger \phi_\alpha, \tag{5.41}$$

where $\mathbf{U}$ is a unitary matrix (that preserves anti-commutation relations of fermionic operators), $\mathbf{E} = \mathrm{diag}(\varepsilon_1, \ldots, \varepsilon_{2L})$ is the diagonal matrix containing the $2L$ eigenvalues ε_α of $\mathbf{T}$, and $\mathbf{\Phi}^\dagger = (\phi_1^\dagger, \ldots, \phi_{2L}^\dagger)$ is defined in terms of the eigenvectors of $\mathbf{T}$:

$$\phi_\alpha^\dagger = \sum_I U_{I,\alpha} d_I^\dagger. \tag{5.42}$$

Now, the many-body state $|\Phi_0\rangle$ can be constructed by occupying the N_e lowest-energy orbitals:

$$|\Phi_0\rangle = \prod_{\alpha=1}^{N_e} \phi_\alpha^\dagger |0\rangle = \left(\sum_I U_{I,1} d_I^\dagger \right) \cdots \left(\sum_I U_{I,N_e} d_I^\dagger \right) |0\rangle \tag{5.43}$$

Following the arguments of section 5.4, the generic configuration, with N_e electrons, which is visited along the Markov process, reads as:

$$|x\rangle = d_{R_1}^\dagger \ldots d_{R_{N_e}}^\dagger |0\rangle, \tag{5.44}$$

where $j = 1, \ldots, N_e$ includes both up and down spins and R_j assumes values from 1 to $2L$: the positions of spin-up electrons coincide with the site number, while the positions of spin-down electrons must be shifted by L. The overlap $\langle x|\Phi_0\rangle$ is then given by:

$$\langle x|\Phi_0\rangle = \langle 0|d_{R_{N_e}}\cdots d_{R_1}\left(\sum_I U_{I,1}d_I^\dagger\right)\cdots\left(\sum_I U_{I,N_e}d_I^\dagger\right)|0\rangle$$

$$= \langle 0|d_{R_{N_e}}\cdots d_{R_1}\left[\sum_p(-1)^p\prod_{\alpha=1}^{N_e}U_{p\{R_j\},\alpha}\right]d_{R_1}^\dagger\cdots d_{R_{N_e}}^\dagger|0\rangle,\quad (5.45)$$

where the sum inside the square bracket is over all the possible permutations of the $\{R_j\}$ in $|x\rangle$; the sign appears because of the anti-commutation relations of fermionic operators. Then, we get:

$$\langle x|\Phi_0\rangle = \det\{U_{R_j,\alpha}\}.\quad (5.46)$$

When constructing the many-body state (5.43), we must pay attention to construct a unique many-body state, i.e., occupy the correct lowest-energy orbitals. When the highest-occupied orbital and the lowest-unoccupied one have different energies (*closed shell* configuration), the choice is unique. Instead, it can also happen that there is a degeneracy that does not allow a unique choice (*open shell* configuration). Whenever the non-interacting Hamiltonian $\mathcal{H}_0$ is diagonalized numerically, the eigenstates provided by standard libraries do not carry definite quantum numbers (like momentum), but are given by generic linear combinations of degenerate orbitals (which erratically depend on the numerical precision used for the computation). This is not a problem whenever *all* the degenerate eigenstates are included in the many-body state (5.43), since taking any linear combination of columns in the matrix $\mathbf{U}$ will not change the value of the determinant in Eq. (5.46). By contrast, not including all degenerate eigenstates will cause a problem, since the determinant will depend on the particular combination of states that is considered. Therefore, whenever a numerical diagonalization is done, we must verify that a closed-shell configuration occurs. Otherwise, the problem of having a vanishing gap can be overcome by constructing suitable orbitals with definite quantum numbers, to obtain a reproducible simulation of the many-body state.

In practice, the diagonalization of the non-interacting Hamiltonian $\mathcal{H}_0$ must be performed at the beginning of the Monte Carlo calculation; then, we need to store the reduced part of the $\mathbf{U}$, obtained by keeping only the N_e columns that correspond to occupied orbitals. This is a $2L \times N_e$ matrix:

$$\mathbf{U} = \begin{pmatrix} U_{1,1} & \cdots & U_{1,N_e} \\ \vdots & \ddots & \vdots \\ U_{2L,1} & \cdots & U_{2L,N_e} \end{pmatrix}.\quad (5.47)$$

Then, the overlap with a generic electronic configuration is given by the determinant of the matrix obtained by considering only the rows of (5.47) corresponding to the electron positions $\{R_j\}$, thus giving a $N_e \times N_e$ matrix.

Remarkably, the same kind of formalism can be used also if the non-interacting Hamiltonian contains an electron pairing that couples up and down spins. Indeed, let us consider a BCS Hamiltonian described by:

$$\mathcal{H}_0 = \sum_{i,j,\sigma} t_{i,j} c^{\dagger}_{i,\sigma} c_{j,\sigma} - \mu_0 \sum_{i,\sigma} c^{\dagger}_{i,\sigma} c_{i,\sigma} + \sum_{i,j} \Delta_{i,j} c^{\dagger}_{i,\uparrow} c^{\dagger}_{j,\downarrow} + \text{h.c.}, \tag{5.48}$$

where we have included a chemical potential μ_0, which fixes, on average, the number of electrons. In this case, the total number of particles is not conserved and the concept of single-particle orbitals is not defined. Indeed, the ground state of the BCS Hamiltonian is naturally written in terms of a pairing function (see Chapter 1). Nevertheless, we can perform a particle-hole transformation on the spin-down electrons:

$$c_{i,\uparrow} \rightarrow f_{i,\uparrow} \equiv d_i \tag{5.49}$$

$$c_{i,\downarrow} \rightarrow f^{\dagger}_{i,\downarrow} \equiv d^{\dagger}_{i+L}. \tag{5.50}$$

Then, apart from constant terms, the transformed BCS Hamiltonian has the form of Eq. (5.38), where the matrix $\mathbf{T}$ is given by (i and $j \leq L$):

$$T_{i,i} = -\mu_0 \tag{5.51}$$

$$T_{i+L,i+L} = \mu_0 \tag{5.52}$$

$$T_{i,j} = t_{i,j}, \tag{5.53}$$

$$T_{i+L,j+L} = -t_{i,j}, \tag{5.54}$$

$$T_{i,j+L} = T^*_{j+L,i} = \Delta_{i,j}. \tag{5.55}$$

Since, after the particle-hole transformation, the number of particles (but not the z component of the spin) is conserved, the eigenstates of the BCS Hamiltonian can be expressed into "orbitals," similarly to the ones of Eq. (5.42), but without having a definite spin component along z.

Notice that the particle-hole transformations (5.49) and (5.50) change a spin down into the vacuum and vice-versa, so that:

$$|0\rangle_i \rightarrow f^{\dagger}_{i,\downarrow} |\tilde{0}\rangle_i, \tag{5.56}$$

$$c^{\dagger}_{i,\uparrow} |0\rangle_i \rightarrow f^{\dagger}_{i,\uparrow} f^{\dagger}_{i,\downarrow} |\tilde{0}\rangle_i, \tag{5.57}$$

$$c^{\dagger}_{i,\downarrow} |0\rangle_i \rightarrow |\tilde{0}\rangle_i, \tag{5.58}$$

$$c^{\dagger}_{i,\uparrow} c^{\dagger}_{i,\downarrow} |0\rangle_i \rightarrow f^{\dagger}_{i,\uparrow} |\tilde{0}\rangle_i, \tag{5.59}$$

where $|\tilde{0}\rangle_i$ is the vacuum of the f electrons, i.e., $f_{i,\uparrow}|\tilde{0}\rangle = f_{i,\downarrow}|\tilde{0}\rangle = 0$. By using the anti-commutation relations of the fermion operators, we have that the occupations $u_{i,\sigma}$ of the new fermionic operators are related to the original ones by:

$$u_{i,\uparrow} = f_{i,\uparrow}^{\dagger} f_{i,\uparrow} = c_{i,\uparrow}^{\dagger} c_{i,\uparrow} = n_{i,\uparrow}, \tag{5.60}$$

$$u_{i,\downarrow} = f_{i,\downarrow}^{\dagger} f_{i,\downarrow} = 1 - c_{i,\downarrow}^{\dagger} c_{i,\downarrow} = 1 - n_{i,\downarrow}, \tag{5.61}$$

which implies that the relations among densities and magnetizations on site i are given by:

$$u_i = u_{i,\uparrow} + u_{i,\downarrow} = n_{i,\uparrow} - n_{i,\downarrow} + 1 = m_i^z + 1, \tag{5.62}$$

$$w_i^z = u_{i,\uparrow} - u_{i,\downarrow} = n_{i,\uparrow} + n_{i,\downarrow} - 1 = n_i - 1, \tag{5.63}$$

where we have introduced u_i and w_i^z, as the density and magnetization for the transformed operators, respectively. The letters u and w are chosen because they look like an upside down n and m, respectively. Therefore, apart from constant terms, upon the particle-hole transformation of Eqs. (5.49) and (5.50), the density changes into the z component of the spin and vice-versa. As a corollary, we obtain that, whenever the total magnetization of the original particles is vanishing, the total number of the new particles is equal to the number of sites L.

5.6.2 Fast Computation of the Determinants

Let us now show how to compute efficiently the ratio of determinants when the two configurations $|x\rangle$ and $|x'\rangle$ differ by one or few electron hoppings. According to Eq. (5.46) the overlap $\langle x|\Phi_0\rangle$ is given by the determinant of $\tilde{U}_{j,\alpha} \equiv U_{R_j,\alpha}$:

$$\tilde{U} = \begin{pmatrix} U_{R_1,1} & \cdots & U_{R_1,N_e} \\ \vdots & \ddots & \vdots \\ U_{R_l,1} & \cdots & U_{R_l,N_e} \\ \vdots & \ddots & \vdots \\ U_{R_{N_e},1} & \cdots & U_{R_{N_e},N_e} \end{pmatrix}, \tag{5.64}$$

which is obtained taking only the rows corresponding to the occupied sites of the matrix U of Eq. (5.47). Notice that the actual order of rows in $\tilde{U}$ is determined by the one of creation operators in $|x\rangle$, see Eq. (5.44). In this regard, the generalization of the vector κ (see section 5.4) for the spinful case, can be used to put the rows of $\tilde{U}$ in the correct order.

Let us start and consider the electronic configurations in which $|x'\rangle$ is obtained from $|x\rangle$ just by hopping the l-th electron from R_l to R'_l, i.e., $|x'\rangle = d_{R'_l}^{\dagger} d_{R_l}|x\rangle$.

The new matrix $\tilde{\mathbf{U}}'$ will be equal to $\tilde{\mathbf{U}}$, except that the elements of the l-th row will be changed from $U_{R_l,\alpha}$ to $U_{R'_l,\alpha}$:

$$\tilde{\mathbf{U}}' = \begin{pmatrix} U_{R_1,1} & \cdots & U_{R_1,N_e} \\ \vdots & \ddots & \vdots \\ U_{R'_l,1} & \cdots & U_{R'_l,N_e} \\ \vdots & \ddots & \vdots \\ U_{R_{N_e},1} & \cdots & U_{R_{N_e},N_e} \end{pmatrix}. \tag{5.65}$$

Then, the ratio between two configurations that differ only by a single fermion hopping is:

$$\frac{\langle x'|\Phi_0\rangle}{\langle x|\Phi_0\rangle} = \frac{\langle x|d^\dagger_{R_l}d_{R'_l}|\Phi_0\rangle}{\langle x|\Phi_0\rangle} = \frac{\det\tilde{\mathbf{U}}'}{\det\tilde{\mathbf{U}}}. \tag{5.66}$$

By denoting with K the new site of the l-th electron (i.e., $K \equiv R'_l$), the updated matrix elements are given by a compact form:

$$\tilde{U}'_{j,\alpha} = \tilde{U}_{j,\alpha} + \delta_{j,l}(U_{K,\alpha} - \tilde{U}_{l,\alpha}) = \tilde{U}_{j,\alpha} + \delta_{j,l}v^{K,l}_\alpha, \tag{5.67}$$

where we have defined $v^{K,l}_\alpha \equiv U_{K,\alpha} - \tilde{U}_{l,\alpha}$; here the indices K and l are fixed, since they specify the site where the electron is hopping and the electron index, respectively. This equation can be rewritten in the following way:

$$\tilde{U}'_{j,\alpha} = \sum_\beta \tilde{U}_{j,\beta}\left(\delta_{\beta,\alpha} + \tilde{U}^{-1}_{\beta,l}v^{K,l}_\alpha\right) = \sum_\beta \tilde{U}_{j,\beta}Q_{\beta,\alpha}, \tag{5.68}$$

where

$$Q_{\beta,\alpha} = \delta_{\beta,\alpha} + \tilde{U}^{-1}_{\beta,l}v^{K,l}_\alpha. \tag{5.69}$$

Therefore, $\tilde{\mathbf{U}}' = \tilde{\mathbf{U}}\mathbf{Q}$, which implies that the calculation of the ratio of the determinants of $\tilde{\mathbf{U}}'$ and $\tilde{\mathbf{U}}$ is equivalent to the calculation of the determinant of $\mathbf{Q}$:

$$\frac{\det\tilde{\mathbf{U}}'}{\det\tilde{\mathbf{U}}} = \det\mathbf{Q}. \tag{5.70}$$

The great simplification comes from the fact that the determinant of $\mathbf{Q}$ can be easily computed. Indeed, $\mathbf{Q}$ has a particularly simple form that can be written as:

$$Q_{\beta,\alpha} = \delta_{\beta,\alpha} + \mathcal{B}_\beta\mathcal{A}_\alpha, \tag{5.71}$$

where $\mathcal{B}_\beta = \tilde{U}^{-1}_{\beta,l}$ and $\mathcal{A}_\alpha = v^{K,l}_\alpha$. Although the matrix is not Hermitian, the eigenvalues of the matrix $\mathbf{Q}$ can be obtained from the secular equation:

$$\sum_\alpha Q_{\beta,\alpha}v_\alpha = \lambda v_\beta; \tag{5.72}$$

by using the explicit form of Eq. (5.71), we obtain:

$$v_\beta + \mathcal{B}_\beta \sum_\alpha \mathcal{A}_\alpha v_\alpha = \lambda v_\beta, \qquad (5.73)$$

which implies that all vectors v_α that are orthogonal to $\mathcal{A}_\alpha$ are eigenvectors with eigenvalue $\lambda = 1$ (there are $N_e - 1$ of such vectors); in addition, $v_\alpha = \mathcal{B}_\alpha$ is also eigenvector with $\lambda = 1 + \sum_\alpha \mathcal{A}_\alpha \mathcal{B}_\alpha = \sum_\alpha U_{K,\alpha} \tilde{U}^{-1}_{\alpha,l}$. Therefore, we have that:

$$\frac{\det \tilde{U}'}{\det \tilde{U}} = \det Q = \sum_\alpha U_{K,\alpha} \tilde{U}^{-1}_{\alpha,l}. \qquad (5.74)$$

Having stored (at the beginning of the simulation) the matrix $\tilde{U}^{-1}$ for the configuration $|x\rangle$, this calculation requires $O(N_e)$ operations, instead of the $O(N_e^3)$ needed to evaluate a determinant. Then, once the new configuration $|x'\rangle$ is accepted along the Markov process, the matrix $\tilde{U}^{-1}$ must be updated. This can be done in $O(N_e^2)$ operations. In fact, we have that $(\tilde{U}')^{-1} = Q^{-1}\tilde{U}^{-1}$, the inverse of the matrix Q being given by (as easily verified):

$$Q^{-1}_{\alpha,\beta} = \delta_{\alpha,\beta} - \frac{1}{\det Q} \mathcal{B}_\alpha \mathcal{A}_\beta. \qquad (5.75)$$

Then, the updated matrix elements of $(\tilde{U}')^{-1}$ are given by:

$$\tilde{U}^{-1'}_{\alpha,j} = \tilde{U}^{-1}_{\alpha,j} - \frac{\tilde{U}^{-1}_{\alpha,l}}{\det Q} \left(\sum_\beta U_{K,\beta} \tilde{U}^{-1}_{\beta,j} - \delta_{l,j} \right). \qquad (5.76)$$

This is a closed equation for updating the matrix $\tilde{U}^{-1}$.

We would like to emphasize that the previous results for the calculation of the ratio of determinants and the updating can be further simplified. Indeed, at the beginning of the calculation, we can compute and store a $2L \times N_e$ matrix W, whose elements are given by:

$$W_{I,j} = \sum_\alpha U_{I,\alpha} \tilde{U}^{-1}_{\alpha,j}; \qquad (5.77)$$

then the ratio of determinants (5.74) costs $O(1)$ operations, since it consists in taking the element corresponding to the new site (row) and the electron performing the hopping process (column):

$$\frac{\det \tilde{U}'}{\det \tilde{U}} = W_{K,l}. \qquad (5.78)$$

The evaluation of W requires the knowledge of the full matrix U, which has been computed and stored once for all at the beginning of the simulation (it does not depend upon the electronic configuration), and $\tilde{\mathbf{U}}^{-1}$, which instead depends upon the configuration $|x\rangle$. Then, a simple updating scheme for $\mathbf{W}$ is possible. In fact, by multiplying both sides of Eq. (5.76) by $U_{I,\alpha}$ and summing over α, we obtain:

$$W'_{I,j} = W_{I,j} - \frac{W_{I,l}}{W_{K,l}} \left(W_{K,j} - \delta_{l,j} \right), \tag{5.79}$$

where, we have used that $\det \mathbf{Q} = W_{K,l}$, according to Eq. (5.78) and the definition of Eq. (5.77). Since each matrix element must be updated with $O(1)$ operations, the total cost is $O(2LN_e)$.

By using a wave function that is constructed from filling single-particle states (also including cases with BCS pairing when a particle-hole transformation is done), the algorithm can be written in terms of the matrix elements of $\mathbf{W}$, which is the equal-time Green's function, see Eq. (5.66):

$$\frac{\langle x|d^\dagger_{R_l} d_K|\Phi_0\rangle}{\langle x|\Phi_0\rangle} = W_{K,l}. \tag{5.80}$$

We would like to mention that Eq. (5.79) has a direct interpretation by using the Wick theorem (with different *bra* and *ket* states) for the updated (equal-time) Green's function:

$$W'_{I,j} = \frac{\langle x'|d^\dagger_{R_I} d_I|\Phi_0\rangle}{\langle x'|\Phi_0\rangle} = \frac{\langle x|d^\dagger_{R_I} d_K d^\dagger_{R_I} d_I|\Phi_0\rangle}{\langle x|d^\dagger_{R_I} d_K|\Phi_0\rangle}; \tag{5.81}$$

in this case, we must consider the contractions without including the anomalous (superconducting) ones, which vanish since both $|x\rangle$ and $|\Phi_0\rangle$ have a fixed number of particles.

Let us finish this part by generalizing the previous formalism to the case where more than one electron hop, i.e., $|x'\rangle = d^\dagger_{R'_{I_1}} d_{R_{I_1}} \ldots d^\dagger_{R'_{I_m}} d_{R_{I_m}} |x\rangle$, thus leading to a modification of m rows of the $\tilde{\mathbf{U}}$ matrix; for example, this could be the case for pair-hopping or spin-flip processes. Then, Eq. (5.67) generalizes into:

$$\tilde{U}'_{j,\alpha} = \tilde{U}_{j,\alpha} + \sum_{r=1}^{m} \delta_{j,l_r} \left(U_{K_r,\alpha} - \tilde{U}_{l_r,\alpha} \right) = \tilde{U}_{j,\alpha} + \sum_{r=1}^{m} \delta_{j,l_r} v^{K_r,l_r}_\alpha; \tag{5.82}$$

as before, the indices K_r and l_r (for $r = 1, \ldots, m$) are fixed, because they specify the sites where the electrons are hopping and the electron indices, respectively. By performing the same algebra as before, we get:

$$\tilde{U}'_{j,\alpha} = \sum_{\beta} \tilde{U}_{j,\beta} \left(\delta_{\beta,\alpha} + \sum_{r=1}^{m} \tilde{U}^{-1}_{\beta,l_r} v^{K_r,l_r}_\alpha \right) = \sum_{\beta} \tilde{U}_{j,\beta} Q_{\beta,\alpha}, \tag{5.83}$$

where now the matrix $\mathbf{Q}$ has the following form:

$$Q_{\beta,\alpha} = \delta_{\beta,\alpha} + \sum_{r=1}^{m} \mathcal{B}_{\beta}^{r} \mathcal{A}_{\alpha}^{r}, \tag{5.84}$$

where $\mathcal{B}_{\beta}^{r} = \tilde{U}_{\beta,l_r}^{-1}$ and $\mathcal{A}_{\alpha}^{r} = v_{\alpha}^{K_r,l_r}$. As before, the determinant of $\mathbf{Q}$ can be easily computed by solving the corresponding eigenvalue problem:

$$\sum_{\alpha} Q_{\beta,\alpha} v_{\alpha} = v_{\beta} + \sum_{r=1}^{m} \mathcal{B}_{\beta}^{r} \sum_{\alpha} \mathcal{A}_{\alpha}^{r} v_{\alpha} = \lambda v_{\beta}, \tag{5.85}$$

which implies that all vectors that are orthogonal to the subspace defined by the $\mathcal{A}_{\alpha}^{r}$'s are eigenvectors with $\lambda = 1$; moreover, $v_{\alpha} = \sum_{r=1}^{m} x_r \mathcal{B}_{\alpha}^{r}$ is an eigenvector provided that the coefficients x_r satisfy:

$$\sum_{s=1}^{m} \left(\delta_{r,s} + \sum_{\alpha} \mathcal{A}_{\alpha}^{r} \mathcal{B}_{\alpha}^{s} \right) x_s = \lambda x_r. \tag{5.86}$$

Therefore, the m non-trivial eigenvalues of $\mathbf{Q}$ are given by the ones of the $m \times m$ matrix:

$$C_{r,s} = \delta_{r,s} + \sum_{\alpha} \mathcal{A}_{\alpha}^{r} \mathcal{B}_{\alpha}^{s} = W_{K_r,l_s}. \tag{5.87}$$

The final expression of the ratio of the two determinants is given by:

$$\frac{\det \tilde{U}'}{\det \tilde{U}} = \det(W_{K_r,l_s}). \tag{5.88}$$

Also in this case, once the move is accepted, we have to update the matrix $(\tilde{U}')^{-1} = \mathbf{Q}^{-1}\tilde{U}^{-1}$. As before, the inverse of the $\mathbf{Q}$ matrix can be obtained:

$$Q_{\alpha,\beta}^{-1} = \delta_{\alpha,\beta} - \sum_{r,s=1}^{m} \mathcal{B}_{\alpha}^{r} C_{r,s}^{-1} \mathcal{A}_{\beta}^{s}. \tag{5.89}$$

Therefore, we obtain:

$$\tilde{U}_{\alpha,j}^{-1'} = \tilde{U}_{\alpha,j}^{-1} - \sum_{r,s=1}^{m} \tilde{U}_{\alpha,l_r}^{-1} C_{r,s}^{-1} \left(\sum_{\beta} U_{K_s,\beta} \tilde{U}_{\beta,j}^{-1} - \delta_{l_s,j} \right). \tag{5.90}$$

Then, the updated $\mathbf{W}'$ is obtained by multiplying both sides of the previous equation by $U_{i,\alpha}$ and summing over α:

$$W_{I,j}' = W_{I,j} + \sum_{r=1}^{m} W_{I,l_r} b_j^{(r)}, \tag{5.91}$$

where

$$b_j^{(r)} = -\sum_{s=1}^{m} C_{r,s}^{-1}\left(W_{K_s,j} - \delta_{l_s,j}\right). \tag{5.92}$$

Similarly to the fast update of the Jastrow factor, it is highly recommended, from time to time, to recompute the matrix $\mathbf{W}$ from scratch. This costs $O(2LN_e^2)$, thus this calculation does not affect the complexity of the algorithm if it is done every $O(N_e)$ electron moves.

5.6.3 Delayed Updates

Let us consider the simple update in which only one electron is displaced. The basic operation in the updating is the so-called rank-1 update of a $2L \times N_e$ matrix, see Eq. (5.79) that can be written as:

$$W'_{I,j} = W_{I,j} + a_I b_j, \tag{5.93}$$

where $\mathbf{a}$ and $\mathbf{b}$ are two vectors of length $2L$ and N_e, respectively:

$$a_I = W_{I,l}, \tag{5.94}$$

$$b_j = -\frac{W_{K,j} - \delta_{l,j}}{W_{K,l}}. \tag{5.95}$$

This updating operation can be computationally inefficient, whenever, for large size, the matrix $\mathbf{W}$ cannot be contained in the cache of the processor. A way to overcome this drawback is to delay the update of the matrix $\mathbf{W}$, without loosing its information. This can be done by storing a set of left and right vectors $\mathbf{a}^{(p)}$ and $\mathbf{b}^{(p)}$ with $p = 1, \ldots, m$, as well as the "initial" matrix (denoted by $\mathbf{W}^0$) from which we begin to delay the updates. Then, the matrix $\mathbf{W}$, after m updates is given by:

$$W_{I,j} = W_{I,j}^0 + \sum_{p=1}^{m} a_I^{(p)} b_j^{(p)}. \tag{5.96}$$

Every time we accept a new configuration, a new pair of vectors $\mathbf{a}^{(m+1)}$ and $\mathbf{b}^{(m+1)}$ can be computed in few operations in term of $\mathbf{W}^0$ and the previous vectors with $p = 1, \ldots, m$, by substituting Eq. (5.96) in Eqs. (5.94) and (5.95). Once the matrix $\mathbf{W}$ is written in the form of Eq. (5.96), the number of operations required to evaluate the factors in the sum is $O[m(2L + N_e)]$, which is negligible compared to the full update for $m \ll L$.

In this way, we can find an optimal $m_{\max}$, for which we can evaluate the full matrix $\mathbf{W}$ by a standard matrix multiplication:

$$\mathbf{W} = \mathbf{W}^0 + \mathbf{A}\mathbf{B}^T, \tag{5.97}$$

where $\mathbf{A}$ and $\mathbf{B}$ are $2L \times m_{\max}$ and $N_e \times m_{\max}$ matrices, which are made of the $p = 1, \ldots, m_{\max}$ vectors $\mathbf{a}^{(p)}$ and $\mathbf{b}^{(p)}$, respectively. After that, we can continue with a new delayed update with a new $\mathbf{W}^0 = \mathbf{W}$, by initializing to zero the integer m. The advantage of this updating procedure is that after a cycle of $m_{\max}$ steps the bulk of the computation is given by the matrix-matrix product in Eq. (5.97), which is much more efficient (and it is not limited by cache memory) than the $m_{\max}$ rank-1 updates of $\mathbf{W}$ given in Eq. (5.79). For large number of electrons, the delayed update procedure, with the optimal value of $m_{\max}$, allows us to improve the speed of the variational Monte Carlo code by about an order of magnitude.

5.6.4 Backflow Correlations

Here, we would like to discuss how to implement the updating of the determinant part in presence of backflow correlations that have been introduced on lattice problem. We notice that, in the lattice case, a continuous change of coordinates, implementing the standard backflow correlation (1.70), is not possible. Therefore, this *Ansatz* requires to find an expression of the orbital in a generic off-lattice backflow coordinate in terms of the allowed lattice positions, consistently to a Taylor expansion of the orbitals (Tocchio et al., 2008 and Tocchio et al., 2011). In practice, we consider a quadratic Hamiltonian (as the ones that have been discussed in section 5.6), to construct the non-interacting orbitals $\{U_{I,\alpha}\}$, see Eq. (5.42). In the simplest approach, in which the backflow correlations act on holon-doublon (nearest-neighbor) pairs (Tocchio et al., 2008), the overlap between the generic configuration $|x\rangle$ and the backflow wave function $|\Phi_0^b\rangle$ is constructed from the "correlated" orbital with backflow correction:

$$U_{I,\alpha}^b = \eta_0 U_{I,\alpha} + \eta_1 \sum_{\langle j \rangle_i} D_i H_j U_{J,\alpha}, \qquad (5.98)$$

where $I = i$ ($I = i + L$) for electrons with spin up (down), and equivalently for J and j; $\langle j \rangle_i$ indicates the sites j that are nearest neighbors of i; D_i (H_i) is the operator that gives 1 if the site i is doubly occupied (empty) and 0 otherwise; finally η_0 and η_1 are variational parameters. Then, the wave function for the given configuration $|x\rangle$ is obtained by taking the determinant of the matrix:

$$\tilde{\mathbf{U}}^b = \begin{pmatrix} U_{R_1,1}^b & \cdots & U_{R_1,N_e}^b \\ \vdots & \ddots & \vdots \\ U_{R_{N_e},1}^b & \cdots & U_{R_{N_e},N_e}^b \end{pmatrix}, \qquad (5.99)$$

in which the rows correspond to the sites occupied by the electrons.

The fast update for a ratio between two determinants with configurations $|x\rangle$ and $|x'\rangle$, which differ by one or few electron hoppings, can be computed as before.

Now, the only difference is that, even if there is only one electron hopping, the new matrix $\tilde{U}^{b'}$ will differ not only for a single row, but for few of them. Indeed, by moving one electron, an empty or doubly occupied site can be created, leading to a modification of the "correlated" orbitals of Eq. (5.98). Therefore, in general, we must use the general update in which several rows are changed; however, in this case, it is convenient to work directly with the inverse matrix $\tilde{U}'$, by updating it according to Eq. (5.90).

In practice, we can consider more general forms for the "correlated" orbitals, by considering further terms (Tocchio et al., 2011), but still remaining in the same spirit of considering a linear combination of non-interacting orbitals depending on the many-body configuration $|x\rangle$ (in the previous case, the linear combination is taken for configurations having holons and doublons at nearest-neighbor sites).

5.7 Fermionic Systems with Pfaffians

In the previous section, we have explained how to construct a correlated wave function starting from non-interacting electrons described by the Hamiltonian of Eq. (5.37), which naturally gives rise to the concept of single-particle orbitals. Here, the uncorrelated part of the many-body state is given by a determinant, see Eq. (5.46). By using a particle-hole transformation on spin-down electrons, the same approach can be also extended to BCS Hamiltonians containing a pairing among electrons with different spins, see Eq. (5.48). By contrast, when the pairing terms couple electrons with the same spin, orbitals cannot be defined, even after the particle-hole transformation and the uncorrelated state is written in terms of a pairing function. In turn, a determinant does not represent the most general uncorrelated anti-symmetric state that, instead, can be defined by means of the Pfaffian of an anti-symmetric matrix. In the following, we will show that, in the most general case, the uncorrelated part of the wave function is given by a Pfaffian, which may describe, as particular cases, states with singlet pairing or Slater determinants. This kind of mean-field wave function was first introduced by Bouchaud et al. (1988) and then used to study small atoms and molecules (Bajdich et al., 2006, 2008). Then, we will describe this kind of state and an efficient algorithm to perform local updates.

5.7.1 Definition of the Pfaffian

Before discussing the non-interacting state, we would like to summarize the definition and few important properties of the Pfaffian. Let us consider a $2n \times 2n$

skew-symmetric matrix $\mathbf{A}$, namely a matrix having $a_{i,j} = -a_{j,i}$. The Pfaffian of $\mathbf{A}$ is defined as the anti-symmetrized product:

$$\text{Pf}\mathbf{A} \equiv \mathcal{A}\left[a_{1,2}\, a_{3,4} \ldots a_{2n-1,2n}\right] = \sum_{\alpha} \text{sign}(\alpha) \prod_{k=1}^{n} a_{i_k j_k}, \tag{5.100}$$

where the sum runs over all the $(2n-1)!!$ pair partitions defined by $\alpha = \{(i_1,j_1),\ldots,(i_n,j_n)\}$ with $i_k < j_k$ and $i_1 < i_2 < \cdots < i_n$; $\text{sign}(\alpha)$ indicates the parity of the permutation corresponding to the partition α. For example, for $n = 2$:

$$\text{Pf}\begin{pmatrix} 0 & a_{1,2} & a_{1,3} & a_{1,4} \\ -a_{1,2} & 0 & a_{2,3} & a_{2,4} \\ -a_{1,3} & -a_{2,3} & 0 & a_{3,4} \\ -a_{1,4} & -a_{2,4} & -a_{3,4} & 0 \end{pmatrix} = a_{1,2}a_{3,4} - a_{1,3}a_{2,4} + a_{1,4}a_{2,3}. \tag{5.101}$$

Notice that $\det\mathbf{A} = (\text{Pf}\mathbf{A})^2$; therefore, *apart form a global sign*, the calculation of the Pfaffian of a matrix can be reduced to the one of the determinant.

The Pfaffian satisfies the following relations:

$$\text{Pf}\begin{pmatrix} \mathbf{A} & 0 \\ 0 & \mathbf{A}' \end{pmatrix} = \text{Pf}\mathbf{A} \times \text{Pf}\mathbf{A}', \tag{5.102}$$

where $\mathbf{A}$ and $\mathbf{A}'$ are $2n \times 2n$ and $2m \times 2m$ skew-symmetric matrices, respectively. Moreover:

$$\text{Pf}\left(\mathbf{B}\mathbf{A}\mathbf{B}^T\right) = \det\mathbf{B} \times \text{Pf}\mathbf{A}, \tag{5.103}$$

where $\mathbf{B}$ is an arbitrary $2n \times 2n$ matrix.

Finally, we report two important identities that will be used to demonstrate the fast updating of the Pfaffian. Let us consider two invertible skew-symmetric matrices $\mathbf{A}$ ($2n \times 2n$) and $\mathbf{C}$ ($2m \times 2m$) and a real matrix $\mathbf{B}$ ($2n \times 2m$). Then, we have that:

$$\begin{pmatrix} \mathbf{A} & \mathbf{B} \\ -\mathbf{B}^T & \mathbf{C}^{-1} \end{pmatrix} = \begin{pmatrix} 1 & \mathbf{BC} \\ 0 & 1 \end{pmatrix}\begin{pmatrix} \mathbf{A}+\mathbf{BCB}^T & 0 \\ 0 & \mathbf{C}^{-1} \end{pmatrix}\begin{pmatrix} 1 & 0 \\ (\mathbf{BC})^T & 1 \end{pmatrix}$$
$$= \begin{pmatrix} 1 & 0 \\ (\mathbf{A}^{-1}\mathbf{B})^T & 1 \end{pmatrix}\begin{pmatrix} \mathbf{A} & 0 \\ 0 & \mathbf{C}^{-1}+\mathbf{B}^T\mathbf{A}^{-1}\mathbf{B} \end{pmatrix}\begin{pmatrix} 1 & \mathbf{A}^{-1}\mathbf{B} \\ 0 & 1 \end{pmatrix}. \tag{5.104}$$

Therefore, by using Eqs. (5.102) and (5.103), we arrive to:

$$\frac{\text{Pf}\left(\mathbf{A}+\mathbf{BCB}^T\right)}{\text{Pf}\mathbf{A}} = \frac{\text{Pf}\left(\mathbf{C}^{-1}+\mathbf{B}^T\mathbf{A}^{-1}\mathbf{B}\right)}{\text{Pf}\mathbf{C}^{-1}}. \tag{5.105}$$

5.7.2 Definition of the Non-Interacting State

A case that cannot be brought back to the previous formalism based upon single-particle orbitals is when the original non-interacting Hamiltonian contains a pairing between the same spins or there are both spin-flip hopping and pairing terms. Indeed, here it is not possible to eliminate the pairing terms by performing particle-hole transformations. The most general quadratic Hamiltonian for a fermionic system is given by a generalized BCS Hamiltonian:

$$\mathcal{H}_0 = \sum_{i,j,\sigma,\tau} t_{i,j}^{\sigma,\tau} c_{i,\sigma}^\dagger c_{j,\tau} - \mu_0 \sum_{i,\sigma} c_{i,\sigma}^\dagger c_{i,\sigma} + \sum_{i,j,\sigma,\tau} \Delta_{i,j}^{\sigma,\tau} c_{i,\sigma}^\dagger c_{j,\tau}^\dagger + \text{h.c.,} \tag{5.106}$$

where $t_{i,j}^{\sigma,\tau} = \left(t_{j,i}^{\tau,\sigma}\right)^*$ and $\Delta_{j,i}^{\tau,\sigma} = -\Delta_{i,j}^{\sigma,\tau}$ are hopping and pairing terms, respectively.

The Hamiltonian $\mathcal{H}_0$ can be diagonalized by a generalized Bogoliubov transformation:

$$\Phi_\alpha = \sum_I \left(u_{I,\alpha} d_I + v_{I,\alpha} d_I^\dagger \right), \tag{5.107}$$

where the $2L \times 2L$ matrices $\mathbf{u}$ and $\mathbf{v}$ are determined by imposing:

$$\left[\mathcal{H}_0, \Phi_\alpha\right] = -\varepsilon_\alpha \Phi_\alpha, \tag{5.108}$$

as well as the conditions of orthogonality (that preserve the anti-commutation relations of fermionic operators):

$$\mathbf{u}^\dagger \mathbf{u} + \mathbf{v}^\dagger \mathbf{v} = \mathbf{1}, \tag{5.109}$$

$$\mathbf{u}^T \mathbf{v} + \mathbf{v}^T \mathbf{u} = \mathbf{0}. \tag{5.110}$$

Notice that, Eq. (5.108) allows pairs of solutions, since if Φ_α has eigenvalue ε_α then $\Phi_\alpha^\dagger$ will also be a solution with eigenvalue $-\varepsilon_\alpha$. Therefore, we can limit ourselves to the eigenvectors with non-negative eigenvalues.

In most cases, the ground state of the non-interacting Hamiltonian (5.106) can be written as a generalized BCS wave function:

$$|\Phi_0\rangle = \exp\left(\frac{1}{2} \sum_{I,J} F_{I,J} d_I^\dagger d_J^\dagger\right) |0\rangle, \tag{5.111}$$

where $F_{I,J}$ is the anti-symmetric pairing function, defined by imposing that $|\Phi_0\rangle$ is annihilated by all the operators Φ_α with $\varepsilon_\alpha \geq 0$, which leads to:

$$\mathbf{Fu} = \mathbf{v}. \tag{5.112}$$

As in the case with determinants, also here we must pay attention in defining a unique many-body wave function. Therefore, only the cases without vanishing eigenvalues ε_α must be considered. Moreover, in order to have a non-singular pairing function $\mathbf{F}$, we must also require that the matrix $\mathbf{u}$ is invertible. Notice that a singular **u** corresponds to the existence of occupied single-particle orbitals. In this case, the many-body state must be constructed including these unpaired orbitals, increasing the computational complexity of the algorithm (Bajdich et al., 2006, 2008). In the following, for the fast update, we will limit ourselves to non-singular cases.

The overlap between $|\Phi_0\rangle$ and a generic configuration of Eq. (5.44) is given by:

$$\langle x|\Phi_0\rangle = \frac{1}{(N_e/2)!}\langle 0|d_{R_{R_e}}\cdots d_{R_1}\left(\frac{1}{2}\sum_{I,J}F_{I,J}d_I^\dagger d_J^\dagger\right)^{N_e/2}|0\rangle, \tag{5.113}$$

then, by expanding the product of the sums, we get a sum over the $(N_e-1)!!$ pair partitions of the fermions in $|x\rangle$ of $(N_e/2)!$ identical contributions. The ordering of the fermion operators gives rise to the sign appearing in the definition of the Pfaffian, see Eq. (5.100), thus leading to:

$$\langle x|\Phi_0\rangle = \mathrm{Pf}\left\{F_{R_i,R_j}\right\}. \tag{5.114}$$

5.7.3 Fast Computation of the Pfaffian

We now show how to compute efficiently the ratio of two Pfaffians when two configurations $|x\rangle$ and $|x'\rangle$ differ by one electron hopping. In particular, suppose that the l-th electron in R_l changes its position to R_l', i.e., $|x'\rangle = d_{R_l'}^\dagger d_{R_l}|x\rangle$. The new matrix $\mathbf{F}'$ that is needed to compute the overlap with $|x'\rangle$ differs from the old one $\mathbf{F}$ only in the l-th row and column. According to Eq. (5.114), the overlap $\langle x|\Phi_0\rangle$ is given by the Pfaffian of the matrix:

$$\tilde{\mathbf{F}} = \begin{pmatrix} F_{R_1,R_1} & \cdots & F_{R_1,R_l} & \cdots & F_{R_1,R_{N_e}} \\ \vdots & \ddots & \vdots & \ddots & \vdots \\ F_{R_l,R_1} & \cdots & F_{R_l,R_l} & \cdots & F_{R_l,R_{N_e}} \\ \vdots & \ddots & \vdots & \ddots & \vdots \\ F_{R_{N_e},R_1} & \cdots & F_{R_{N_e},R_l} & \cdots & F_{R_{N_e},R_{N_e}} \end{pmatrix}, \tag{5.115}$$

which can be constructed from the original matrix $\mathbf{F}$. As for the caste with determinants, the actual order of rows (an columns) is determined by the one of creation operators in the configuration $|x\rangle$, see Eq. (5.44). Then, a generalization of the vector κ (see section 5.4) for the spinful case can be used to select the correct order of columns and rows. Then, the overlap $\langle x'|\Phi_0\rangle$ is given by the Pfaffian of:

$$\tilde{\mathbf{F}}' = \begin{pmatrix} F_{R_1,R_1} & \cdots & F_{R_1,R_l'} & \cdots & F_{R_1,R_{N_e}} \\ \vdots & \ddots & \vdots & \ddots & \vdots \\ F_{R_l',R_1} & \cdots & F_{R_l',R_l'} & \cdots & F_{R_l',R_{N_e}} \\ \vdots & \ddots & \vdots & \ddots & \vdots \\ F_{R_{N_e},R_1} & \cdots & F_{R_{N_e},R_l'} & \cdots & F_{R_{N_e},R_{N_e}} \end{pmatrix}. \tag{5.116}$$

In a compact form, we have:

$$\tilde{F}'_{i,j} = \tilde{F}_{i,j} + \delta_{i,l}\left(F_{R'_l,R_j} - \tilde{F}_{l,j}\right) - \delta_{j,l}\left(F_{R'_l,R_i} - \tilde{F}_{l,i}\right), \tag{5.117}$$

which can be cast in the form of $\tilde{\mathbf{F}}' = \tilde{\mathbf{F}} + \mathbf{B}\mathbf{C}\mathbf{B}^T$ where $\mathbf{B}$ is a $N_e \times 2$ matrix:

$$\mathbf{B} = \begin{pmatrix} F_{R_1,R'_l} - \tilde{F}_{1,l} & \delta_{1,l} \\ \vdots & \vdots \\ F_{R_{N_e},R'_l} - \tilde{F}_{N_e,l} & \delta_{N_e,l} \end{pmatrix}, \tag{5.118}$$

and $\mathbf{C}$ is a 2×2 matrix:

$$\mathbf{C} = \begin{pmatrix} 0 & 1 \\ -1 & 0 \end{pmatrix}. \tag{5.119}$$

Therefore, by using Eq. (5.105), we get a simple expression for the ratio of the Pfaffians:

$$\frac{\langle x'|\Phi_0\rangle}{\langle x|\Phi_0\rangle} = \frac{\langle x|d^\dagger_{R_l}d_{R'_l}|\Phi_0\rangle}{\langle x|\Phi_0\rangle} = \frac{\mathrm{Pf}\tilde{\mathbf{F}}'}{\mathrm{Pf}\tilde{\mathbf{F}}} = \frac{\mathrm{Pf}(\mathbf{C}^{-1} + \mathbf{B}^T\tilde{\mathbf{F}}^{-1}\mathbf{B})}{\mathrm{Pf}\mathbf{C}}, \tag{5.120}$$

which can be easily computed since it requires the computation of the Pfaffian of a 2×2 matrix, being $\mathrm{Pf}\mathbf{C}^{-1} = -1$. The explicit form is worked out by using Eq. (5.118) for the matrix $\mathbf{B}$:

$$\frac{\mathrm{Pf}\tilde{\mathbf{F}}'}{\mathrm{Pf}\tilde{\mathbf{F}}} = \sum_j F_{R'_l,R_j}\tilde{F}^{-1}_{j,l} \equiv G_{R'_l,l}. \tag{5.121}$$

Having computed the matrix $\tilde{\mathbf{F}}^{-1}$ at the beginning of the simulation, this calculation requires $O(N_e)$ operations. As before, once the new configuration $|x'\rangle$ is accepted along the Markov chain, the matrix $\tilde{\mathbf{F}}^{-1}$ must be updated. From Eq. (5.117), we have that:

$$\tilde{F}'_{i,j} = \sum_m \tilde{F}_{i,m}\left(\delta_{m,j} + \tilde{F}^{-1}_{m,l}v_j^{R'_l,R_l} - w_m^{R'_l,R_l}\delta_{j,l}\right), \tag{5.122}$$

where

$$v_j^{R'_l,R_l} = F_{R'_l,R_j} - \tilde{F}_{l,j}, \tag{5.123}$$

$$w_m^{R'_l,R_l} = \sum_j \tilde{F}^{-1}_{m,j}(F_{R'_l,R_j} - \tilde{F}_{l,j}) = \delta_{m,l} - G_{R'_l,m}. \tag{5.124}$$

Then, the inverse $\tilde{\mathbf{F}}^{-1}$ can be easily found by following the same strategy as in section 5.6:

$$\tilde{F}^{-1'}_{i,j} = \tilde{F}^{-1}_{i,j} + \frac{1}{G_{R'_l,l}}\left(\tilde{F}^{-1}_{i,l}w_j^{R'_l,R_l} + w_i^{R'_l,R_l}\tilde{F}^{-1}_{l,j}\right). \tag{5.125}$$

The important difference with the case of section 5.6 is that the updating cannot be written by using the matrix $\mathbf{G}$ alone, but $\tilde{\mathbf{F}}^{-1}$ must be also kept. In fact, by multiplying both sides of Eq. (5.125) by F_{R,R_i} and summing over i, we have:

$$
G'_{R,j} - (F_{R,R'_l} - F_{R,R_l})\tilde{F}^{-1'}_{l,j} = G_{R,j}
$$
$$
+ \frac{1}{G_{R'_l,l}}\left(G_{R,l}w_j^{R'_l,R_l} + \sum_i F_{R,R_i}w_i^{R'_l,R_l}\tilde{F}^{-1}_{l,j} \right), \quad (5.126)
$$

where we have used the fact that:

$$
G'_{R,j} = \sum_i F_{R,R'_i}\tilde{F}^{-1'}_{i,j} = \sum_i \left[F_{R,R_i} + \delta_{i,l}\left(F_{R,R'_l} - F_{R,R_l} \right) \right] \tilde{F}^{-1'}_{i,j}. \quad (5.127)
$$

Then, by using Eq. (5.125) to express $\tilde{F}^{-1'}_{l,j}$ in terms of $\tilde{F}^{-1}_{l,j}$, we finally get:

$$
G'_{R,j} = G_{R,j} + \frac{1}{G_{R'_l,l}}\left[G_{R,l}w_j^{R'_l,R_l} + \left(F_{R,R'_l} + \sum_k G_{R,k}F_{R'_l,R_k} \right)\tilde{F}^{-1}_{l,j} \right]. \quad (5.128)
$$

Therefore, a fast updating can be performed by considering the matrices $\mathbf{G}$ and $\tilde{\mathbf{F}}^{-1}$ and using Eqs. (5.125) and (5.128) every time the proposed new configuration is accepted.

Within the Pfaffian wave function, in addition to the standard (equal-time) Green's function:

$$
\frac{\langle x|d^\dagger_{R_l}d_R|\Phi_0\rangle}{\langle x|\Phi_0\rangle} = G_{R,l}, \quad (5.129)
$$

also the anomalous (superconducting) ones are present:

$$
\frac{\langle x|d^\dagger_{R_l}d^\dagger_{R_k}|\Phi_0\rangle}{\langle x|\Phi_0\rangle} = \tilde{F}^{-1}_{k,l}, \quad (5.130)
$$

$$
\frac{\langle x|d_R d_{R'}|\Phi_0\rangle}{\langle x|\Phi_0\rangle} = F_{R,R'} + \sum_k G_{R,k}F_{R',R_k}. \quad (5.131)
$$

As before, the interpretation of Eqs. (5.125) and (5.128) is given by the Wick theorem (when *bra* and *ket* states are different) for the updated Green's functions $G'_{R,j}$ and $\tilde{F}^{-1'}_{i,j}$; in this case, the anomalous contractions are also present, since $|\Phi_0\rangle$ has not a fixed number of particles.

The generalization to the case where more than one electron hop is straightforward by using Eq. (5.105) but it is very cumbersome. However, we can bypass the m-electron update by applying successively single-electron updates.

5.8 Energy and Correlation Functions

The variational energy or any other observable, including correlation functions, can be easily computed by using the same tricks of the fast updates that we have discussed in the previous sections. Indeed, the local energy or other local observables are written in terms of a sum of ratios of wave functions times the matrix elements of the operator $\langle x|\mathcal{O}|x'\rangle$, according to Eq. (5.13). Let us consider a generic n-body operator:

$$\mathcal{O} = \sum_{I_1,\dots,I_n} \sum_{J_1,\dots,J_n} \mathcal{O}_{I_1,\dots,I_n;J_1,\dots,J_n} d^{\dagger}_{I_1} d_{J_1} \dots d^{\dagger}_{I_n} d_{J_n}, \tag{5.132}$$

where $d^{\dagger}_I$ (d_I) creates (destroys) an electron on the site $I = 1,\dots,2L$ (a similar definition holds for a bosonic system). Then, the local estimator of Eq. (5.5) is given by:

$$\mathcal{O}_L(x) = \sum_{I_1,\dots,I_n} \sum_{J_1,\dots,J_n} \mathcal{O}_{I_1,\dots,I_n;J_1,\dots,J_n} \frac{\langle x|d^{\dagger}_{I_1} d_{J_1} \dots d^{\dagger}_{I_n} d_{J_n}|\Psi\rangle}{\langle x|\Psi\rangle}. \tag{5.133}$$

The ratio in the r.h.s. of this equation can be easily computed by using the results discussed in the sections dedicated to the fast-update algorithms.

5.9 Practical Implementation

Here, we would like to sketch the important steps in a practical implementation of the variational Monte Carlo algorithm.

1. **Initialization** at the beginning of the calculation.
   - Generate a random state of the basis set $|x\rangle$ that is stored into a vector `iconf(L)`, whose elements give the local configuration on the site $i = 1,\dots,L$. For example, for the fermionic (single-band) Hubbard model, `iconf(i)` can assume the values 0 (empty site), ± 1 (one electron with spin up or down), or 2 (doubly-occupied site). In order to change the correct row of the matrix $\tilde{U}$ (for the determinant) or row and column of $\tilde{F}$ (for the Pfaffian), it is also important to store a vector `kel(2L)`, whose non-zero elements give, for each site, the position of the creation operators in the string defining the sampled configuration $|x\rangle$ of Eq. (5.44). For the bosonic Hubbard model, `iconf(i)` can assume all the non-negative integer values and there is no need to store `kel(2L)`.
   - Verify that the initial configuration is not singular, i.e., $\langle x|\Psi_J\rangle \neq 0$. The Jastrow factor usually does not give rise to any problem in this sense and we must only check whether $\langle x|\Phi_0\rangle$ is vanishing or not. Indeed, by working with

fermionic states, it may happen that $\langle x|\Phi_0\rangle = 0$ for some $|x\rangle$. Although this is not a problem along the simulation, since if one of these singular configurations is proposed it will not be accepted by the Metropolis algorithm, it would be a problem to initialize the Markov chain with a singular configuration (because the first acceptance probability in the Metropolis algorithm would have a vanishing denominator). Notice that the typical value for the overlap $\langle x|\Phi_0\rangle$ is exponentially small with the size of the system; however, within the fast update algorithms, we never need to compute the actual values of determinants or Pfaffians but only ratios of them. Therefore, at the beginning of the simulation, we must only require that the matrix $\tilde{U}$ is invertible without numerical roundoff, in order to construct the equal-time Green's functions.

- Compute the table of Eq. (5.33) to perform the fast update of the Jastrow factor and all the Green's functions that are necessary to perform the fast update. For the determinant case, only the Green's function of Eq. (5.80) is necessary, while for the Pfaffian case both the standard Green's function (5.129) and the anomalous ones (5.130) and (5.131) are needed.

2. **Markov process** with the Metropolis algorithm.
   - Propose a new (random) configuration $|x'\rangle$ by moving one or few particles.
   - Compute the ratio between the new and the old wave functions, see Eqs. (5.78), (5.88), or (5.121).
   - Accept or reject the proposed configuration according to the Metropolis algorithm with Eq. (5.12).
   - If the new configuration is accepted, update the table for the Jastrow factor (5.34) and all the Green's functions, i.e., Eq. (5.79) for determinants or Eqs. (5.125) and (5.128) for Pfaffians.

3. **Computation of observables**.

   Observables (e.g., the variational energy) can be computed every $O(L)$ steps (one step corresponds to propose a move in which one or few particles are moved) in order to have uncorrelated configurations. Of course, the frequency at which the observables are computed depends upon the acceptance ratio, e.g., if the latter one is small, we need to perform more Markov steps before recomputing the observable, in order to decorrelate subsequent measurements. A further binning technique (see section 3.11) can be applied to reduce correlation.

   It is always recommended to write the observables on the hard-disk and then perform the statistical analysis, since the correlation time is not usually known. Indeed, the calculation of observables during the Monte Carlo simulation would require the knowledge of this quantity before starting the simulation. In the unfortunate case in which this is not correctly estimated, we must restart the simulation, which may have a large computational cost. Instead, the post-processing analysis is usually much less expensive, since it only requires simple operations.

6

Optimization of Variational Wave Functions

6.1 Introduction

In the first work that used the variational Monte Carlo approach to study a many-body system, a relatively simple wave function has been considered, containing only a couple of parameters; thus, it was possible to obtain their optimal values by a simple fitting procedure of few energy values (McMillan, 1965). The possibility to include several variational parameters in a Monte Carlo framework is a relatively recent achievement. Indeed, until the beginning of the nineties, people limited the optimization to very few (e.g., at most two or three) variational parameters, which were determined with trial and error calculations of energies or by fitting methods. Nowadays, there is a growing interest in attaining fast and accurate schemes to optimize a large number of parameters to describe correlated many-body wave functions.

Hereafter, we denote by the vector $\boldsymbol{\alpha} = (\alpha_1, \ldots, \alpha_p)$ the set of all p variational parameters describing the wave function:

$$\Psi_\alpha(x) = \langle x | \Psi_\alpha \rangle, \tag{6.1}$$

where $\{|x\rangle\}$ denotes a complete basis set in the Hilbert space (which is taken to be orthogonal and normalized). The first breakthrough in the optimization has been done by realizing that, once a sampling for $\Psi_\alpha(x)$ has been obtained, the reweighting technique can be used to compute the energy or the variance for a wave function with different parameters, i.e., $\Psi_{\alpha+\delta\alpha}(x)$ (Umrigar et al., 1988). This method was shown to be very efficient and stable for a small number of particles (e.g., for few atoms and molecules), especially when the variance minimization was employed, while it has considerable drawbacks for large number of particles. Few years after the reweighting technique was applied, a further progress in the stochastic optimization techniques of several parameters was achieved. The basic idea is to compute energy differences for an arbitrary small variation of the parameters, namely evaluating the energy derivatives. It turns out that these quantities correspond to well-defined estimators with variance that increases only polynomially (e.g., at most quadratically) with the number of particles. The simplest approach is based upon the steepest-descent minimization (Harju et al., 1997). More elaborated algorithms include the so-called *stochastic reconfiguration* (Sorella, 1998, 2001), approximated versions of the Newton-Raphson approach (Umrigar and Filippi, 2005; Sorella, 2005), and the so-called *linear method* (Nightingale and Melik-Alaverdian, 2001; Umrigar and Filippi, 2005; Umrigar et al., 2007; Toulouse and Umrigar, 2007). The key point in all these techniques is to devise a scheme that is stable even in the presence of the statistical noise and may converge quickly to the global minimum. These iterative schemes are based on energy derivatives, which must be calculated as accurately as possible within a given Monte Carlo simulation.

We emphasize that a robust and efficient method for the wave function optimization will also improve the convergence of projection techniques, which crucially depend on the quality of the guiding wave function that is used (see Chapters 8 and 9). In particular, the fixed-node approach benefits from an optimized wave function, since this approximation becomes exact by systematically improving the trial state (see Chapter 10).

6.2 Reweighting Techniques for the Optimization of Wave Functions

Here, we briefly discuss how the reweighting technique, described in section 3.2, can be used to get, with a single simulation, the energy or variance for several sets of variational parameters α. In particular, we consider the method based on the variance optimization, which has been widely used for several years after being introduced by Umrigar et al. (1988). Suppose that a set of configurations $\{x_i\}$ with $i = 1, \ldots, N$ have been sampled from the square of the variational state $\Psi_\alpha(x)$. Of course, the energy and variance corresponding to this "initial" guess with α are given by:

$$E_\alpha = \frac{\langle \Psi_\alpha | \mathcal{H} | \Psi_\alpha \rangle}{\langle \Psi_\alpha | \Psi_\alpha \rangle} \approx \frac{1}{N} \sum_{i=1}^{N} e_{L,\alpha}(x_i), \tag{6.2}$$

$$\sigma_{\Psi_\alpha}^2 = \frac{\langle \Psi_\alpha | (\mathcal{H} - E_\alpha)^2 | \Psi_\alpha \rangle}{\langle \Psi_\alpha | \Psi_\alpha \rangle} \approx \frac{1}{N} \sum_{i=1}^{N} \left[e_{L,\alpha}(x_i) - E_\alpha \right]^2, \tag{6.3}$$

where the local energy is given by, see Eq. (5.7):

$$e_{L,\alpha}(x_i) = \frac{\langle x_i | \mathcal{H} | \Psi_\alpha \rangle}{\langle x_i | \Psi_\alpha \rangle}. \tag{6.4}$$

Then, by using Eq. (3.10), with:

$$R_{\alpha+\delta\alpha_k}(x) = \left| \frac{\Psi_{\alpha+\delta\alpha_k}(x)}{\Psi_\alpha(x)} \right|^2, \tag{6.5}$$

we can estimate (with the same set of configurations $\{x_i\}$) also the energy and the variance that correspond to a wave function with (slightly) different variational parameters $\alpha + \delta\alpha$:

$$E_{\alpha+\delta\alpha} = \frac{\sum_{i=1}^N e_{L,\alpha+\delta\alpha}(x_i) R_{\alpha+\delta\alpha}(x_i)}{\sum_{i=1}^N R_{\alpha+\delta\alpha}(x_i)}, \tag{6.6}$$

$$\sigma^2_{\Psi_{\alpha+\delta\alpha}} = \frac{\sum_{i=1}^N \left[e_{L,\alpha+\delta\alpha}(x_i) - E_{\alpha+\delta\alpha} \right]^2 R_{\alpha+\delta\alpha}(x_i)}{\sum_{i=1}^N R_{\alpha+\delta\alpha}(x_i)}. \tag{6.7}$$

Therefore, by performing a single simulation with variational parameters α, it would be possible to "reconstruct" the energy and variance also for different values in the neighborhood of α. In order to find the optimal set of parameters, we can minimize either the energy of Eq. (6.6) or the variance of Eq. (6.7) with respect to $\delta\alpha$, by applying the steepest-descent approach (for the given set of configurations $\{x_i\}$, these are non-linear functions of the variational parameters). Within the reweighting method, the advantage of the variance minimization is that this is a sum of positive terms and, therefore, a minimum is always present (in other words, the variance is always a positive quantity, even for a finite number of samples). Instead, by performing the energy minimization, it is not guaranteed to find a result that is bounded from below; indeed, it may happen that a configuration x_i, which has been sampled according to $|\Psi_\alpha(x)|^2$ (corresponding to a finite local energy $e_{L,\alpha}(x_i)$), gives a very large and negative local energy for the parameters $\alpha + \delta\alpha$; for example, x_i may be arbitrary close to the nodal surface of $\Psi_{\alpha+\delta\alpha}(x)$, such that $e_{L,\alpha+\delta\alpha}(x_i) \to -\infty$.

The reweighting technique is unstable when the number of particles is large. Indeed, let us consider the case where only one component α_k of the vector α has been incremented by $\delta\alpha_k$; then, the reweighting factors are determined by the ratio of two many-body wave functions that are exponentially large or small, leading to:

$$R_{\alpha+\delta\alpha_k}(x) \approx \exp(\pm\delta\alpha_k N_p), \tag{6.8}$$

where N_p is the number of particles. Therefore, for large N_p, even a tiny change of the variational parameters would imply an exponential increase of the statistical noise. This fact makes it extremely hard (if not impossible) to find out the optimal set of parameters in a system with several particles, because the statistical noise in the variation of the energy or variance is very large. Therefore, it is necessary to consider alternative optimization approaches: one possibility is to consider methods that are based upon the calculation of energy derivatives. Indeed, although approaches based upon the variance minimization are quite satisfactory, there are good motivations to develop energy-based minimization methods: for example, an energy-optimized wave function usually provides better expectation values with respect to variance-optimized ones, since the former approach is more sensitive to low-energy variation than the latter one.

6.3 Energy Derivatives

In this section, we discuss the basics ingredients that are necessary to compute the derivatives of the variational energy with respect to a given variational parameter α_k:

$$f_k = -\frac{\partial E_\alpha}{\partial \alpha_k} = -\frac{\partial}{\partial \alpha_k} \frac{\langle \Psi_\alpha | \mathcal{H} | \Psi_\alpha \rangle}{\langle \Psi_\alpha | \Psi_\alpha \rangle}. \tag{6.9}$$

The dependence of E_α on the variational parameters is just a consequence of the fact that the wave function $|\Psi_\alpha\rangle$ depends upon $\boldsymbol{\alpha}$. Thus, in order to differentiate E_α, it is convenient to expand $|\Psi_\alpha\rangle$ for small changes $\alpha_k \to \alpha_k + \delta\alpha_k$. For a given configuration $|x\rangle$, where $\Psi_\alpha(x)$ is a complex number (in case of complex parameters, we can assume that all the α_k are real, once we consider their real and imaginary parts separately), we have that:

$$\Psi_{\alpha+\delta\alpha_k}(x) = \Psi_\alpha(x) + \delta\alpha_k \frac{\partial \Psi_\alpha(x)}{\partial \alpha_k} + O(\delta\alpha_k^2), \tag{6.10}$$

where the notation $\Psi_{\alpha+\delta\alpha_k}(x)$ means that only the component α_k of the vector $\boldsymbol{\alpha}$ has been incremented by $\delta\alpha_k$. In the following, for simplicity, we assume that $\Psi_\alpha(x) \neq 0$ for all the configurations. For fermionic systems in the continuous space, the nodal region $\Psi_\alpha(x) = 0$ represents a negligible (i.e., with zero measure) integration domain. On the lattice, accidental configurations with $\Psi_\alpha(x) = 0$ can be removed by considering a tiny perturbation of the variational *Ansatz* (e.g., by adding a small noisy part) and considering the limit of vanishing perturbation. Then, Eq. (6.10) can be formally written in terms of a local operator $\mathcal{O}_k$, corresponding to the parameter α_k and defined by diagonal matrix elements $\mathcal{O}_k(x)$:

$$\langle x | \mathcal{O}_k | x' \rangle = \delta_{x,x'} \mathcal{O}_k(x), \tag{6.11}$$

$$\mathcal{O}_k(x) = \frac{\partial \ln \Psi_\alpha(x)}{\partial \alpha_k} = \frac{1}{\Psi_\alpha(x)} \frac{\partial \Psi_\alpha(x)}{\partial \alpha_k}; \tag{6.12}$$

here, in principle, $\mathcal{O}_k(x)$ may depend upon the variational parameters $\boldsymbol{\alpha}$, however, to keep the notation simple, we prefer not to put the label α in the local operators.

The important point is that $\mathcal{O}_k(x)$ can be usually computed for the given *Ansatz* of the variational state (see section 6.7 for the Jastrow-Slater wave functions). In this way, we can write a formal expansion of the many-body state as:

$$|\Psi_{\alpha+\delta\alpha_k}\rangle = (1 + \delta\alpha_k\mathcal{O}_k)|\Psi_\alpha\rangle, \tag{6.13}$$

which can be readily verified by taking the overlap of both sides of the above equation with $|x\rangle$ and using Eqs. (6.11) and (6.12). Notice that the diagonal operator $\mathcal{O}_k$ is not necessarily Hermitian, as its diagonal elements are not necessarily real, for a generic complex case.

Let us now show how to obtain the explicit form of the energy derivative with respect to a given variational parameter. It is clear that the variational energy E_α (as well as any other correlation function) does not depend on the overall normalization (and global phase) of the wave function. In other words, by scaling the wave function by an arbitrary complex constant c, i.e., $|\Psi_\alpha\rangle \to c|\Psi_\alpha\rangle$, E_α remains unchanged. In order to exploit this property, it is better to consider explicitly normalized wave functions. First of all we define:

$$|v_{0,\alpha}\rangle \equiv \frac{|\Psi_\alpha\rangle}{||\Psi_\alpha||}, \tag{6.14}$$

where $||\Psi_\alpha||$ indicates the norm of the state $|\Psi_\alpha\rangle$. Then, we define a set of states (one for each value of $k = 1, \ldots, p$):

$$|v_{k,\alpha}\rangle \equiv (\mathcal{O}_k - \overline{\mathcal{O}}_k)|v_{0,\alpha}\rangle, \tag{6.15}$$

where

$$\overline{\mathcal{O}}_k = \langle v_{0,\alpha}|\mathcal{O}_k|v_{0,\alpha}\rangle = \frac{\langle\Psi_\alpha|\mathcal{O}_k|\Psi_\alpha\rangle}{\langle\Psi_\alpha|\Psi_\alpha\rangle}. \tag{6.16}$$

The states $|v_{k,\alpha}\rangle$ are orthogonal to $|v_{0,\alpha}\rangle$, as easily verified when using Eq. (6.16); however, they are neither normalized nor orthogonal to each other, i.e., in general $\langle v_{k,\alpha}|v_{k',\alpha}\rangle \neq \delta_{k,k'}$ for $k, k' \neq 0$. Therefore, the set of states $|v_{0,\alpha}\rangle$ and $\{|v_{k,\alpha}\rangle\}$ defines a *semi-orthogonal* basis.

In order to compute the normalized wave function when the parameter α_k is changed, we first compute the norm of $|\Psi_{\alpha+\delta\alpha_k}\rangle$:

$$||\Psi_{\alpha+\delta\alpha_k}||^2 = \langle\Psi_\alpha|(1 + \delta\alpha_k\mathcal{O}_k)^*(1 + \delta\alpha_k\mathcal{O}_k)|\Psi_\alpha\rangle$$
$$= ||\Psi_\alpha||^2 \left[1 + 2\Re(\delta\alpha_k\overline{\mathcal{O}}_k) + O(\delta\alpha_k^2)\right]. \tag{6.17}$$

Then, we have that:

$$|v_{0,\alpha+\delta\alpha_k}\rangle = \frac{|\Psi_{\alpha+\delta\alpha_k}\rangle}{||\Psi_{\alpha+\delta\alpha_k}||} = |v_{0,\alpha}\rangle + \left[\delta\alpha_k\mathcal{O}_k - \Re(\delta\alpha_k\overline{\mathcal{O}}_k)\right]|v_{0,\alpha}\rangle + O(\delta\alpha_k^2)$$
$$= \left[1 + i\Im(\delta\alpha_k\overline{\mathcal{O}}_k)\right]|v_{0,\alpha}\rangle + \delta\alpha_k|v_{k,\alpha}\rangle + O(\delta\alpha_k^2), \tag{6.18}$$

which can be finally recast as:

$$|v_{0,\alpha+\delta\alpha_k}\rangle = \exp(i\delta\phi)\left[|v_{0,\alpha}\rangle + \delta\alpha_k|v_{k,\alpha}\rangle\right] + O(\delta\alpha_k^2), \qquad (6.19)$$

where $\delta\phi = \Im(\delta\alpha_k\overline{\mathcal{O}}_k)$.

By using the above expression, it is immediate to work out the derivative of the variational energy E_α with respect to a given variational parameter α_k:

$$\frac{\partial E_\alpha}{\partial\alpha_k} = \lim_{\delta\alpha_k\to 0}\frac{\langle v_{0,\alpha+\delta\alpha_k}|\mathcal{H}|v_{0,\alpha+\delta\alpha_k}\rangle - \langle v_{0,\alpha}|\mathcal{H}|v_{0,\alpha}\rangle}{\delta\alpha_k}$$

$$= \langle v_{k,\alpha}|\mathcal{H}|v_{0,\alpha}\rangle + \langle v_{0,\alpha}|\mathcal{H}|v_{k,\alpha}\rangle = 2\Re\left[\frac{\langle\Psi_\alpha|\mathcal{H}(\mathcal{O}_k - \overline{\mathcal{O}}_k)|\Psi_\alpha\rangle}{\langle\Psi_\alpha|\Psi_\alpha\rangle}\right]. \quad (6.20)$$

Notice also that, as expected, the phase factor $\delta\phi$ does not enter in the above expression.

In order to evaluate the force f_k by a standard Monte Carlo sampling, we introduce a completeness relation to have:

$$f_k = -2\Re\left[\frac{\sum_x\langle\Psi_\alpha|\mathcal{H}|x\rangle\langle x|\left(\mathcal{O}_k - \overline{\mathcal{O}}_k\right)|\Psi_\alpha\rangle}{\sum_x\langle\Psi_\alpha|x\rangle\langle x|\Psi_\alpha\rangle}\right]$$

$$= -2\Re\left[\sum_x\frac{e_L^*(x)\left(\mathcal{O}_k(x) - \overline{\mathcal{O}}_k\right)|\Psi_\alpha(x)|^2}{\sum_x|\Psi_\alpha(x)|^2}\right], \qquad (6.21)$$

where $e_L^*(x)$ is the complex conjugate of the local energy (here, we omitted the index α, in harmony with the notation adopted for the local operators $\mathcal{O}_k$), as it is generally a complex-valued function. Then, f_k can be evaluated by considering:

$$f_k \approx -2\Re\left[\frac{1}{N}\sum_{i=1}^N e_L^*(x_i)\left(\mathcal{O}_k(x_i) - \overline{\mathcal{O}}_k\right)\right], \qquad (6.22)$$

$$\overline{\mathcal{O}}_k \approx \frac{1}{N}\sum_{i=1}^N\mathcal{O}_k(x_i). \qquad (6.23)$$

We remark that for the evaluation of the errorbars, it is important to take advantage of the correlation between the expectation value $\overline{\mathcal{O}}_k$ and the actual derivatives f_k, which must be computed using the same sample $\{x_i\}$.

Finally, we emphasize two main properties of the above expression of derivatives, which represent the basis of any efficient stochastic minimization method:

1. From Eq. (6.22), we have that if the wave function is an exact eigenstate of $\mathcal{H}$, the local energy coincides with the corresponding exact eigenvalue E, regardless the sample $\{x_i\}$. This implies that f_k identically vanishes without statistical fluctuations and thus we recover the zero-variance property for energy derivatives.

2. In the general case, the variance is always bounded by the square of the number of particles N_p^2, because the local energy is an extensive random variable that is multiplied by a derivative estimator, i.e., $O_k(x)$, that is also at most extensive. Thus the variance of the product of two extensive quantity is at most $O(N_p^2)$. This implies that these expressions can be used to optimize wave functions with a large number of particles, since, at variance with methods based on the reweighting technique, the statistical fluctuations to estimate energy derivatives increases at most polynomially with the number of particles.

6.4 The Stochastic Reconfiguration

The knowledge of energy derivatives of Eq. (6.9) allows us to employ the steepest-descent method (Press et al., 2007) to change the variational parameters $\boldsymbol{\alpha} = (\alpha_1, \ldots, \alpha_p)$, even when p is very large:

$$\alpha_k' = \alpha_k + \delta\alpha_k, \tag{6.24}$$

$$\delta\alpha_k = \Delta f_k, \tag{6.25}$$

where Δ is an arbitrary (small) constant. In principle, its value can be optimized to reach the lowest possible energy at each iteration; however, in most applications, it is a common practice to keep Δ constant along the minimization procedure. Then, the variational parameters are iteratively improved along a Markov chain procedure. In absence of noise, the steepest-descent method always converges to a minimum, where the Euler conditions $f_k = 0$ are satisfied. Indeed, let us suppose that $f_k \neq 0$, then the energy for $\boldsymbol{\alpha}'$ is given by a Taylor expansion to linear order in Δ:

$$E_{\alpha'} = E_\alpha + \sum_k \frac{\partial E_\alpha}{\partial \alpha_k} \delta\alpha_k + O(\Delta^2) = E_\alpha - \Delta \sum_k f_k^2 + O(\Delta^2), \tag{6.26}$$

where we used that $\partial E_\alpha / \partial \alpha_k = -f_k$ and $\delta\alpha_k = \Delta f_k$. Therefore, for small Δ, when the linear truncation is accurate enough in the Taylor expansion, we obtain that:

$$\Delta E \equiv E_{\alpha'} - E_\alpha = -\Delta \sum_k f_k^2 \leq 0; \tag{6.27}$$

here, the equality sign holds only when $f_k = 0$. Thus, the method converges to a minimum for a large number of iterations just because the energy monotonically decreases with the number of iterations. Within the steepest-descent method only the first derivative of the energy is computed and it is certain that a small change of the parameters $\delta\boldsymbol{\alpha} = (\delta\alpha_1, \ldots, \delta\alpha_p)$ parallel to the force $\mathbf{f} = (f_1, \ldots, f_p)$ will decrease the energy; the only issue concerns the size of Δ, which must be taken sufficiently small to make the quadratic term in Eq. (6.26) negligible.

Let us discuss a simple case where all the parameters are "equivalent" and consider a wave function that is a linear combination of p orthonormal Slater determinants $|D_k\rangle$ (i.e., $\langle D_k|D_{k'}\rangle = \delta_{k,k'}$):

$$|\Psi_\alpha\rangle = \sum_k \alpha_k |D_k\rangle; \qquad (6.28)$$

then, it is reasonable to make a search with the condition that

$$\delta s^2 = \sum_k \delta \alpha_k^2 \qquad (6.29)$$

is small enough along the optimization procedure, where it is assumed that for small δs the Taylor expansion of the energy to first order in $\delta \alpha$ is accurate enough. In order to enforce the constraint of Eq. (6.29), the usual scheme is to introduce a Lagrange multiplier μ and minimize the quadratic form:

$$\Delta E + \mu \delta s^2 = \sum_k \left(-f_k \delta \alpha_k + \mu \delta \alpha_k^2 \right); \qquad (6.30)$$

this approach yields a minimum condition:

$$\delta \alpha_k = \frac{f_k}{2\mu}, \qquad (6.31)$$

which is compatible with the steepest-descent approach of Eq. (6.25) with $\Delta = 1/2\mu$. Although the above argument makes sense for simple wave functions, as for example the ones with linear dependence upon the parameters, serious difficulties arise in strongly-correlated states, where the dependence on the variational parameters is highly non-linear (e.g., in the Jastrow factors). Here, a small change of a given variational parameter can produce very different wave functions and physical quantities, whereas another parameter may weakly affect the wave function. In order to overcome these difficulties, it is important to introduce a more appropriate metric δs^2 that is used to estimate the "proximity" of two normalized (complex) wave functions $|v_{0,\alpha}\rangle$ and $|v_{0,\alpha+\delta\alpha}\rangle$:

$$\delta s^2 = \text{Min}_{\delta\theta} || \exp(-i\delta\theta) v_{0,\alpha+\delta\alpha} - v_{0,\alpha} ||^2. \qquad (6.32)$$

In the previous definition of δs^2, the minimization on the phase factor $\delta\theta$ is necessary because we do not want to distinguish between two wave functions that differ only by an overall phase factor, as they produce the same correlation functions. In other words, we want to define a distance δs^2 that vanishes when we have physically equivalent wave functions. Then, we replace in Eq. (6.32) the expression for $|v_{0,\alpha+\delta\alpha}\rangle$ that is obtained by generalizing Eq. (6.19) to the case where several parameters are changed:

$$|v_{0,\alpha+\delta\alpha}\rangle = \exp(i\delta\phi)\left[|v_{0,\alpha}\rangle + \sum_k \delta\alpha_k|v_{k,\alpha}\rangle\right] + O(|\delta\alpha|^2), \qquad (6.33)$$

where $\delta\phi = \sum_k \Im(\delta\alpha_k\overline{\mathcal{O}}_k)$. Now, the minimization over $\delta\theta$ gives $\delta\theta = \delta\phi$, thus leading to:

$$\delta s^2 = \sum_{k,k'}\langle v_{k,\alpha}|v_{k',\alpha}\rangle\delta\alpha_k\delta\alpha_{k'} + o\left(|\delta\alpha|^2\right). \qquad (6.34)$$

Since all increments $\delta\alpha_k$ are assumed real (as discussed previously, here we assume that all parameters are real), we can symmetrize the previous expression with respect to the indices k and k' and neglect the terms that are $o(|\delta\alpha|^2)$, obtaining:

$$\delta s^2 = \sum_{k,k'}\left(\frac{\langle v_{k,\alpha}|v_{k',\alpha}\rangle + \langle v_{k',\alpha}|v_{k,\alpha}\rangle}{2}\right)\delta\alpha_k\delta\alpha_{k'}. \qquad (6.35)$$

In this way, we can finally identify a matrix $\mathbf{S}$ that fully determines the metric in the space of normalized wave functions:

$$S_{k,k'} = \Re\left(\langle v_{k,\alpha}|v_{k',\alpha}\rangle\right), \qquad (6.36)$$

which implies that the distance between two wave functions reads:

$$\delta s^2 = \sum_{k,k'}S_{k,k'}\delta\alpha_k\delta\alpha_{k'}. \qquad (6.37)$$

At this point, it is natural to improve the steepest-descent method by using the metric given by $\mathbf{S}$. The minimization of $\Delta E + \mu\delta s^2$ with the metric δs^2 given in Eq. (6.37) improves the convergence to the minimum of the variational energy with respect to the simple steepest-descend approach, as non-equivalent parameters can be appropriately changed with a different scale. This approach is called *stochastic reconfiguration* (Sorella, 1998 and 2001). The minimization of $\Delta E + \mu\delta s^2$ gives:

$$\sum_{k'}S_{k,k'}\delta\alpha_{k'} = \frac{f_k}{2\mu}, \qquad (6.38)$$

which is a set of linear equations for the unknown vector $\delta\alpha$. After having solved this linear system, we can update the variational parameters until convergence is reached; as in the steepest-descent method, we can set $\Delta = 1/(2\mu)$ small enough, which may be kept fixed during the optimization. We would like to stress the fact that, since the matrix $\mathbf{S}$ is strictly positive definite, the energy is monotonically decreasing along the optimization as:

$$\Delta E = -\Delta \sum_{k,k'} S_{k,k'}^{-1} f_k f_{k'} < 0. \tag{6.39}$$

Within a Monte Carlo procedure, the matrix $\mathbf{S}$ is evaluated by a finite sampling of N configurations $\{x_i\}$ as:

$$S_{k,k'} \approx \Re \left[\frac{1}{N} \sum_{i=1}^{N} \left(\mathcal{O}_k(x_i) - \overline{\mathcal{O}}_k \right) \left(\mathcal{O}_{k'}(x_i) - \overline{\mathcal{O}}_{k'} \right) \right]; \tag{6.40}$$

the forces are also computed in a similar way, see Eq. (6.22). Therefore, the solution of the linear system (6.38) is affected by statistical errors, yielding statistical fluctuations of the final variational parameters $\{\alpha_k\}$, even when convergence has been reached. In this regime, it is convenient to perform several iterations, in order to obtain accurate values for the variational parameters by averaging them after equilibration.

By considering the limit of $\Delta \to 0$, the optimization procedure reduces to a set of stochastic differential equations that are approximately described by a Langevin dynamics (see Chapter 4):

$$\frac{d\boldsymbol{\alpha}(t)}{dt} = \mathbf{S}^{-1}\mathbf{f}[\boldsymbol{\alpha}(t)] + \boldsymbol{\eta}(t), \tag{6.41}$$

where we have explicitly indicated that the forces depend upon the variational parameters. Here, we neglect the correlation between the force components in the statistical noise, since this aspect does not alter the qualitative discussion. The noise is described by:

$$\langle \eta_k(t) \rangle = 0, \tag{6.42}$$

$$\langle \eta_k(t)\eta_{k'}(t') \rangle = 2T_{\text{noise}}\delta_{k,k'}\delta(t-t'). \tag{6.43}$$

Within a Monte Carlo optimization, the parameter T_{noise} depends on the fact that both $\mathbf{f}$ and $\mathbf{S}$ are evaluated by using a finite sampling with N configurations. We would like to emphasize that the inversion of the matrix $\mathbf{S}$ introduces a controllable bias, which vanishes for $N \to \infty$. This is due to the fact that the matrix elements $S_{k,k'}$, obtained from Eq. (6.40), are random variables distributed around the corresponding exact values with fluctuations that are $O(1/\sqrt{N})$. Therefore, the noise can be tuned to zero by increasing N as $T_{\text{noise}} \propto 1/N$, for the central limit theorem. The number of iterations necessary to reach convergence is weakly dependent on the "temperature" T_{noise} (see Chapter 4); however, it crucially depends on the energy landscape of the system. The optimal value for N is the smallest one that provides the fluctuations of the parameters within the desired accuracy. It should be noted

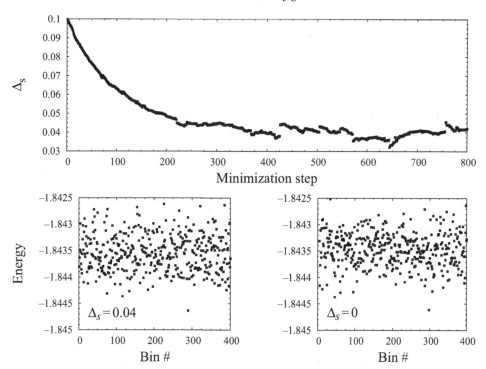

Figure 6.1 Results for the attractive Hubbard model with $U/t = -1$ for $L = 162$ sites at half filling (i.e., $N_e = L$). The wave function is obtained by applying a density-density Jastrow factor (1.65) to a BCS wave function (1.91) with on-site s-wave pairing $\Delta_k = \Delta_s$; the number of Jastrow pseudo-potentials is 29. Upper panel: by using the stochastic reconfiguration, the optimization of Δ_s as a function of the Monte Carlo steps, here $N = 5000$. Lower panels: the variational energy calculated by fixing all the variational parameters, for the optimal value $\Delta_s = 0.04$ (left panel) and for $\Delta_s = 0$ (right panel).

that the variational parameters, averaged over the Langevin simulation, are close to the true energy minimum, but may be affected by a bias that scales to zero with T_{noise}, due to the presence of non-quadratic terms in the energy landscape. Therefore, the best strategy to obtain accurate results is to average the variational parameters by systematically increasing the value of N.

As an example for the convergence of the stochastic reconfiguration, we report in Fig. 6.1 the case of the attractive Hubbard model with $U/t = -1$ at half filling, with $N_e = L = 162$. The variational wave function is given by the BCS state (1.91) with on-site s-wave pairing $\Delta_k = \Delta_s$, supplemented with the density-density Jastrow factor (1.65). For this value of the electron-electron interaction, a small value of Δ_s is stabilized, i.e., $\Delta_s \approx 0.04$. We would like to emphasize that the energy gain due to this term is $\Delta E = 0.00007(2)$, which is much smaller than the energy fluctuations that are present in the simulation; nevertheless, an accurate determination of this variational parameter is possible by using the stochastic optimization, demonstrating the power of such Monte Carlo methods.

6.4.1 Covariance Property

One important property of the stochastic reconfiguration optimization is that it has the same efficiency when the parametrization of the many-body state is changed without changing its form. This fact represents an important improvement with respect to the simple steepest-descent method. For example, let us consider the case in which the wave function contains the Gutzwiller factor g and a fugacity term f (suitable whenever the number of fermions is not fixed along the simulation); for the fermionic case, in which $n_i = n_{i,\uparrow} + n_{i,\downarrow}$ and $n_{i\sigma}^2 = n_{i\sigma}$, the correlation term can be written in two equivalent ways:

$$\exp\left(-\sum_i \frac{g}{2}n_i^2 - \sum_i f n_i\right) = \exp\left(-\sum_i f'n_i\right)\prod_i\left[1 + (g' - 1)n_{i\uparrow}n_{i\downarrow}\right], \quad (6.44)$$

where $g' = \exp(-g)$ and $f' = f + g/2$. Choosing one parametrization (with g and f) or the other (with g' and f') does not necessarily give the same efficiency in the optimization procedure (while it must give the same energy at equilibrium, apart from problems related to local minima). Here, we show that, within the stochastic reconfiguration technique, the minimization procedure has the same effectiveness when changing the way of parameterizing the quantum state.

In the general case, let us consider an arbitrary transformation of the variational parameters:

$$\alpha_k' = \alpha_k'(\alpha_1, \ldots, \alpha_p), \quad (6.45)$$

which gives the same quantum state, i.e., $\Psi_{\alpha'}(x) = \Psi_\alpha(x)$. Then, we want to impose that, after one step of optimization, the same improvement is obtained when considering the parametrization with α or α'. The condition $\Psi_{\alpha'+\delta\alpha'}(x) = \Psi_{\alpha+\delta\alpha}(x)$ would require that:

$$\delta\alpha_k' = \sum_{k'} \frac{\partial\alpha_k'}{\partial\alpha_{k'}}\delta\alpha_{k'} = (\mathbf{J}\delta\alpha)_k, \quad (6.46)$$

where we have defined the Jacobian matrix by $J_{k,k'} = \frac{\partial\alpha_k'}{\partial\alpha_{k'}}$. The previous relation is indeed satisfied within the stochastic reconfiguration technique. In fact, the local operator $\mathcal{O}_k'(x)$, which enters into the calculation of the energy derivative, is given by:

$$\mathcal{O}_k'(x) = \frac{1}{\Psi_{\alpha'}(x)}\frac{\partial\Psi_{\alpha'}(x)}{\partial\alpha_k'} = \sum_{k'}\frac{\partial\alpha_{k'}}{\partial\alpha_k'}\left(\frac{1}{\Psi_\alpha(x)}\frac{\partial\Psi_\alpha(x)}{\partial\alpha_{k'}}\right) = \sum_{k'}J_{k,k'}^{-1}\mathcal{O}_{k'}(x), \quad (6.47)$$

where we have used the fact that $\Psi_{\alpha'}(x) = \Psi_{\alpha}(x)$ and the property that the Jacobian of the inverse transformation is just the inverse of the Jacobian of the direct transformation. In this way, the new forces $\mathbf{f}'$ and the new matrix $\mathbf{S}'$ are:

$$\mathbf{f}' = \mathbf{J}^{-1}\mathbf{f}, \tag{6.48}$$

$$\mathbf{S}' = \mathbf{J}^{-1}\mathbf{S}\mathbf{J}^{-1}. \tag{6.49}$$

Therefore, the iteration step with the new parametrization α' is given by:

$$\delta\alpha' = \Delta(\mathbf{S}')^{-1}\mathbf{f}' = \Delta\left(\mathbf{J}^{-1}\mathbf{S}\mathbf{J}^{-1}\right)^{-1}\mathbf{J}^{-1}\mathbf{f} = \mathbf{J}\delta\alpha, \tag{6.50}$$

which is exactly what we required at the beginning. This fact can be named *covariance property* of the method, because (exactly as in general relativity) the matrix $\mathbf{S}$ represents the metric of our non-linear space (of wave functions) and the direction of the optimization is covariant and does not depend on the way in which the wave function is represented. This property is very important since it implies that there is no way to improve the method by performing any non-linear transformation among the parameters.

6.4.2 A Scale-Invariant Regularization

A fundamental ingredient along the stochastic optimization is the inversion of the matrix $\mathbf{S}$, which is a non-negative matrix, even when considering a finite sampling of N configurations. However, it may happen that it has eigenvalues that differ by several order of magnitudes and, therefore, it is always worth doing a *pre-conditioning* that stabilizes the inversion, namely considering the scaling:

$$S_{k,k'}^{\text{pc}} = \frac{S_{k,k'}}{\sqrt{S_{k,k}S_{k',k'}}}. \tag{6.51}$$

In this way, we first find the solutions of the linear equations:

$$\sum_{k'} S_{k,k'}^{\text{pc}} \delta\alpha_{k'}^{\text{pc}} = \frac{f_k^{\text{pc}}}{2\mu}, \tag{6.52}$$

where the scaled unknowns are $\delta\alpha_{k'}^{\text{pc}} = \sqrt{S_{k',k'}}\delta\alpha_{k'}$ and the scaled forces are $f_k^{\text{pc}} = f_k/\sqrt{S_{k,k}}$; then, we obtain the original $\delta\alpha$ by a simple rescaling $\delta\alpha_k = \delta\alpha_k^{\text{pc}}/\sqrt{S_{k,k}}$.

A most serious problem arises whenever the matrix $\mathbf{S}^{\text{pc}}$ acquires very small and even negligible eigenvalues, implying that the application of $(\mathbf{S}^{\text{pc}})^{-1}$ to the force $\mathbf{f}^{\text{pc}}$ can be unstable, especially if we consider that both quantities are evaluated within a stochastic procedure. For example, if there is a redundancy in the parametrization of the wave function, one particular linear combination of the forces will vanish, as well as the matrix $\mathbf{S}^{pc}$ will be singular. For example, by diagonalizing $\mathbf{S}^{pc}$, we have that:

$$\lambda_i \, (\delta\alpha^{pc})_i^U = (f^{pc})_i^U, \tag{6.53}$$

where λ_i are the eigenvalues of $\mathbf{S}^{pc} = \mathbf{U}^\dagger \Lambda \mathbf{U}$, $(\delta\boldsymbol{\alpha}^{pc})^U = \mathbf{U}\delta\boldsymbol{\alpha}^{pc}$, and $(\mathbf{f}^{pc})^U = \mathbf{U}\mathbf{f}^{pc}$. For a redundant parameter, $(f^{pc})_k^U = \lambda_k = 0$, which makes the solution of the problem undetermined. Therefore, it is clear that some form of regularization is necessary. A simple but efficient possibility is to modify the diagonal elements only:

$$S_{k,k'}^{pc,\epsilon} = S_{k,k'}^{pc} + \epsilon\delta_{k,k'}. \tag{6.54}$$

This defines a regular matrix that can be safely inverted for any $\epsilon > 0$; in this way, we get rid of a possible dependence among the variational parameters. Indeed, Eq. (6.53) is modified into:

$$(\lambda_i + \epsilon)(\delta\alpha^{pc})_i^U = (f^{pc})_i^U, \tag{6.55}$$

which implies that $(\delta\alpha^{pc})_i^U \approx 0$ for the cases in which $(f^{pc})_i^U \approx 0$ and $\lambda_i \ll \epsilon$. Therefore, the effect of introducing a finite ϵ is to keep fixed the parameter direction along which $(f^{pc})_i^U \approx 0$. Since $(f^{pc})_i^U \approx \sqrt{\lambda_i}$, neglecting this direction implies an error in the optimization of the order of $\sqrt{\epsilon}$.

Although this regularization does not maintain the covariance property discussed above, it preserves a subset of transformations obtained when scaling the parameters by arbitrary constants ξ_k:

$$\alpha_k' = \xi_k\alpha_k; \tag{6.56}$$

indeed, we can immediately verify this property by using that, in this case, $\mathbf{J}$ is diagonal and, therefore, $\mathbf{S}^{pc}$ of Eq. (6.51) remains unchanged when performing the transformation (6.49) on the matrix $\mathbf{S}$. Therefore, such a regularization is *scale invariant*. The smaller is the value of ϵ the faster will be the approach to the lowest variational energy, but the probability of an instability will be higher (if the matrix $\mathbf{S}$ is singular); a common practice is to use $\epsilon \approx 10^{-3}$, for a reasonable efficiency and accuracy.

6.4.3 Optimization Using the Signal to Noise Ratio

At this point, we want to discuss a very simple argument that explains the reason why the method described above represents an efficient stochastic optimization procedure. The generalized force component f_k, defined in Eq. (6.22) is a random variable with a corresponding noise, whose value can be estimated statistically, as well as its correlation with the other components. Indeed, the covariance matrix can be evaluated by using jackknife or bootstrap methods described in section 3.10:

$$\sigma_{k,k'}^2 \approx \langle\langle f_k f_{k'} \rangle\rangle - \langle\langle f_k \rangle\rangle \langle\langle f_{k'} \rangle\rangle. \tag{6.57}$$

Let us now consider a particular direction in the parameter space that is described by taking $\boldsymbol{\alpha}(\Delta) = \boldsymbol{\alpha}_0 + \Delta \boldsymbol{\tau}$, where Δ parametrizes a line change of the parameters in the direction of the vector $\boldsymbol{\tau}$, starting from $\boldsymbol{\alpha}_0$. This implies that the generalized force f_τ along the direction $\boldsymbol{\tau}$ is given by:

$$f_\tau = -\frac{\partial E_\alpha}{\partial \Delta} = -\sum_k \frac{\partial E_\alpha}{\partial \alpha_k} \frac{\partial \alpha_k}{\partial \Delta} = \mathbf{f} \cdot \boldsymbol{\tau}. \tag{6.58}$$

Its standard deviation is:

$$\sigma_\tau = \sqrt{\sum_{k,k'} \tau_k \sigma_{k,k'}^2 \tau_{k'}}. \tag{6.59}$$

Now, we can look at the direction where the signal (i.e., the actual value of the derivative) is the largest possible compared to its standard deviation. A change of parameters along this direction guarantees a lowering of the energy if the squared signal to noise ratio

$$\Sigma^2(\boldsymbol{\tau}) = \frac{\sum_{k,k'} \tau_k f_k f_{k'} \tau_{k'}}{\sum_{k,k'} \tau_k \sigma_{k,k'}^2 \tau_{k'}} \tag{6.60}$$

is much larger than one (e.g., if the derivative along a given direction is non-zero by more than 3 standard deviations). In this case, the energy will be lowered by changing $\boldsymbol{\alpha}_0 \to \boldsymbol{\alpha}_0 + \Delta \boldsymbol{\tau}$, with Δ sufficiently small. In this sense, a simple standard maximization yields that the "best direction" is given by:

$$\tau_k = \Delta_\tau \sum_{k'} (\sigma^2)_{k,k'}^{-1} f_{k'}, \tag{6.61}$$

where Δ_τ is an arbitrary constant, as the maximum of $\Sigma^2(\boldsymbol{\tau})$ does not change when scaling the solution by an arbitrary constant. In practice, for accurate variational wave functions, the covariance matrix is expected to be close to the matrix $\mathbf{S}$, since the fluctuations of the force component f_k are mainly given by the fluctuations of $(\mathcal{O}_k - \overline{\mathcal{O}}_k)$, and the fluctuations of the local energy can be neglected, leading to:

$$\sigma_{k,k'}^2 \simeq S_{k,k'}. \tag{6.62}$$

Therefore, Eqs. (6.61) and (6.62) explain the reason why the stochastic reconfiguration approach gives a particularly efficient way to perform Monte Carlo optimizations. In fact, the minimization according to Eq. (6.38) follows (approximately) the one in which there is a maximal signal to noise ratio, which is fundamental for a stable stochastic optimization.

6.5 Stochastic Reconfiguration as a Projection Technique

Here, we show that the stochastic reconfiguration method is equivalent, in the case of real wave functions, to an imaginary-time projection technique, which is restricted in the subspace defined by $|v_{0,\alpha}\rangle$ and $\{|v_{k,\alpha}\rangle\}$, as defined in Eqs. (6.14) and (6.15). In order to prove this statement, let us define a wave function that is obtained from the variational one $|\Psi_\alpha\rangle$ by applying one step of the power method (see section 1.7):

$$|\Psi_\Delta\rangle = (1 - \Delta\mathcal{H})|\Psi_\alpha\rangle, \tag{6.63}$$

where Δ is a small constant, such to improve the energy with respect to $|\Psi_\alpha\rangle$. Here, we consider an imaginary-time evolution (i.e., real Δ), but all the results are also valid for a real-time propagation (i.e., $\Delta \to i\Delta$), which allows the basis for simulating quantum dynamics within variational Monte Carlo (Carleo et al., 2012). However, in this case, we cannot assume that the variational parameters are real because they will be turned to complex already after the first steps of dynamics. Therefore, we are led to consider general complex quantum states, that are analytic functions of the variational parameters . Then, it is easy to show that Eqs. (6.14), (6.15), and (6.19) remain unchanged, even when considering complex variations $\{\delta\alpha_k\}$. In general, for any finite value of Δ, the state $|\Psi_\Delta\rangle$ will be different from any state that can be parametrized by using the original variational form, namely $|\Psi_\Delta\rangle \neq |\Psi_{\alpha'}\rangle$ for any choice of the parameters α'. Nevertheless, for small enough Δ, $|\Psi_\Delta\rangle$ is very close to $|\Psi_\alpha\rangle$ and we may ask whether a suitable parametrization with $\alpha' = \alpha + \delta\alpha$ can give an accurate representation of $|\Psi_\Delta\rangle$. In particular, we can minimize the "distance" between $|\Psi_\Delta\rangle$ and $e^{i\delta\theta}|\Psi_{\alpha+\delta\alpha}\rangle$ (where $\delta\theta$ is a generic angle that has to be taken into account for the real-time propagation). By using the notations of Eqs (6.14) and (6.15), we have:

$$|\Psi_\Delta\rangle = ||\Psi_\alpha||(1 - \Delta\mathcal{H})|v_{0,\alpha}\rangle, \tag{6.64}$$

$$|\Psi_{\alpha+\delta\alpha}\rangle = e^{i\delta\phi}||\Psi_{\alpha+\delta\alpha}|| \left[|v_{0,\alpha}\rangle + \sum_k \delta\alpha_k |v_{k,\alpha}\rangle \right], \tag{6.65}$$

where $||\Psi_{\alpha+\delta\alpha}|| = C||\Psi_\alpha||$ with $C \approx 1$, see Eq. (6.17). The minimization of the distance can be achieved by projecting $|\Psi_\Delta\rangle$ into the subspace spanned by the semi-orthogonal basis set:

$$\langle v_{0,\alpha}|\Psi_\Delta\rangle = e^{i\delta\theta}\langle v_{0,\alpha}|\Psi_{\alpha+\delta\alpha}\rangle, \tag{6.66}$$

$$\langle v_{k,\alpha}|\Psi_\Delta\rangle = e^{i\delta\theta}\langle v_{k,\alpha}|\Psi_{\alpha+\delta\alpha}\rangle, \tag{6.67}$$

which lead to the following set of linear equations for the $\delta\alpha_k$'s:

$$1 - \Delta\langle v_{0,\alpha}|\mathcal{H}|v_{0,\alpha}\rangle = Ce^{i(\delta\theta+\delta\phi)}, \tag{6.68}$$

$$-\Delta\langle v_{k,\alpha}|\mathcal{H}|v_{0,\alpha}\rangle = Ce^{i(\delta\theta+\delta\phi)} \sum_{k'}\langle v_{k,\alpha}|v_{k',\alpha}\rangle\delta\alpha_{k'}. \tag{6.69}$$

Here, the first equation defines the normalization constant $C = 1 + O(\Delta)$, which is irrelevant, ($\delta\theta = -\delta\phi$ for real Δ), while the second one can be recast as:

$$\sum_{k'}\tilde{S}_{k,k'}\delta\alpha_{k'} = \frac{\tilde{f}_k}{2\mu}, \tag{6.70}$$

where $\mu = Ce^{i\theta}/\Delta$ and both the force component $\tilde{f}_k$ and the matrix $\tilde{S}_{k,k'}$ are generally complex and given by:

$$\tilde{f}_k = -2\langle v_{k,\alpha}|\mathcal{H}|v_{0,\alpha}\rangle = -2\frac{\langle\Psi_\alpha|\left(\mathcal{O}_k - \overline{\mathcal{O}}_k\right)^*\mathcal{H}|\Psi_\alpha\rangle}{\langle\Psi_\alpha|\Psi_\alpha\rangle}, \tag{6.71}$$

$$\tilde{S}_{k,k'} = \langle v_{k,\alpha}|v_{k',\alpha}\rangle = \frac{\langle\Psi_\alpha|\left(\mathcal{O}_k - \overline{\mathcal{O}}_k\right)^*\left(\mathcal{O}_{k'} - \overline{\mathcal{O}}_{k'}\right)|\Psi_\alpha\rangle}{\langle\Psi_\alpha|\Psi_\alpha\rangle}. \tag{6.72}$$

We would like to emphasize that the projection approach of Eq. (6.70) coincides with the minimization algorithm we have derived in Eq. (6.38) only when the variational wave function $\Psi_\alpha(x)$ is real.

We finally remark that, also within the projection technique, the matrix $\tilde{S}$ requires some regularization to avoid too small eigenvalues in its inversion. In the case where we want to integrate the short time dynamics with an accuracy Δ^2, then the regularization cutoff has to be scaled according to $\epsilon \propto \Delta^4$. This is not a problem in principle, but makes prohibitive a highly accurate time integration, since we have to increase the number of samples $N \propto 1/\epsilon^2$, in order to obtain a statistical accuracy that is $O(\epsilon)$; otherwise, the error is larger than the target accuracy. Fortunately, this issue is not a problem if we are interested only to the optimization of the variational wave function and not to the accurate description of the exact time propagation, as in real time quantum dynamics.

6.6 The Linear Method

The stochastic reconfiguration method performs very well whenever there are not very different energy scales in the optimization problem, as in the case of lattice models or in *ab-initio* electronic calculations where the high-energy scales of the core electrons are eliminated by using pseudo-potentials. However, in particular cases, it may happen that the speed of convergence to the minimum remains extremely slow, also within the stochastic reconfiguration scheme. An established way to speed-up any optimization scheme would be to take into account the exact second derivatives of the energy with respect to the variational parameters; this can be done by using the Newton-Raphson scheme, which is extremely efficient, since it requires very few steps to converge to an energy minimum. Unfortunately, the exact evaluation of the Hessian matrix:

$$H_{k,k'} = \frac{\partial^2 E_\alpha}{\partial \alpha_k \partial \alpha_{k'}} \tag{6.73}$$

is quite cumbersome in practice. Indeed, this would require to compute the Taylor expansion of the variational wave function up to the second order in the variation of the parameters α:

$$\Psi_{\alpha+\delta\alpha}(x) \approx \Psi_\alpha(x) + \sum_k \delta\alpha_k \frac{\partial \Psi_\alpha(x)}{\partial \alpha_k} + \frac{1}{2} \sum_{k,k'} \delta\alpha_k \delta\alpha_{k'} \frac{\partial^2 \Psi_\alpha(x)}{\partial \alpha_k \partial \alpha_{k'}}; \tag{6.74}$$

as before, see Eqs. (6.11) and (6.12), we can express the second derivative in terms of local operators $\mathcal{O}_{k,k'}$:

$$\langle x | \mathcal{O}_{k,k'} | x' \rangle = \delta_{x,x'} \mathcal{O}_{k,k'}(x), \tag{6.75}$$

$$\mathcal{O}_{k,k'}(x) = \frac{1}{\Psi_\alpha(x)} \frac{\partial^2 \Psi_\alpha(x)}{\partial \alpha_k \partial \alpha_{k'}}; \tag{6.76}$$

however, the actual evaluation of these operators is not easy in general (while it can be afforded for Jastrow parameters).

Instead of using the full Hessian approach, a great simplification arises when using the expansion of the wave function up to the linear order (i.e., by considering the first derivatives only) and then compute the expectation value of the Hamiltonian, even beyond the linear regime:

$$E_{\alpha+\delta\alpha} = \frac{\sum_{k,k'=0}^{p} z_k^* z_{k'} \langle v_{k,\alpha} | \mathcal{H} | v_{k',\alpha} \rangle}{\sum_{k,k'=0}^{p} z_k^* z_{k'} \langle v_{k,\alpha} | v_{k',\alpha} \rangle}, \tag{6.77}$$

where we have explicitly indicated that the sum includes the terms with $k = 0$ and $k' = 0$; then $z_k = \delta\alpha_k$ for $k > 0$ and $z_0 = 1$. This approach has been dubbed as *linear method* (Umrigar et al., 2007; Toulouse and Umrigar, 2007). Within this simplified approach, we have to compute the matrix elements:

$$H_{k,k'} = \langle v_{k,\alpha} | \mathcal{H} | v_{k',\alpha} \rangle, \tag{6.78}$$

$$\overline{S}_{k,k'} = \langle v_{k,\alpha} | v_{k',\alpha} \rangle, \tag{6.79}$$

where $\bar{S}_{0,0} = 1$, $\bar{S}_{k,0} = \bar{S}_{0,k} = 0$ for $k > 0$, and $\bar{S}_{k,k'} = \langle v_{k,\alpha} | v_{k',\alpha} \rangle$ for k and $k' > 0$. Then, the change $\delta\alpha_k$ can be obtained by minimizing the energy with respect to each z_k. This procedure leads to a generalized eigenvalue equation (since the states $\{|v_k\rangle\}$ are not orthogonal to each other):

$$\sum_{k'=0}^{p} H_{k,k'} z_{k'} = E \sum_{k'=0}^{p} \bar{S}_{k,k'} z_{k'}. \tag{6.80}$$

Among all possible right eigenvectors of the above generalized eigenvalue equation, it is not always convenient to take the one corresponding to the lowest (real) eigenvalue E. Indeed, whenever the matrices are particularly large, or the simulation is particularly noisy, it is better to take the eigenvector with maximum overlap $|z_0|^2$ with $|v_{0,\alpha}\rangle$; in this way, the correction $\delta\alpha_k = z_k/z_0$ is small and the expansion of $|\Psi_{\alpha+\delta\alpha}\rangle$ remains under control.

The matrix elements $\bar{S}_{k,k'}$ and $H_{k,k'}$ can be evaluated by a Monte Carlo sampling:

$$\bar{S}_{k,k'} \approx \frac{1}{N} \sum_{i=1}^{N} \left(\mathcal{O}_k(x_i) - \bar{\mathcal{O}}_k \right)^* \left(\mathcal{O}_{k'}(x_i) - \bar{\mathcal{O}}_{k'} \right), \tag{6.81}$$

$$H_{k,k'} \approx \frac{1}{N} \sum_{i=1}^{N} \left(\mathcal{O}_k(x_i) - \bar{\mathcal{O}}_k \right)^* \frac{\langle x_i | \mathcal{H} \left(\mathcal{O}_{k'} - \bar{\mathcal{O}}_{k'} \right) | \Psi_\alpha \rangle}{\langle x_i | \Psi_\alpha \rangle}, \tag{6.82}$$

in which we defined $\mathcal{O}_0 = \mathbb{I}$ and $\bar{\mathcal{O}}_0 = 0$ in order to include the cases where $k = 0$ or $k' = 0$. The factor in the r.h.s. of Eq. (6.82) can be expressed in terms of derivatives of the local energy, which can be efficiently computed within the variational Monte Carlo technique. Notice that the Hamiltonian matrix **H** is Hermitian only on average, since its Hermitian character is spoiled by statistical fluctuations that are present in samples with a finite number of configurations. Thus, it would be tempting to symmetrize the matrix to restore its Hermitian character, e.g., by considering $(H_{k,k'} + H_{k',k}^*)/2$. Instead, our experience indicates that it is convenient to solve the above equations without performing such symmetrization, in order to maintain the so-called "strong" zero-variance property. Indeed, if a particular linear combination of states in the semi-orthogonal basis

$$|\varphi\rangle = \sum_{k=0}^{p} z_k \left(\mathcal{O}_k - \bar{\mathcal{O}}_k \right) | \Psi_\alpha \rangle \tag{6.83}$$

is a right eigenstate of the system (6.80) with eigenvalue E_φ, we obtain that:

$$\sum_{k'=0}^{p} H_{k,k'} z_{k'} \approx \frac{1}{N} \sum_{i=1}^{N} (\mathcal{O}_k(x_i) - \overline{\mathcal{O}}_k)^* \frac{\langle x_i | \mathcal{H} \sum_{k'=0}^{p} z_{k'} (\mathcal{O}_{k'} - \overline{\mathcal{O}}_{k'}) | \Psi_\alpha \rangle}{\langle x_i | \Psi_\alpha \rangle}$$

$$= E_\varphi \frac{1}{N} \sum_{i=1}^{N} (\mathcal{O}_k(x_i) - \overline{\mathcal{O}}_k)^* \frac{\langle x_i | \sum_{k'=0}^{p} z_{k'} (\mathcal{O}_{k'} - \overline{\mathcal{O}}_{k'}) | \Psi_\alpha \rangle}{\langle x_i | \Psi_\alpha \rangle}$$

$$\approx E_\varphi \sum_{k'=0}^{p} \overline{S}_{k,k'} z_{k'}, \tag{6.84}$$

which implies that Eq. (6.80) is satisfied with zero variance: by estimating both $H_{k,k'}$ and $\overline{S}_{k,k'}$ with the *same* configurations, we obtain a right eigenvector of the linear system corresponding to the exact eigenstate of $\mathcal{H}$ given by the assumption of Eq. (6.83). This is a "strong" zero-variance principle because it can be satisfied even when $|\Psi_\alpha\rangle$ is not an exact eigenstate of the Hamiltonian.

We would like to make some final considerations on the linear method, which may lead, in several cases, to a substantial improvement with respect to the stochastic reconfiguration approach. Even if its implementation is more difficult than the one of the stochastic reconfiguration technique and requires some extra computational time for each iteration, the total number of iterations needed to reach a converged result are usually drastically reduced as compared with the stochastic reconfiguration technique. Unfortunately, it is not possible to quantify the actual gain in computational time that is expected when using the linear method. In fact, in quantum Monte Carlo simulations, a stable, accurate, and fast optimization can be obtained even with a slowly convergent method, since each iteration may require a small number of samples; in this way, performing thousands of iterations with the stochastic reconfiguration technique does not represent a serious problem, especially when the number of variational parameters is small enough.

In all these optimization techniques, the number of samples that are necessary to achieve a meaningful accuracy of the matrices involved in the optimization should be much larger than the leading dimension of the matrix (i.e., the number of parameters), otherwise the sampling will produce rank-deficient singular (i.e., "dirty") matrices. A practical rule of thumb is to take the number of samples N such that:

$$N \geq 10 \times p, \tag{6.85}$$

where p is the number of parameters that must be optimized.

As an example of the linear method, we show the case of the repulsive Hubbard model with $U/t = 4$, on a cluster with $L = 98$ at half filling ($N_e = L$). In this case, the variational state is given by a Jastrow-Slater wave function, where the Jastrow term contains density-density terms, see Eq. (1.65), and the Slater determinant is constructed from free-electron orbitals. The variational parameters to be optimized are the Jastrow pseudo-potentials, for each independent distances

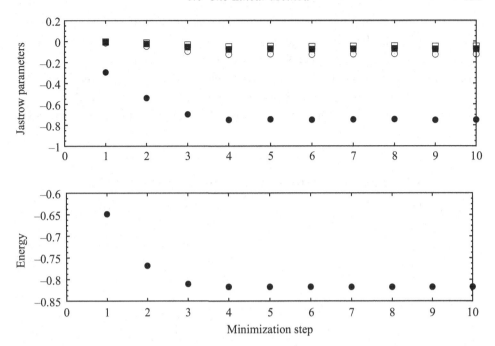

Figure 6.2 Results for the repulsive Hubbard model with $U/t = 4$ for $L = 98$ sites at half filling (i.e., $N_e = L$). The wave function is obtained by applying a density-density Jastrow factor (1.65) to a Slater determinant obtained by taking free-electron orbitals; the total number of Jastrow pseudo-potentials is 19 (including the on-site Gutzwiller term). By using the linear method, the optimization of the Jastrow pseudo-potentials (Gutzwiller term and the ones corresponding to the smallest 3 distances) is reported (upper panel). The variational energy along the minimization technique is also shown (lower panel). The number of independent Monte Carlo attempts for each optimization step is $N = 80000$.

(including the on-site Gutzwiller factor). In Fig. 6.2, we report the energy and few Jastrow parameters (the Gutzwiller term and the ones corresponding to the smallest four distances) for the first ten iterations (each one corresponding to $N = 80000$ steps). The initial parameters are set to zero. We would like to emphasize that, with the linear method, a very small number of steps $N \approx 4$ is sufficient to converge to the equilibrated values, also for the Jastrow terms that correspond to long-range distances.

Finally, we list a number of points that are still a subject of research, which have not been solved within the linear method yet:

- The "strong" zero-variance property does not hold whenever some parameters are restricted to be real (for example, usually we want to work with a real Jastrow factor). Instead, this method does not have this drawback if the calculation is restricted to a real *Ansatz* and a real Hamiltonian, since, in this case, both $H_{k,k'}$ and $\overline{S}_{k,k'}$ are real and the corresponding eigenvectors are also real.

- Far from the minimum, the change of the variational parameters given by $\delta \alpha_k = z_k/z_0$ can be rather large and sometimes may even increase the energy (because the change may exceed the regime where the linear expansion is valid). In this case, it is probably better to start the optimization with the stochastic reconfiguration technique and then apply the linear method only close to the minimum. Notice that this problem can be more and more relevant when increasing the system size.

6.7 Calculations of Derivatives in the Jastrow-Slater Case

Here, we show how derivatives of the wave function, see Eq. (6.12), can be computed in the case of a Jastrow-Slater *Ansatz*:

$$|\Psi_J\rangle = \mathcal{J}|\Phi_0\rangle, \tag{6.86}$$

where

$$\mathcal{J} = \exp\left(-\frac{1}{2}\sum_{i,j} v_{i,j} n_i n_j\right), \tag{6.87}$$

and $|\Phi_0\rangle$ is the ground state of a non-interacting (BCS) Hamiltonian:

$$\mathcal{H}_0 = \sum_{I,J} t_{I,J} d_I^\dagger d_J. \tag{6.88}$$

In the following, we consider the translationally invariant case, where both the Jastrow pseudo-potential and the parameters in $\mathcal{H}_0$ only depends upon the distance between the sites, i.e., $v_{i,j} \rightarrow v_k$ and $t_{I,J} \rightarrow t_k$. The generalization to a case where translational symmetry is broken is straightforward. A different approach can be devised for the Slater part if $|\Phi_0\rangle$ is defined without passing through an auxiliary Hamiltonian, namely defined by given single-particle orbitals that are suitably parametrized.

Let us start from the case of the Jastrow factor, which is very simple. Given the translational invariance, we have that:

$$\mathcal{J} = \exp\left(-\frac{1}{2}\sum_k v_k \sum_i n_i \sum_{s_i(k)} n_{s_i(k)}\right), \tag{6.89}$$

where $s_i(k)$ indicates all the sites at distance k from the site i, e.g., in one spatial dimension $s_i(k)$ includes the two sites on the left and right (for an even number of sites, the largest distance corresponds to a single site); in two dimensions, the number of connected sites depends upon the punctual group of the lattice, but also on the distance (e.g., on the square lattice, it can be 8, 4, 2, and 1). From Eq. (6.12), we have that:

$$\mathcal{O}_k(x) = -\frac{1}{2}\frac{\langle x|\left(\sum_i n_i \sum_{s_i(k)} n_{s_i(k)}\right)|\Psi_J\rangle}{\langle x|\Psi_J\rangle} = -\frac{1}{2}\langle x|\left(\sum_i n_i \sum_{s_i(k)} n_{s_i(k)}\right)|x\rangle, \quad (6.90)$$

where the last equality is due to the fact that n_i is diagonal in the basis $\{|x\rangle\}$ where electrons have definite positions (and spin) in the lattice, see section 5.4. Therefore, the calculation of the derivatives corresponds to the evaluation of a simple set of diagonal correlation functions.

Then, we consider the derivatives of the variational *Ansatz* with respect to the parameters contained in the non-interacting state $|\Phi_0\rangle$ for the case with determinants (see section 5.6). Then, a small change $t_k \rightarrow t_k + \delta t_k$ corresponds to a modification of the auxiliary Hamiltonian, i.e., $\mathcal{H}_0 \rightarrow \mathcal{H}_0 + \delta t_k \mathcal{V}^{(k)}$, where $\mathcal{V}^{(k)}$ is a two-body operator that includes all the hopping terms associated to t_k:

$$\mathcal{V}^{(k)} = \sum_{I,J} V_{I,J}^{(k)} d_I^\dagger d_J, \quad (6.91)$$

where the matrix $V_{I,J}^{(k)}$ is equal to 1 (0) if the hopping t_k connects (does not connect) the sites I and J. In the following, we will assume that the ground state of $\mathcal{H}_0$ is non-degenerate, implying that all orbitals corresponding to eigenvalues $\varepsilon_\alpha \leq \varepsilon_F$ are occupied, while all the others are empty. Then, the change in the non-interacting wave function can be computed in perturbation theory:

$$|\Phi_0(t_k + \delta t_k)\rangle = |\Phi_0(t_k)\rangle - \frac{\delta t_k}{\mathcal{H}_0 - E_0}\left(\mathcal{V}^{(k)} - \langle\mathcal{V}^{(k)}\rangle_0\right)|\Phi_0(t_k)\rangle + O(\delta t_k^2), \quad (6.92)$$

where $\langle\mathcal{V}^{(k)}\rangle_0 = \langle\Phi_0(t_k)|\mathcal{V}^{(k)}|\Phi_0(t_k)\rangle$ and E_0 is the non-interacting ground-state energy. The derivative can be easily worked out by expressing the perturbation $\mathcal{V}^{(k)}$ in the basis that diagonalizes the Hamiltonian $\mathcal{H}_0$, see Eq. (5.42):

$$\mathcal{V}^{(k)} = \sum_{\alpha,\beta} \overline{V}_{\alpha,\beta}^{(k)} \phi_\alpha^\dagger \phi_\beta, \quad (6.93)$$

where

$$\overline{V}_{\alpha,\beta}^{(k)} = \sum_{I,J} U_{I,\alpha}^* V_{I,J}^{(k)} U_{J,\beta}, \quad (6.94)$$

and $\mathbf{U}$ is the unitary matrix used to diagonalize $\mathcal{H}_0$, see Eq. (5.42). Thus, we obtain that:

$$|\Phi_0(t_k + \delta t_k)\rangle = |\Phi_0(t_k)\rangle - \delta t_k \sum_{\alpha\neq\beta} \frac{\overline{V}_{\alpha,\beta}^{(k)}}{\varepsilon_\alpha - \varepsilon_\beta} \phi_\alpha^\dagger \phi_\beta |\Phi_0(t_k)\rangle + O(\delta t_k^2); \quad (6.95)$$

here, we have used the fact that the state $\phi_\alpha^\dagger \phi_\beta |\Phi_0(t_k)\rangle$ is an eigenstate of $\mathcal{H}_0$ with eigenvalue $E_{\alpha,\beta} = E_0 + \varepsilon_\alpha - \varepsilon_\beta$. Therefore, within perturbation theory:

$$\frac{|\Phi_0(t_k + \delta t_k)\rangle - |\Phi_0(t_k)\rangle}{\delta t_k} = -\sum_{\alpha \neq \beta} \frac{\overline{V}_{\alpha,\beta}^{(k)}}{\varepsilon_\alpha - \varepsilon_\beta} \phi_\alpha^\dagger \phi_\beta |\Phi_0(t_k)\rangle + O(\delta t_k); \qquad (6.96)$$

the local operator $\mathcal{O}_k(x)$ is obtained by taking the limit of $\delta t_k \to 0$:

$$\mathcal{O}_k(x) = -\sum_{\alpha \neq \beta} \frac{\overline{V}_{\alpha,\beta}^{(k)}}{\varepsilon_\alpha - \varepsilon_\beta} \frac{\langle x|\phi_\alpha^\dagger \phi_\beta |\Phi_0(t_k)\rangle}{\langle x|\Phi_0(t_k)\rangle}; \qquad (6.97)$$

the last term can be computed by re-expressing the operators $\phi_\alpha^\dagger$ (and ϕ_β) in terms of $d_I^\dagger$ (and d_J):

$$\mathcal{O}_k(x) = -\sum_{I,J} W_{I,J}^{(k)} \frac{\langle x|d_I^\dagger d_J |\Phi_0(t_k)\rangle}{\langle x|\Phi_0(t_k)\rangle}, \qquad (6.98)$$

where

$$W_{I,J}^{(k)} = \sum_{\alpha,\beta}' U_{I,\alpha} \frac{\overline{V}_{\alpha,\beta}^{(k)}}{\varepsilon_\alpha - \varepsilon_\beta} U_{J,\beta}^*; \qquad (6.99)$$

here, the primed sum is carried over occupied orbitals ($\varepsilon_\beta \leq \varepsilon_F$) and unoccupied ones ($\varepsilon_\alpha > \varepsilon_F$).

The case with Pfaffians, see Eq. (5.106), can be obtained by a straightforward generalization of the previous steps. Here, we have to include also perturbations containing two creation or annihilation operators:

$$\mathcal{V}^{(k)} = \sum_{I,J} W_{I,J}^{(k)} d_I^\dagger d_J^\dagger + h.c., \qquad (6.100)$$

which is associated to a small change in the pairing term (for a translationally invariant case, $\Delta_{j,i}^{\tau,\sigma} \to \Delta_k^{\tau,\sigma}$). Then, by inverting the Bogoliubov transformation of Eq. (5.107):

$$d_I = \sum_\alpha \left(u_{I,\alpha}^* \Phi_\alpha + v_{I,\alpha} \Phi_\alpha^\dagger \right), \qquad (6.101)$$

the perturbation will possess both standard and anomalous terms, i.e., $\Phi_\alpha^\dagger \Phi_\beta$, which are similar to the ones in Eq. (6.93), but also $\Phi_\alpha^\dagger \Phi_\beta^\dagger$ and $\Phi_\alpha \Phi_\beta$. The rest of the derivation will follow the one shown for the determinant case.

Finally, we show how it is possible to compute the derivatives of the variational wave function with respect to the parameters η_k that define the backflow correlations. In our formulation, the "correlated" orbitals depend linearly upon these parameters, e.g., see the simplest case of Eq. (5.98). The logarithmic derivative can be computed by using the general formula for a matrix $\mathbf{M}$:

$$\frac{\partial}{\partial \eta_k} \ln \det \mathbf{M} = \frac{\partial}{\partial \eta_k} \mathrm{Tr} \ln \mathbf{M} = \mathrm{Tr} \left(\mathbf{M}^{-1} \frac{\partial \mathbf{M}}{\partial \eta_k} \right), \qquad (6.102)$$

where $\partial \mathbf{M}/\partial \eta_k$ can be easily computed by differentiating every element of the matrix $\mathbf{M}$ with respect to the parameter η_k. Then, the local operator $\mathcal{O}_k(x)$ is obtained by taking the matrix $\tilde{\mathbf{U}}^b$ of Eq. (5.99), which defines the state with backflow correlations, as $\mathbf{M}$ in Eq. (6.102). Notice that the inverse matrix does not cost an extra computation time, since it is already calculated for the Metropolis algorithm.

7

Time-Dependent Variational Monte Carlo

7.1 Introduction

Variational calculations are usually limited to study the ground-state and low-energy states, as discussed in Chapter 1. Here, we present a straightforward generalization of the variational principle to consider non-equilibrium properties of strongly correlated systems. In this regard, the simplest example is given by the time-dependent Hartree-Fock approach, which has been introduced by Dirac (1930) at the dawn of quantum mechanics (Ring and Schuck, 2004). In the following, we present the formalism for a real-time variational approach that is able to incorporate the Jastrow factor (Carleo et al., 2012, 2014); the method has been further generalized to incorporate a time-dependent fermionic part (Ido et al., 2015).

In principle, a time-dependent approach would require solving the full many-body Schrödinger equation:

$$i\frac{d}{dt}|\Phi(t)\rangle = \mathcal{H}|\Phi(t)\rangle, \tag{7.1}$$

where $|\Phi(t)\rangle$ is the wave function at time t. The formal solution (here, we assume that the Hamiltonian does not depend upon time) is given by:

$$|\Phi(t)\rangle = e^{-i\mathcal{H}t}|\Phi(0)\rangle. \tag{7.2}$$

However, an exact treatment of the time evolution is possible only for small systems, since the Hilbert space grows exponentially as the system size increases and, therefore, approximated techniques are needed. In this regard, the time-dependent Hartree-Fock approximation is not accurate for interacting many-body systems because it drastically underestimate correlation effects. In the recent past, few numerical methods have been developed to treat large systems accurately, as time-dependent density matrix renormalization group (White and Feiguin, 2004; Daley et al., 2004) and non-equilibrium dynamical mean-field theory (Aoki et al., 2014). However, both these methods have severe difficulties in treating large systems in two or three dimensions. Therefore, alternative techniques are needed and the time-dependent extension of the variational Monte Carlo approach represents a suitable way to follow the exact dynamics of correlated wave functions.

7.2 Real-Time Evolution of the Variational Parameters

The heart of our variational approach is the definition of the time-dependent wave function. Here, we will consider bosonic problems only, where the quantum state has the form of a time-dependent Jastrow factor, but the generalization to fermionic systems is straightforward and can be done by considering a time-dependent Slater determinant (Ido et al., 2015). Therefore, we take:

$$\Psi(x,t) = \langle x|\Psi(t)\rangle = \exp\left[\sum_k \alpha_k(t)\mathcal{O}_k(x)\right]\zeta(x), \tag{7.3}$$

where $\zeta(x)$ is a time-independent bosonic state and $\alpha_k(t) = \alpha_k^R(t) + i\alpha_k^I(t)$ are *complex* variational parameters coupled to the real local excitation operators $\mathcal{O}_k$, which are taken to be diagonal in the chosen basis, i.e., $\langle x|\mathcal{O}_k|x'\rangle = \delta_{x,x'}\mathcal{O}_k(x)$. Among the set of the operators $\{\mathcal{O}_k\}$, we include the identity $\mathcal{O}_{k=0} = \mathbb{I}$, thus $\alpha_0^R(t)$ allows us to fulfill the normalization condition of $\Psi(x,t)$ and $\alpha_0^I(t)$ to pick up an arbitrary phase factor during the real time dynamics.

The time evolution of this quantum state is governed by $\alpha_k(t)$, whose trajectories in time are fixed by a set of differential equations such to minimize the "distance" between the approximate and the exact states in the Hilbert space; equivalently, the same set of equations may be also obtained from the principle of stationary action as it will be described in the following.

7.2.1 Minimal Hilbert Space Distance

From the one hand, starting at a given time t, the *exact* infinitesimal real-time evolution of $\Psi(x,t)$ is given by:

$$\Phi(x,t+\epsilon) = \Psi(x,t)\left[1 - i\epsilon e_L(x,t)\right] + O(\epsilon^2), \tag{7.4}$$

where

$$e_L(x,t) = e_L^R(x,t) + ie_L^I(x,t) = \frac{\langle x|\mathcal{H}|\Psi(t)\rangle}{\langle x|\Psi(t)\rangle} \tag{7.5}$$

is the complex-valued local energy for a given set of variational parameters and for a given many-body configuration $|x\rangle$. On the other hand, an infinitesimal variation of the variational wave function due to the change of parameters is given by:

$$\Psi(x,t+\epsilon) = \Psi(x,t)\left[1 + \sum_k \delta\alpha_k(t)\mathcal{O}_k(x)\right] + O(\epsilon^2), \tag{7.6}$$

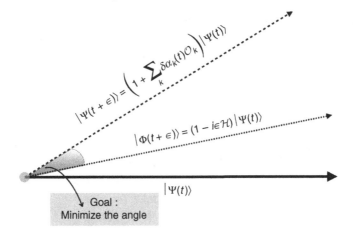

Figure 7.1 Pictorial representation of the infinitesimal real-time evolution of $|\Psi(t)\rangle$ (represented by a vector in the Hilbert space, solid arrow). The exact time evolved state is given by $|\Phi(t+\epsilon)\rangle$ (dotted arrow) and the approximated one is given by $|\Psi(t+\epsilon)\rangle$ (dashed arrow). The best "variational" state is obtained by minimizing the distance between these two states.

where $\delta\alpha_k(t)$ are small, i.e., $O(\epsilon)$, complex variables. Therefore, in order to obtain the optimal changes $\delta\alpha_k(t)$ of the variational parameters in the small-time interval considered, we can minimize the Euclidean distance $\Delta_\epsilon(t)$ between the exact time evolved state $\Phi(x, t+\epsilon)$ and our approximate *Ansatz* $\Psi(x, t+\epsilon)$, see Fig. 7.1:

$$\Delta_\epsilon^2(t) = \sum_x |\Psi(x, t+\epsilon) - \Phi(x, t+\epsilon)|^2. \tag{7.7}$$

This quantity is conveniently written as an expectation value over the square-modulus of the wave function as:

$$\Delta_\epsilon^2(t) = \sum_x |\Psi(x, t)|^2 \left| i\epsilon e_L(x, t) + \sum_k \delta\alpha_k(t)\mathcal{O}_k(x) \right|^2. \tag{7.8}$$

By using the Euler minimum condition:

$$\frac{d}{d\alpha_k^*(t)} \Delta_\epsilon^2(t) = 0, \tag{7.9}$$

we obtain:

$$\sum_{k'} \langle \mathcal{O}_{k'}\mathcal{O}_k \rangle_t \delta\alpha_{k'}(t) = -i\epsilon \langle \mathcal{O}_k e_L(t) \rangle_t, \tag{7.10}$$

which is correct up to $O(\epsilon^2)$; here, the symbol $\langle \ldots \rangle_t$ indicates the average over square-modulus $|\Psi(x, t)|^2$:

$$\langle \mathcal{O}_k\mathcal{O}_{k'} \rangle_t = \sum_x |\Psi(x, t)|^2 \mathcal{O}_k(x)\mathcal{O}_{k'}(x) = \langle \Psi(t)|\mathcal{O}_k\mathcal{O}_{k'}|\Psi(t)\rangle, \tag{7.11}$$

$$\langle \mathcal{O}_k e_L(t) \rangle_t = \sum_x |\Psi(x,t)|^2 \mathcal{O}_k(x) e_L(x,t) = \langle \Psi(t)|\mathcal{O}_k \mathcal{H}|\Psi(t)\rangle, \qquad (7.12)$$

which can be easily implemented numerically by the standard variational Monte Carlo technique.

At this stage we can take the limit $\epsilon \to 0$ by introducing time derivatives of the variational parameters:

$$\dot{\alpha}_k(t) = \lim_{\epsilon \to 0} \frac{\delta \alpha_k(t)}{\epsilon}. \qquad (7.13)$$

Furthermore, by taking the explicit dependence upon the real and imaginary parts of $\delta \alpha_k(t)$ in Eq. (7.10), we obtain closed differential equations for $\dot{\alpha}_k^R(t)$ and $\dot{\alpha}_k^I(t)$:

$$\sum_{k'} \langle \mathcal{O}_k \mathcal{O}_{k'} \rangle_t \dot{\alpha}_{k'}^R(t) = \langle \mathcal{O}_k e_L^I(t) \rangle_t, \qquad (7.14)$$

$$\sum_{k'} \langle \mathcal{O}_k \mathcal{O}_{k'} \rangle_t \dot{\alpha}_{k'}^I(t) = -\langle \mathcal{O}_k e_L^R(t) \rangle_t; \qquad (7.15)$$

these differential equations minimize $\Delta_\epsilon^2(t)$ in the limit $\epsilon \to 0$.

Numerically it is very important to be as close as possible to the continuous limit, because only in this case the linearized evolution in Eq. (7.4) becomes a proper propagation, and non-unitary terms consistently vanish for $\epsilon \to 0$. It should be noted that the solution of Eq. (7.10) also guarantees that in the short-time propagation the expectation values of the operators $\mathcal{O}_k$ remain close (i.e., up to $O(\epsilon^2)$) to the exact dynamics:

$$\langle \Psi(t+\epsilon)|\mathcal{O}_k|\Psi(t+\epsilon) \rangle \approx \langle \Phi(t+\epsilon)|\mathcal{O}_k|\Phi(t+\epsilon) \rangle. \qquad (7.16)$$

We want to finish this part by showing a more explicit form of the differential equations (7.14) and (7.15) that highlights the part related to the norm and the global phase of the time-dependent wave function. Given that $\mathcal{O}_{k=0} = \mathbb{I}$, we can simplify Eqs. (7.14) and (7.15) and decouple the equations for $\alpha_0(t)$ from the others:

$$\sum_{k'>0} S_{k,k'} \dot{\alpha}_{k'}^R = \langle \mathcal{O}_k e_L^I(t) \rangle_t, \qquad (7.17)$$

$$\sum_{k'>0} S_{k,k'} \dot{\alpha}_{k'}^I = -\langle \mathcal{O}_k e_L^R(t) \rangle_t + \langle e_L^R(t) \rangle_t \langle \mathcal{O}_k \rangle_t, \qquad (7.18)$$

where $S_{k,k'} = \langle \mathcal{O}_k \mathcal{O}_{k'} \rangle_t - \langle \mathcal{O}_k \rangle_t \langle \mathcal{O}_{k'} \rangle_t$ and $\langle e_L^I(t) \rangle_t = 0$. In addition, the equations for $\alpha_0(t)$ read:

$$\dot{\alpha}_0^R(t) = -\sum_{k'>0} \langle \mathcal{O}_{k'} \rangle_t \dot{\alpha}_{k'}^R(t), \qquad (7.19)$$

$$\dot{\alpha}_0^I(t) = -\langle e_L^R(t) \rangle_t - \sum_{k'>0} \langle \mathcal{O}_{k'} \rangle_t \dot{\alpha}_{k'}^I(t). \qquad (7.20)$$

7.2.2 Principle of Stationary Action

The equations of motion for the variational parameters can be also derived from an alternative approach, based on the principle of stationary action. In order to proceed, we introduce the action:

$$S = \int dt \langle \Psi(t)| \left(i\frac{\partial}{\partial t} - \mathcal{H} \right) |\Psi(t)\rangle, \tag{7.21}$$

which is a functional of the variational parameters $\alpha_k(t)$. Notice that the assumption of a normalized wave function $\Psi(x,t)$ implies that the action S is real. As well known, the real exact dynamics can be obtained by taking the stationary solution of S, namely:

$$\frac{\delta S}{\delta \Psi^*(x,t)} = 0, \tag{7.22}$$

among all possible variations of $\Psi(x,t)$. In the following, we will show that the principle of stationary action can be extended to a variational *Ansatz* and allows the optimal choice for the time evolution of the variational parameters derived in the previous section. When the variational wave function is restricted to the form of Eq. (7.3), the action can be easily expressed in terms of $\dot{\alpha}_k(t)$ and reads:

$$S = \int dt \langle \Psi(t)| \left(i\sum_{k'} \dot{\alpha}_{k'}(t)\mathcal{O}_{k'} - \mathcal{H} \right) |\Psi(t)\rangle. \tag{7.23}$$

The stationary condition for a normalized wave function parametrized by a set of complex parameters $\alpha_k(t)$ is given by:

$$\frac{\delta S}{\delta \alpha_k^*(t)} = 0. \tag{7.24}$$

By using the definition of $|\Psi(t)\rangle$ given in Eq. (7.3), we notice that the dependence on $\alpha_k^*(t)$ in S is given only in $\langle \Psi(t)|$ and, therefore, we immediately get the stationary condition:

$$\langle \Psi(t)| \left(i\sum_{k'} \mathcal{O}_k \mathcal{O}_{k'} \dot{\alpha}_{k'} - \mathcal{O}_k \mathcal{H} \right) |\Psi(t)\rangle = 0, \tag{7.25}$$

which is clearly equivalent to Eq. (7.10) in the continuous limit $\epsilon \to 0$.

7.2.3 Norm and Energy Conservation

The differential equations (7.14) and (7.15) define a real-time dynamics of the variational states that preserves both the norm and the energy. As far as the norm of the wave function is concerned, we have that:

$$N(t) = \sum_x |\Psi(x,t)|^2, \tag{7.26}$$

whose time-derivative is given by:

$$\dot{N}(t) = 2\left[\dot{\alpha}_0^R(t) + \sum_{k>0}\langle\mathcal{O}_k\rangle_t\dot{\alpha}_k^R(t)\right],$$ (7.27)

which vanishes, as a direct consequence of Eq. (7.19).

Moreover, the energy is given by:

$$E(t) = \sum_x |\Psi(x,t)|^2 e_L(x,t).$$ (7.28)

The time-derivative can be easily found:

$$\dot{E}(t) = 2\sum_k \left(\dot{\alpha}_k^R(t)\langle\mathcal{O}_k e_L^R\rangle_t + \dot{\alpha}_k^I(t)\langle\mathcal{O}_k e_L^I\rangle_t\right),$$ (7.29)

which also vanishes, given Eqs. (7.14) and (7.15).

7.2.4 Real-Time Variational Monte Carlo

Given the correlated nature of the variational wave function, a crucial point is to provide a reliable solution of the differential equations (7.14) and (7.15). The variational Monte Carlo method allows us for a numerically exact solution for the variational trajectories. Indeed, at each time t for a set of variational parameters $\{\alpha_k(t)\}$, the square modulus of the wave function $|\Psi(x,t)|^2$ can be straightforwardly interpreted as a probability distribution over the Hilbert space spanned by the configurations $|x\rangle$ and a Markov process can be devised, whose stationary equilibrium distribution coincides with the desired measure. All expectation values that enter into Eqs. (7.17) and (7.18) are computed as statistical averages over the random walk. Then, after a suitable number of Monte Carlo steps, the linear system of differential equations can be solved in order to obtain the first-order derivatives of the variational parameters, which can be in turn integrated by means of standard algorithms for first-order differential equations. To give the new set of parameters $\{\alpha_k(t+\epsilon)\}$. This procedure is iterated from the initial time until the final one is reached.

7.3 An Example for the Quantum Quench in One Dimension

Here, we would like to present a simple application of the time-dependent variational methodology that we have described in the previous sections. In particular, we show the spreading of density-density correlations in the Bose-Hubbard model after a sudden quench of the interaction strength U, from $U = U_{\text{init}}$ to $U = U_{\text{fin}}$. In particular, we consider the one-dimensional model defined by:

$$\mathcal{H} = -J\sum_i b_i^\dagger b_{i+1} + \text{h.c.} + \frac{U}{2}\sum_i n_i(n_i - 1),$$ (7.30)

where $b_i^\dagger$ (b_i) creates (destroys) a boson on site i, and $n_i = b_i^\dagger b_i$ is the density of bosons on site i; periodic-boundary conditions are assumed on the chain with L sites, such that $b_{L+1}^\dagger \equiv b_1^\dagger$. Notice that, here, the hopping amplitude is denoted by J, in order not to confuse it with the time t.

The initial state at $t = 0$, is the best variational Jastrow wave function for $U = U_{init}$, then the real-time evolution of this state is performed according to the Hamiltonian (7.30) with $U = U_{fin}$, according to Eq. (7.10). In the numerical calculations, we use a sufficiently small time-step $\epsilon = 0.01$ and a fourth-order Runge-Kutta integration scheme, which conserves the energy with a very small systematic error of the order of one part in a thousand, for times up to $t = 100$ (Carleo et al., 2014). We consider the case with one boson per site in average, i.e., $\sum_i n_i = L$ and examine the evolution of the density-density correlation function:

$$N(r, t) = \frac{1}{L} \sum_i \left(\langle \Psi(t)|n_{i+r}n_i|\Psi(t)\rangle - \langle \Psi(0)|n_{i+r}n_i|\Psi(0)\rangle \right). \tag{7.31}$$

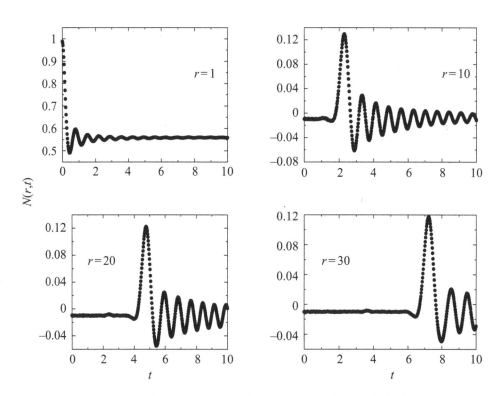

Figure 7.2 Time evolution of the density-density correlations $N(r, t)$, given by Eq. (7.31), for $r = 1, 10, 20$, and 30. The calculations are performed for the Bose-Hubbard model with $L = 100$ sites with periodic-boundary conditions with one boson per site in average. The initial state is the ground state for $U_{init} = 0$, the time evolution is done for $U_{fin}/J = 3$.

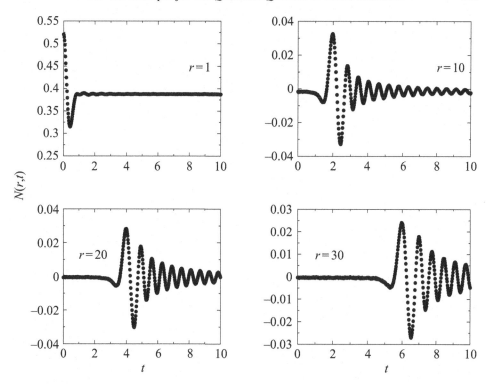

Figure 7.3 The same as in Fig. 7.2 for $U_{\text{init}}/J = 2$ and $U_{\text{fin}}/J = 4$. In this case, the initial state is given by the best Jastrow wave function for the initial interaction strength.

The results of the density-density correlation $N(r, t)$ for a chain with $L = 100$ are shown in Figs. 7.2 and 7.3 for $r = 1$, 10, 20, and 30. In the former case, we set $U_{\text{init}} = 0$ and $U_{\text{fin}}/J = 3$ (here, the initial state is the exact ground state of the non-interacting Bose-Hubbard Hamiltonian), while, in the latter one, we fixed $U_{\text{init}}/J = 2$ and $U_{\text{fin}}/J = 4$ (here, the initial state is just a variational *Ansatz* for the exact ground state). The existence of the so-called light cone is clearly visible: $N(r, t)$ is unaffected at short times, then develops a maximum at a finite time $t^{\star}(r)$ and finally undergoes damped oscillations. For large enough separation r, the activation time $t^{\star}(r)$ depends linearly on the separation, $t^{\star}(r) \approx v_{\text{lc}} \times r$, which defines the light-cone velocity v_{lc}. Remarkably, the time-dependent variational method allows us to simulate very long times in the ballistic regime without any instability. Indeed, we can easily reach times as large as $t \approx 100$, which are much larger than what is possible by using other numerical methods, as time-dependent density matrix renormalization group (White and Feiguin, 2004; Daley et al., 2004) and non-equilibrium dynamical mean-field theory (Aoki et al., 2014).

Part IV

Projection Techniques

8

Green's Function Monte Carlo

8.1 Basic Notions and Formal Derivations

In this Chapter, we describe both the basic principles and the more advanced details that are necessary to implement the so-called Green's function Monte Carlo (GFMC) technique. Within this approach, the ground-state wave function of a given Hamiltonian is stochastically sampled by using the power method that has been discussed in section 1.7. As already done in Chapter 5 for the variational Monte Carlo technique, we fix a complete basis set $\{|x\rangle\}$ in the Hilbert space:

$$\sum_x |x\rangle\langle x| = \mathbb{I}, \qquad (8.1)$$

in which the states are taken to be *orthogonal* and *normalized*. Moreover, we assume that, given the Hamiltonian $\mathcal{H}$, the matrix elements $\mathcal{H}_{x'x} = \langle x'|\mathcal{H}|x\rangle$ can be computed efficiently for each $|x\rangle$ and $|x'\rangle$. For local Hamiltonians, although the dimension of the Hilbert space increases exponentially with the system size L, for each $|x\rangle$, the number of the non-zero elements $\mathcal{H}_{x'x}$ scales with L; therefore, given the state of the basis $|x\rangle$, all matrix elements $\mathcal{H}_{x'x}$ can be computed with a reasonable computational effort. By using this property, it is possible to define a stochastic algorithm that allows us to perform the power method (see section 1.7) in a statistical way (Ceperley and Alder, 1980; Reynolds et al., 1982; Foulkes et al., 2001). Here, the ground-state wave function $|\Upsilon_0\rangle$ is filtered out from an initial state $|\Psi_0\rangle$, by using a suitable projection operator:

$$\lim_{n\to\infty} (\Lambda - \mathcal{H})^n|\Psi_0\rangle \propto |\Upsilon_0\rangle, \qquad (8.2)$$

where Λ is a diagonal operator with $\Lambda_{x,x} = \lambda$. The necessary requirements to approach the ground state $|\Upsilon_0\rangle$ is to take a sufficiently large value of λ and choose an initial state $|\Psi_0\rangle$ such that $\langle\Upsilon_0|\Psi_0\rangle \neq 0$. In practice, Eq. (8.2) can be implemented iteratively:

$$|\Psi_{n+1}\rangle = (\Lambda - \mathcal{H})|\Psi_n\rangle. \tag{8.3}$$

By expanding over the given basis set, we have:

$$\Psi_{n+1}(x') = \sum_x \mathcal{G}_{x',x}\Psi_n(x), \tag{8.4}$$

where $\Psi_n(x) = \langle x|\Psi_n\rangle$ and

$$\mathcal{G}_{x',x} = \langle x'|(\Lambda - \mathcal{H})|x\rangle \tag{8.5}$$

is the so-called Green's function (even though it does not correspond to a Green's function). On the lattice, the GFMC technique is particularly simple and has been used for the first time to study the ground-state properties of the Heisenberg model on the square lattice (Trivedi and Ceperley, 1989, 1990).

Within a statistical implementation of the power method, we would be tempted to interpret Eq. (8.4) as a Master equation for a stochastic variable where $\Psi_n(x)$ represents the probability distribution at the iteration n and $\mathcal{G}_{x',x}$ is the transition probability. However, some important points must be elucidated in order to reach a final statistical interpretation of this quantum evolution. First of all, within this approach, *all* the matrix elements of the Green's function $\mathcal{G}_{x',x}$ have to be non-negative, as required for defining a transition probability. As far as the diagonal elements $\mathcal{G}_{x,x}$ are concerned, there is no problem, since we can always define a sufficiently large and positive value of λ that gives $\mathcal{G}_{x,x} \geq 0$. By contrast, the requirement that off-diagonal elements are non-negative is highly non trivial and, indeed, is satisfied only by certain models; if $\mathcal{G}_{x',x} < 0$ for some couples (x',x), we say that we are in presence of the *sign problem*, which will be discussed in Chapter 10. In the following, we will assume that the Hamiltonian is such that $\mathcal{G}_{x',x} \geq 0$ for all the couples (x,x'). Furthermore, even when the Green's function is non-negative, we generically have that:

$$\sum_{x'} \mathcal{G}_{x',x} \neq 1, \tag{8.6}$$

and thus $\mathcal{G}_{x',x}$ cannot be interpreted as a transition probability, since it is not normalized. Notice that Eq. (8.6) implies that the normalization of the wave function is not conserved along the projection technique of Eq. (8.4). Nevertheless, we can always split up the Green's function into the product of two factors: a stochastic matrix $p_{x',x}$ that represents a *bona fide* transition probability, and a real number b_x, which is the normalization of the Green's function:

$$\mathcal{G}_{x',x} = p_{x',x}b_x, \tag{8.7}$$

where b_x is given by:

$$b_x = \sum_{x'} \mathcal{G}_{x',x}, \qquad (8.8)$$

and then the matrix $p_{x',x}$ can be obtained by:

$$p_{x',x} = \frac{\mathcal{G}_{x',x}}{b_x}. \qquad (8.9)$$

We now want to devise a simple Markov process from which the evolution of the wave function of Eq. (8.4) can be obtained. Since the Green's function cannot be directly interpreted as a transition probability, a description with only the configuration x_n (that identifies the quantum state $|x\rangle$ at the iteration step n) is not sufficient. The simplest way of considering the presence of the scale factor b_x is to add a weight w_n in the statistical description of the projection method. Therefore, the Markov process is defined by a dyad (x_n, w_n); its evolution, according to the decomposition of Eqs. (8.8) and (8.9) is described by:

1) generate $x_{n+1} = x'$ with probability p_{x',x_n},   (8.10)

2) update the weight with $w_{n+1} = w_n b_x$.   (8.11)

Notice that here the new configuration x_{n+1} can be the same as x_n, because it is selected among all the ones that have $p_{x',x_n} \neq 0$, including also x_n when $p_{x_n,x_n} \neq 0$ (this is the case when $\mathcal{G}_{x,x} = \lambda - \mathcal{H}_{x,x} \neq 0$). The Markov process describes a *diffusion* of the configuration x_n (as well as its weight w_n) with the following transition probability:

$$K(x', w' | x, w) = p_{x',x}\, \delta(w' - w b_x), \qquad (8.12)$$

which defines the new dyad (x', w') given the old one (x, w). Thus, the Master equation corresponding to the probability density $\mathcal{P}_n(x, w)$ is:

$$\mathcal{P}_{n+1}(x', w') = \sum_x \int dw\, K(x', w' | x, w)\, \mathcal{P}_n(x, w). \qquad (8.13)$$

Hereafter, the integration limits over the variable w are assumed to run from $-\infty$ to $+\infty$, the probability density $\mathcal{P}_n(x, w)$ being zero for the values of w that are not reached along the Markov chain (e.g., for $w < 0$). Within this formalism, the wave function $\Psi_n(x)$ can be obtained from integrating over the weight the probability density $\mathcal{P}_n(x, w)$ multiplied by w. Indeed, from the Master equation (8.13) and the definition of the transition probability of Eq. (8.12), we have that:

$$\mathcal{P}_{n+1}(x', w') = \sum_x p_{x',x} \int dw\, \delta(w' - w b_x)\, \mathcal{P}_n(x, w); \qquad (8.14)$$

by multiplying both sides by w' and then integrating over w', we have:

$$\int dw' \, w' \, \mathcal{P}_{n+1}(x', w') = \sum_{x} p_{x'x} \, b_x \int dw \, w \, \mathcal{P}_n(x, w)$$

$$= \sum_{x} \mathcal{G}_{x'x} \int dw \, w \, \mathcal{P}_n(x, w), \qquad (8.15)$$

which is just the original iterative procedure of Eq. (8.4), once the following identification is performed:

$$\Psi_n(x) \equiv \int dw \, w \, \mathcal{P}_n(x, w). \qquad (8.16)$$

Therefore, after a large number n of iterations, $\mathcal{P}_n(x, w)$ converges to an equilibrium distribution $\mathcal{P}_{eq}(x, w)$, which determines the ground-state wave function $\Upsilon_0(x)$:

$$\Upsilon_0(x) \equiv \int dw \, w \, \mathcal{P}_{eq}(x, w). \qquad (8.17)$$

In the following, we will describe a practical implementation of the Markov process, which allows us to compute observables over the ground state $|\Upsilon_0\rangle$, once the high-energy components present in $|\Psi_0\rangle$ are filtered out according to the iterative procedure of Eq. (8.4).

8.2 Single Walker Technique

Here, we explain how to implement the Markov process of Eqs. (8.10) and (8.11) within the simplest approach. To this purpose, we define the basic element of this stochastic process, the so-called *walker*. A walker is determined by an index x, labelling the configuration $|x\rangle$, and a weight w, which is associated to the amplitude of the wave function, as given by Eq. (8.16). The walker changes its configuration and weight by performing a Markovian process with a discrete iteration time n: the dyad (x_n, w_n), which denotes the walker at the time n, is distributed according to $\mathcal{P}_n(x, w)$. Most importantly, the walker determines, in a statistical sense, the quantum state $\Psi_n(x)$:

$$\Psi_n(x) \equiv \int dw \, w \, \mathcal{P}_n(x, w) \approx \langle\langle w_n \delta_{x,x_n} \rangle\rangle, \qquad (8.18)$$

where $\langle\langle \ldots \rangle\rangle$ denotes the statistical average over (infinitely) many independent realizations of the Markov chain with configurations (x_n, w_n). We would like to mention that the actual determination of the wave function $\Psi_n(x)$ is rarely pursued in the actual practice, as in a many-body system the Hilbert space is exponentially large and the information on the wave function cannot be even stored in the computer memory. Moreover, in order to have an adequate errorbar on the configurations, each of them must be visited many times, implying an exponentially large computation time. Remarkably, quantum Monte Carlo approaches are based on the fact that it is not necessary to have an accurate statistical information on the wave function to obtain a reliable estimation of its energy or correlation functions.

In practice, we can use a single walker and start from an initial condition with x_0, corresponding to a given many-body configuration $|x_0\rangle$, and a weight $w_0 = 1$, so to have:

$$\mathcal{P}_0(x, w) = \delta(w - 1)\delta_{x,x_0}. \tag{8.19}$$

Then, we evolve the walker by using Eq. (8.10) and (8.11). The Markov process can be very easily implemented for generic Hamiltonians on a lattice, since the number of non-zero entries in the stochastic matrix p_{x',x_n}, for given x_n, is small, and moderately growing with the number of lattice sites L. Thus, in order to define x_{n+1}, it is enough to divide the interval $[0, 1)$ into smaller intervals for all possible $\{x'\}$ connected to x_n with non-zero probability p_{x',x_n} and generate a random number ξ between 0 and 1. As discussed in section 3.5, ξ will lie in one of the above defined intervals, with a probability of hitting the interval corresponding to a certain x' exactly equal to p_{x',x_n}, see Fig. 8.1. The weight w_{n+1} is just obtained by multiplying w_n by b_{x_n}.

Let us finally discuss the evolution of the marginal probability of the configuration x alone:

$$\Pi_n(x) = \int dw \, \mathcal{P}_n(x, w). \tag{8.20}$$

Since, within the single-walker implementation, the evolution of the configuration x_n is independent from its weight, as given by Eqs. (8.10) and (8.11), it is easy to write down the Master equation for $\Pi_n(x)$:

$$\Pi_{n+1}(x') = \sum_x p_{x',x} \Pi_n(x). \tag{8.21}$$

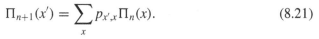

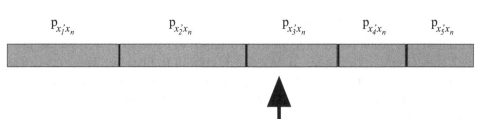

Figure 8.1 The interval $[0, 1)$ is divided into sub-intervals of length equal to the probabilities $p_{x'_i,x_n}$, for all i's labeling the possible configurations x' with non-zero $p_{x'_i,x_n}$ (here, we show the case with 5 entries). Then, a random number $0 \le \xi < 1$ is generated and the new configuration x_{n+1} is selected (here $i = 3$), as ξ (here denoted by the arrow) lies in the corresponding interval, see section 3.5.

Therefore, at equilibrium, the marginal probability is given by the right eigenvector of the stochastic matrix $p_{x'x}$ with eigenvalue $\lambda_0 = 1$, which is given by:

$$\Pi_{eq}(x) = \frac{b_x}{\sum_{x'} b_{x'}}. \tag{8.22}$$

Indeed, we have that:

$$\sum_x p_{x'x} b_x = \sum_x \mathcal{G}_{x'x} = \sum_x \mathcal{G}_{xx'} = b_{x'}, \tag{8.23}$$

where we have just used the fact that the matrix $\mathcal{G}_{x'x}$ is symmetric. Then, for an ergodic Hamiltonian, which implies an ergodic matrix $p_{x'x}$, the probability $\Pi_n(x)$ converges (with an exponential rate) to $\Pi_{eq}(x)$, no matter what is the initial condition, see section 3.8.

8.2.1 Ground-State Energy

Let us now discuss how to compute physical observables and start by considering the ground-state energy E_0. Since for large enough n, $\mathcal{H}\mathcal{G}^n|\Psi_0\rangle \propto E_0|\Upsilon_0\rangle$, we have that:

$$E_0 \approx \frac{\sum_{x'} \langle x'|\mathcal{H}\mathcal{G}^n|\Psi_0\rangle}{\sum_{x'} \langle x'|\mathcal{G}^n|\Psi_0\rangle} = \frac{\sum_{x,x'} \mathcal{H}_{x'x} \langle x|\mathcal{G}^n|\Psi_0\rangle}{\sum_{x'} \langle x'|\mathcal{G}^n|\Psi_0\rangle}, \tag{8.24}$$

which, by using Eq. (8.18) for $\Psi_n(x) = \langle x|\mathcal{G}^n|\Psi_0\rangle$, gives:

$$E_0 \approx \frac{\langle\langle e_L(x_n) w_n \rangle\rangle}{\langle\langle w_n \rangle\rangle}, \tag{8.25}$$

where we have defined the local energy as:

$$e_L(x) = \sum_{x'} H_{x'x} = \lambda - b_x. \tag{8.26}$$

Therefore, the ground-state energy can be obtained by considering several independent calculations with n steps (n being large enough to filter out the ground-state wave function from the initial state $|\Psi_0\rangle$) and then averaging them together.

A more efficient and straightforward way of performing the numerical calculation is to do a single simulation of length $N_{sim} \gg n$; after a thermalization time, the configurations x along the Markov process will be equilibrated according to $\Pi_{eq}(x)$ of Eq. (8.22); then, we can imagine to start, from *each* step $n - p$, a projection technique of length p. In this sense, the initial probability is no longer given by Eq. (8.19), but instead by:

$$P_0(x, w) = \delta(w - 1) \Pi_{eq}(x). \tag{8.27}$$

After p steps, the accumulated weight for each projection process is given by:

$$G_n^p = \prod_{j=1}^{p} b_{x_{n-j}}.$$ (8.28)

Then, the ground-state energy can be estimated by taking a sufficiently large value of p in the accumulated weight:

$$E_0 \approx \frac{\sum_n G_n^p e_L(x_n)}{\sum_n G_n^p}.$$ (8.29)

In practice, in order to avoid numerical overflows/underflows, it is useful to store (e.g., in the hard disk) the factor b_{x_n} for each Markov step, or even better $b_{x_n}/\bar{b}$ (where $\bar{b}$ is an estimation of the average value of b_{x_n}), without performing many multiplications. In the post-processing analysis, we can compute G_n^p for the smallest possible value of p that gives a converged energy and evaluate E_0 through Eq. (8.29).

This simple procedure gives an exact estimation of the ground-state energy. However, few remarks should be done. First of all, when computing the error-bars, we must keep in mind that, by taking each iteration as a starting point for the projection technique, we have correlated measures, which require a binning approach, as described in section 3.11. Most importantly, according to Eq. (8.26), the calculation of the energy does not satisfy the zero-variance property because the random quantity $e_L(x)$ does not depend on any variational guess. In this sense, the inclusion of the importance sampling is fundamental to reduce the statistical fluctuations.

8.3 Importance Sampling

Let us now describe how it is possible to implement importance sampling within the GFMC technique. This can be done by introducing the so-called *guiding function* $\Psi_G(x)$, which must be real (see below) and non-vanishing for all configurations $|x\rangle$. Usually, it is chosen to be the best variational *Ansatz* or very close to it. By multiplying both sides of Eq. (8.4) by $\Psi_G(x')$, we obtain:

$$\Psi_G(x')\Psi_{n+1}(x') = \sum_x \mathcal{G}_{x',x} \frac{\Psi_G(x')}{\Psi_G(x)} \Psi_G(x)\Psi_n(x).$$ (8.30)

Then, by defining the Green's function with importance sampling:

$$\tilde{\mathcal{G}}_{x',x} = \mathcal{G}_{x',x} \frac{\Psi_G(x')}{\Psi_G(x)},$$ (8.31)

we have that the wave function

$$\tilde{\Psi}_n(x) = \Psi_G(x)\Psi_n(x) \tag{8.32}$$

satisfies the same evolution as Eq. (8.4) with $\tilde{\mathcal{G}}$ replacing $\mathcal{G}$:

$$\tilde{\Psi}_{n+1}(x') = \sum_x \tilde{\mathcal{G}}_{x'x}\tilde{\Psi}_n(x). \tag{8.33}$$

We would like to mention the important fact that Eq. (8.31) represents a similarity transformation (also used when discussing the Markov chains in section 3.8). Such transformation does not modify the spectrum of the matrix, i.e., $\mathcal{G}$ and $\tilde{\mathcal{G}}$ have the same eigenvalues, while it trivially changes the eigenvectors, i.e., if $\upsilon(x)$ is an eigenvector of $\mathcal{G}$, then $\tilde{\upsilon}(x) = \Psi_G(x)\upsilon(x)$ is a (right) eigenvector of $\tilde{\mathcal{G}}$. Notice that, in general, $\tilde{\mathcal{G}}_{x'x}$ is no longer symmetric.

Whenever $\tilde{\mathcal{G}}_{x'x}$ is real and non-negative for all couples (x',x), it is possible to apply the same decomposition of Eq. (8.7) in terms of a conditional probability and a real coefficient (the fact of having a real Green's function implies that the guiding function must be also real):

$$\tilde{\mathcal{G}}_{x'x} = \tilde{p}_{x'x}\tilde{b}_x, \tag{8.34}$$

where:

$$\tilde{b}_x = \sum_{x'} \tilde{\mathcal{G}}_{x'x}, \tag{8.35}$$

$$\tilde{p}_{x'x} = \frac{\tilde{\mathcal{G}}_{x'x}}{\tilde{b}_x}. \tag{8.36}$$

Then, the corresponding Markov process can be devised, in analogy to Eqs. (8.10) and (8.11).

Two important facts can be achieved by using the importance sampling transformation. First of all, the exact value of the ground-state energy E_0 is obtained without statistical fluctuations whenever the guiding wave function $\Psi_G(x)$ is the correct ground-state wave function (see below). Moreover, the guiding function can be used as a remedy to the sign problem: there are cases where the original Green's function $\mathcal{G}_{x'x}$ does not satisfy the non-negativity condition for all off-diagonal matrix elements, and this fact can be adjusted by including a properly chosen $\Psi_G(x)$. The antiferromagnetic nearest-neighbor Heisenberg model on the square lattice represents the simplest example in which the Green's function can be negative, i.e., $\mathcal{G}_{x'x} = -J/2$ for configurations $|x\rangle$ and $|x'\rangle$ having two nearest-neighbor spins with opposite orientations. Once considering a guiding function with the so-called Marshall sign (Marshall, 1955):

$$\mathrm{Sign}[\Psi_G(x)] = (-1)^{N_{A,\uparrow}(x)}, \tag{8.37}$$

where $N_{A,\uparrow}(x)$ is the number of up spin in one sub-lattice in the configuration $|x\rangle$, the Green's function with importance sampling is non-negative for all couples of configurations.

In presence of importance sampling, all the derivations of section 8.2 can be repeated. For example, Eq. (8.14) is replaced by:

$$\tilde{P}_{n+1}(x',\tilde{w}') = \sum_x \tilde{p}_{x'x} \int d\tilde{w}\, \delta(\tilde{w}' - \tilde{w}\tilde{b}_x)\, \tilde{P}_n(x,\tilde{w}); \qquad (8.38)$$

moreover, Eq. (8.18) becomes:

$$\tilde{\Psi}_n(x) = \Psi_G(x)\Psi_n(x) \equiv \int d\tilde{w}\, \tilde{w}\, \tilde{P}_n(x,\tilde{w}) \approx \langle\langle\tilde{w}_n\delta_{x,x_n}\rangle\rangle. \qquad (8.39)$$

Finally, we can show that the right eigenvector of the stochastic matrix $\tilde{p}_{x'x}$ (with unit eigenvalue) is given by:

$$\tilde{\Pi}_{eq}(x) = \frac{\tilde{b}_x\Psi_G^2(x)}{\sum_{x'}\tilde{b}_{x'}\Psi_G^2(x')}. \qquad (8.40)$$

8.3.1 Ground-State Energy and Correlation Functions

We are now in the position to compute the ground-state energy in presence of the importance sampling. Again, for large enough n, we have that:

$$E_0 \approx \frac{\langle\Psi_G|\mathcal{H}\mathcal{G}^n|\Psi_0\rangle}{\langle\Psi_G|\mathcal{G}^n|\Psi_0\rangle} = \frac{\sum_x\langle\Psi_G|\mathcal{H}|x\rangle\langle x|\mathcal{G}^n|\Psi_0\rangle}{\sum_x\langle\Psi_G|x\rangle\langle x|\mathcal{G}^n|\Psi_0\rangle}. \qquad (8.41)$$

By using the fact that:

$$\tilde{\Psi}_n(x) = \langle x|\tilde{\mathcal{G}}^n|\Psi_0\rangle = \Psi_G(x)\Psi_n(x) = \langle x|\Psi_G\rangle\langle x|\mathcal{G}^n|\Psi_0\rangle, \qquad (8.42)$$

together with Eq. (8.39) and considering that the guiding function is real, we have:

$$E_0 \approx \frac{\langle\langle\tilde{e}_L(x_n)\tilde{w}_n\rangle\rangle}{\langle\langle\tilde{w}_n\rangle\rangle}, \qquad (8.43)$$

where we have defined the local energy with importance sampling:

$$\tilde{e}_L(x) = \frac{\langle\Psi_G|\mathcal{H}|x\rangle}{\langle\Psi_G|x\rangle} = \sum_{x'}\mathcal{H}_{x'x}\frac{\Psi_G(x')}{\Psi_G(x)}. \qquad (8.44)$$

Notice that if $|\Psi_G\rangle$ is an eigenvector of the Hamiltonian $\mathcal{H}$ with eigenvalue E, then $\tilde{e}_L(x) = E$ for all the configurations $|x\rangle$, which implies that there are no statistical fluctuations. Of course, this is a very unrealistic situation for a generic correlated Hamiltonian; however, the advantage of considering importance sampling is that, whenever the guiding function is close to an eigenstate (e.g., the exact ground state), the statistical fluctuations are strongly reduced with respect to the case without importance sampling.

In practice, as before, we can perform a single simulation of length $N_{\text{sim}} \gg n$ and, after a thermalization time, imagine to start the projection technique of p steps from the configurations that are distributed according to $\tilde{\Pi}_{\text{eq}}(x)$ of Eq. (8.40). In this case, the accumulated weight from iteration $n - p$ for p steps is:

$$\tilde{G}_n^p = \prod_{j=1}^{p} \tilde{b}_{x_{n-j}}, \tag{8.45}$$

and the ground-state energy is given by:

$$E_0 \approx \frac{\sum_n \tilde{G}_n^p \tilde{e}_L(x_n)}{\sum_i \tilde{G}_n^p}. \tag{8.46}$$

By using the so-called *forward-walking* technique, the GFMC method can also efficiently compute expectation values of local operators $\mathcal{O}$, which are diagonal in the basis set $\{|x\rangle\}$:

$$\mathcal{O}|x\rangle = \mathcal{O}(x)|x\rangle, \tag{8.47}$$

where $\mathcal{O}(x)$ is the eigenvalue corresponding to the configuration $|x\rangle$. Indeed, for large n and m, the true expectation value over the ground state can be written as:

$$\frac{\langle \Upsilon_0|\mathcal{O}|\Upsilon_0\rangle}{\langle \Upsilon_0|\Upsilon_0\rangle} \approx \frac{\langle \Psi_0|\tilde{\mathcal{G}}^m \mathcal{O}\tilde{\mathcal{G}}^{n-m}|\Psi_0\rangle}{\langle \Psi_0|\tilde{\mathcal{G}}^n|\Psi_0\rangle}. \tag{8.48}$$

For these operators, the Markov chain can be easily modified in order to account the application of the operator $\mathcal{O}$ at a selected iteration $n - m$. In fact, this is exactly equivalent, in the statistical sense, to modify the weight:

$$\tilde{w}_{n-m} \to \mathcal{O}(x_{n-m})\tilde{w}_{n-m}, \tag{8.49}$$

in such a way that:

$$\frac{\langle \Upsilon_0|\mathcal{O}|\Upsilon_0\rangle}{\langle \Upsilon_0|\Upsilon_0\rangle} \approx \frac{\langle\langle \mathcal{O}(x_{n-m})\tilde{w}_n\rangle\rangle}{\langle\langle \tilde{w}_n\rangle\rangle}. \tag{8.50}$$

As before, the quantity on the r.h.s. can be computed in a single run by using the accumulated weight of Eq. (8.45):

$$\frac{\langle \Upsilon_0|\mathcal{O}|\Upsilon_0\rangle}{\langle \Upsilon_0|\Upsilon_0\rangle} \approx \frac{\sum_n \tilde{G}_n^p \mathcal{O}(x_{n-m})}{\sum_n \tilde{G}_n^p}, \tag{8.51}$$

which gives the ground-state expectation value for $m = p/2$ and large enough p. The important point here is that both numerator and denominator of Eq. (8.48) are evaluated in a single simulation, i.e., by using the same Markov chain.

An alternative method to compute averages of general operators is obtained by performing two independent simulations for the numerator and the denominator of Eq. (8.48). The advantage of this approach is the possibility to measure not only diagonal operators but also general ones with $\mathcal{O}_{x'x} = \langle x'|\mathcal{O}|x\rangle \neq 0$ for $|x'\rangle \neq |x\rangle$. In this case, first of all, we carry out a simulation for evaluating the denominator. Then, we take the equilibrated walker configurations $(x_n, \tilde{w}_n)$ and apply the operator $\mathcal{O}$ at a given iteration. Whenever the operator has off-diagonal elements, this is not simply equivalent to scale the weight as in Eq. (8.49). Nevertheless, we can use a stochastic approach to select only one configuration among the ones that are connected to x with non-zero matrix elements $\mathcal{O}_{x'x}$. Finally, starting from the selected configuration, we perform the propagation for further m steps.

Whenever $\mathcal{O}_{x'x} > 0$, this stochastic approach can be implemented in few steps:

1. As far as the denominator of Eq. (8.51) is concerned, the original Markov chain with n steps is used to compute the accumulated weight $\tilde{G}_n^p$.
2. As far as the numerator of Eq. (8.51) is concerned:
   - Consider the original Markov chain at the iteration $n - m$ with the corresponding configuration $x_{n-m} \equiv x$ and weight $\tilde{G}_{n-m}^p$.
   - Compute $\mathcal{O}(x_{n-m}) \equiv \mathcal{O}(x)$:

$$\mathcal{O}(x) = \sum_{x'} \mathcal{O}_{x'x} \frac{\Psi_G(x')}{\Psi_G(x)}. \tag{8.52}$$

   - Select a random new configuration x' with a probability proportional to $\mathcal{O}_{x'x}\Psi_G(x')/\Psi_G(x)$.
   - Finally, start an independent Markov chain from x' for m steps and keep accumulating the weight from $\tilde{G}_{n-m}^p$ to $(\tilde{G}_n^p)'$. Notice that in this case the weights on numerator and denominator are different, given the forward propagation with m steps.

We would like to stress the important point that the general operators $\mathcal{O}$, with arbitrary signs in the matrix elements, can be always cast as a difference $\mathcal{O} = \mathcal{O}^+ - \mathcal{O}^-$ of two operators with positive definite matrix elements and, therefore, the method just described can be applied to $\mathcal{O}^+$ and $\mathcal{O}^-$ separately.

Finally, we mention that, by using Eq. (8.48) for $m = 0$, the so-called *mixed average* $\mathcal{O}_{MA}$ is obtained, which is a biased estimator of the exact quantum average:

$$\mathcal{O}_{MA} = \frac{\langle\Psi_G|\mathcal{O}|\Upsilon_0\rangle}{\langle\Psi_G|\Upsilon_0\rangle}. \tag{8.53}$$

The calculation of this mixed average is possible for any type of operator, not only the local ones. For operators that are defined on the ground state, i.e., $\mathcal{O}|\Upsilon_0\rangle = \mathcal{O}_0|\Upsilon_0\rangle$ ($\mathcal{O}_0$ being the eigenvalue), the mixed average estimator is exact (as we have seen for the ground-state energy). For all other operators, an approximated scheme to evaluate the ground-state expectation value can be used, which is valid whenever the state $|\Psi_G\rangle$ is close to $|\Upsilon_0\rangle$. Indeed, if $|\Psi_G\rangle = |\Upsilon_0\rangle + \epsilon|\Upsilon'\rangle$ (where $|\Upsilon'\rangle$ is normalized and orthogonal to $|\Upsilon_0\rangle$), the following relation holds up to $O(\epsilon^2)$:

$$\frac{\langle\Upsilon_0|\mathcal{O}|\Upsilon_0\rangle}{\langle\Upsilon_0|\Upsilon_0\rangle} \approx 2\frac{\langle\Psi_G|\mathcal{O}|\Upsilon_0\rangle}{\langle\Psi_G|\Upsilon_0\rangle} - \frac{\langle\Psi_G|\mathcal{O}|\Psi_G\rangle}{\langle\Psi_G|\Psi_G\rangle}. \tag{8.54}$$

8.4 The Continuous-Time Limit

Up to now, we have seen a statistical implementation of the power method, which is based upon the Markov process of Eqs. (8.10) and (8.11). As we discussed in section 8.1, the Green's function must be non-negative, given the probabilistic nature of the GFMC approach. Therefore, the constant λ in Eq. (8.2) has to be taken large enough to have all the diagonal elements $\mathcal{G}_{x,x} \geq 0$. Here, to simplify the notation, we do not put a tilde over the various quantities, assuming that we are using importance sampling. The requirement of having non-negative diagonal elements, often needs a very large shift λ, which also increases with the size of the cluster; in addition, in some cases, this shift cannot be given *a priori* (as for example when the fixed-node approximation is considered, see Chapter 10). If, for the chosen λ, a negative diagonal element $\mathcal{G}_{x,x}$ is found, we need to start a new simulation from scratch with a larger value for λ, with a considerable waste of computational time. In order to avoid this problem, we could work with an exceedingly large value of λ; however, this choice slows down the efficiency of the algorithm, since the probability to remain in the same configuration

$$p_d = \frac{p_{x,x}}{b_x} = \frac{\lambda - \mathcal{H}_{x,x}}{\lambda - e_L(x)} \tag{8.55}$$

becomes very close to one, thus implying a very large correlation time.

In the following, we show that the problem of working with a large λ can be solved without loss of efficiency. Indeed, given p_d, the probability $\mathcal{Q}(k)$ that the walker remains in the same configuration $|x\rangle$ for k steps (i.e., it makes k diagonal moves) and then changes into $|x'\rangle \neq |x\rangle$ is simply given by:

$$\mathcal{Q}(k) = p_d^k(1 - p_d). \tag{8.56}$$

As we described in section 3.3, we can sample this discrete distribution by extracting a random number ξ that is uniformly distributed in $[0, 1)$; then the number of diagonal moves k is determined by the inequalities:

$$(1 - p_d)\sum_{i=0}^{k-1}p_d^i \leq \xi < (1 - p_d)\sum_{i=0}^{k}p_d^i, \tag{8.57}$$

which leads to

$$k \leq \frac{\ln(1 - \xi)}{\ln p_d} < k + 1. \tag{8.58}$$

These relations are equivalent to take:

$$k = \left[\frac{\ln(1 - \xi)}{\ln p_d} \right], \tag{8.59}$$

where $[\dots]$ indicates the integer part. Here, we assume that $0 \leq \xi < 1$, such that k is always finite.

The idea is to consider a set of M elementary iterations as a single Markov step, for which $\mathcal{Q}(k)$ is given by Eq. (8.56) for $k = 0, \dots, M - 1$ and $\mathcal{Q}(M) = p_d^M$ (denoting the possibility that there are no off-diagonal moves during M trials). Then, along the M iterations, we can sample $\mathcal{Q}(k)$ with:

$$k = \min \left(M_{\text{left}}, \left[\frac{\ln(1 - \xi)}{\ln p_d} \right] \right), \tag{8.60}$$

where M_{left} is the number of iterations that are left to complete the Markov step. At the beginning, $M_{\text{left}} = M$, then we can iteratively apply Eq. (8.60) and bookkeep M_{left}. The weight w of the walker is updated according to k diagonal moves:

$$w \rightarrow w b_x^k; \tag{8.61}$$

if $k < M_{\text{left}}$ a new configuration is extracted randomly according to the transition probability $t_{x',x}$ defined by:

$$t_{x',x} = \begin{cases} \frac{p_{x'x}}{1 - p_{x,x}} & \text{for } x \neq x', \\ 0 & \text{for } x = x'. \end{cases} \tag{8.62}$$

Finally, if $k < M_{\text{left}}$, M_{left} is changed into $M_{\text{left}} - (k + 1)$ (k diagonal moves and the off-diagonal one), so that we can continue to use Eq. (8.60) until $M_{\text{left}} = 0$, where the Markov step is concluded, see Fig. 8.2. We stress the fact that the Markov step is composed of M elementary steps; therefore, in this case, the appropriate factors to compute G_n^p in Eq. (8.45) are replaced by the accumulated weights along the M steps.

It is interesting to observe that this method can be readily generalized for $\lambda \rightarrow \infty$ by increasing M with λ, namely for $M = \lambda\beta$, where β represents the imaginary-time evolution of the exact propagator $\exp(-\beta\mathcal{H})$ applied statistically:

$$(\Lambda - \mathcal{H})^M = \lambda^M \left(\mathbb{I} - \frac{\mathcal{H}}{\lambda} \right)^{\lambda\beta} \approx \exp(-\beta\mathcal{H}). \tag{8.63}$$

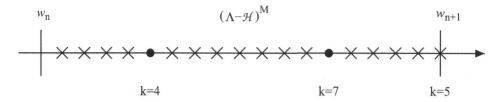

Figure 8.2 Picture of the Markov process based upon sampling the number of diagonal moves of Eq. (8.60). Here, the number of elementary steps is $M = 18$: the steps in which the configuration does not change are marked by crosses, while the (two) steps in which the configuration is changed are marked by full dots. In order to compute observables, only the weights at the end of the Markov step (i.e., after 18 elementary processes) are written in the file, to construct the accumulated weight G_n^p of Eq. (8.45). When $\lambda \to \infty$, a similar picture holds, where the weights are stored only at the end of the Markov step described by the evolution in imaginary time τ.

To this purpose, it is enough to bookkeep the corresponding time β_{left} remaining to complete the imaginary-time propagation of length β. For $\lambda \to \infty$, p_d becomes:

$$p_d \approx 1 + \frac{e_L(x) - \mathcal{H}_{x,x}}{\lambda}. \tag{8.64}$$

Then, Eq. (8.60) becomes:

$$\frac{k}{\lambda} \equiv \tau = \min\left(\beta_{\text{left}}, \frac{\ln(1 - \xi)}{e_L(x) - \mathcal{H}_{x,x}}\right), \tag{8.65}$$

where we have used that $\ln p_d \approx [e_L(x) - \mathcal{H}_{x,x}]/\lambda$. Along the imaginary time β, we iteratively apply this formula and bookkeep the remaining time β_{left} that is left to complete the Markov step of length β. At the beginning, $\beta_{\text{left}} = \beta$, then τ is extracted using Eq. (8.65); the weight is updated according to:

$$w \to w \exp\left[-\tau e_L(x)\right]; \tag{8.66}$$

if $\tau < \beta_{\text{left}}$ the new configuration is taken according to Eq. (8.62) and the remaining time is changed into $\beta_{\text{left}} - \tau$. The Markov step is completed when $\beta_{\text{left}} = 0$.

The important aspect of this approach is that the evolution with the imaginary-time propagator $\exp(-\beta\mathcal{H})$ is done *without* any Trotter error (Trotter, 1959; Suzuki, 1976a,b), which is possible on the lattice.

8.5 Many Walkers Formulation

In practice, the single-walker formulation described in the previous sections is unstable when the number of projection p in Eq. (8.45) is too large. Indeed, the accumulated weight G_n^p is a product of p random variables, thus having huge fluctuations as seen in section 2.7:

$$\frac{\sqrt{\langle (G_n^p)^2 \rangle - \langle G_n^p \rangle^2}}{\langle G_n^p \rangle} \approx \exp\left(\frac{p}{2}\sigma^2\right), \tag{8.67}$$

where σ^2 is the variance of $\ln b_x$. (Also here we prefer to use a simplified notation without a tilde over the different quantities, but still assuming that we are using importance sampling.)

In order to overcome the problem of large fluctuations, we discuss a way to propagate a set of N_w walkers simultaneously, defined by their weights $w_{\alpha,n}$ and configurations $x_{\alpha,n}$, for each Markov iteration n and $\alpha = 1, \ldots, N_w$. Given N_w walkers, we indicate the corresponding configurations and weights with a dyad of vectors $(\mathbf{x}_n, \mathbf{w}_n)$. By evolving them independently, no improvement is obtained for the aforementioned large fluctuations, (e.g., we would just reduce the variance of the accumulated weight by a factor $1/N_w$):

$$\langle (G_n^p)^2 \rangle - \langle G_n^p \rangle^2 \rightarrow \frac{1}{N_w}\left[\langle (G_n^p)^2 \rangle - \langle G_n^p \rangle^2\right]. \tag{8.68}$$

Instead, by performing an iterative *reconfiguration* of the walkers, before that the variance of the weights becomes too large, it is possible to keep the weights of all the walkers approximately equal during the simulation and then obtain a considerable reduction of statistical errors of each single factor appearing in the evaluation of G_n^p, implying that:

$$\sigma^2 \rightarrow \frac{\sigma^2}{N_w}, \tag{8.69}$$

which leads an exponential reduction of the fluctuations in Eq. (8.67). Therefore, by taking $N_w \approx p$, this procedure allows us a stable propagation for a projection of p steps.

Let us now discuss the reconfiguration scheme, which goes under the name of *branching*. The Master equation (8.38) can be easily generalized to the case of many *independent* walkers. Indeed, if the evolution of the probability distribution $\mathcal{P}_n(\mathbf{x}, \mathbf{w})$ is done without any restriction, each walker is uncorrelated from any other one, and we have:

$$\mathcal{P}_n(\mathbf{x}, \mathbf{w}) = \prod_\alpha \mathcal{P}_n(x_\alpha, w_\alpha). \tag{8.70}$$

Similarly to the previous case of Eq. (8.39), we can define the state evolved at iteration n by:

$$\Psi_G(x)\Psi_n(x) \equiv \int d\mathbf{w} \sum_{\mathbf{x}} \left(\frac{\sum_\alpha w_\alpha \delta_{x,x_\alpha}}{N_w}\right) \mathcal{P}_n(\mathbf{x}, \mathbf{w}) \approx \left\langle\!\!\left\langle \frac{1}{N_w} \sum_{\alpha=1}^{N_w} w_{\alpha,n} \delta_{x,x_{\alpha,n}} \right\rangle\!\!\right\rangle. \tag{8.71}$$

Table 8.1. *Reconfiguration scheme. Here $\overline{w}_n$ is defined in Eq. (8.72) and the new walkers are chosen according to the probability $p_{\alpha,n}$ of Eq. (8.73), see Fig. 8.3.*

Old Walkers		New Walkers	
$(x_{1,n}, w_{1,n})$	$\longrightarrow$	$x'_{1,n} = x_{j(1),n}$	$w'_{1,n} = \overline{w}_n$
$(x_{2,n}, w_{2,n})$	$\longrightarrow$	$x'_{2,n} = x_{j(2),n}$	$w'_{2,n} = \overline{w}_n$
$(x_{3,n}, w_{3,n})$	$\longrightarrow$	$x'_{3,n} = x_{j(3),n}$	$w'_{2,n} = \overline{w}_n$
$\vdots$	$\vdots$	$\vdots$	
$(x_{N_w,n}, w_{N_w,n})$	$\longrightarrow$	$x'_{N_w,n} = x_{j(N_w),n}$	$w'_{N_w,n} = \overline{w}_n$

The crucial point is that the physical information is contained in the first moment (with respect to the weight) of the distribution $\mathcal{P}_n(\mathbf{x}, \mathbf{w})$, which gives the wave function (e.g., $\Psi_G(x)\Psi_n(x)$ with importance sampling). Therefore, we can define a reconfiguration process that changes the probability distribution *without* changing the statistical average of Eq. (8.71). This can be obtained by a Markovian step that changes the configurations $(\mathbf{x}_n, \mathbf{w}_n)$ into $(\mathbf{x}'_n, \mathbf{w}'_n)$. In particular, the following reconfiguration does a proper job (Calandra Buonaura and Sorella, 1998):

1. Take the new weights equal to the average of the old ones:

$$w'_{\alpha,n} = \overline{w}_n \equiv \frac{1}{N_w} \sum_\beta w_{\beta,n}. \tag{8.72}$$

2. Choose the new configurations among the old ones with a probability $p_{\alpha,n}$ that is proportional to the weight of the old configurations:

$$p_{\alpha,n} \equiv \frac{w_{\alpha,n}}{\sum_\beta w_{\beta,n}}. \tag{8.73}$$

The branching procedure is schematically shown in Table 8.1 and Fig. 8.3. In principle, the selection of the new configurations can be done by extracting N_w independent random numbers z_α, uniformly distributed in $[0, 1)$, and apply the strategy of section 3.5. However, a better reconfiguration scheme is obtained by extracting N_w *correlated* random numbers $z_\alpha = (\alpha + \xi - 1)/N_w$, where ξ is a single random number uniformly distributed in $[0, 1)$ (and $\alpha = 1, \ldots, N_w$). This set of numbers is then used to select the new configurations, giving an unbiased result despite their correlation. Notice that, when all weights $p_{\alpha,n}$ are equal, this improved scheme does not modify the population of walkers, while, in the general case, the number of surviving walkers is optimal. After this reconfiguration, the new set of N_w walkers have the same weights $\overline{w}_n$, and most of the irrelevant walkers with small weights have dropped out, since they will not be selected according to $p_{\alpha,n}$.

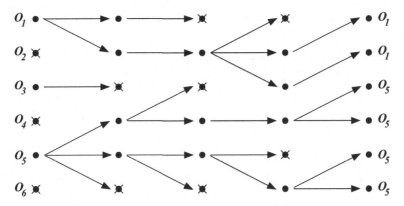

Figure 8.3 Schematic picture of the branching process for $N_{\mathrm{w}} = 6$ walkers. Here, every configuration $x_{\alpha,n}$ (denoted with a full dot) may be killed (denoted by crosses) or continue its evolution (denoted by arrows). In the latter case, the configuration may give rise to one or more copies of it, according to the probability of Eq. (8.73). Reconfigurations are done every n_{bra} Markov steps (here, the intermediate steps are not shown). The forward-walking technique is also shown: at the leftmost step, a (diagonal) operator is applied, whose values are $\mathcal{O}_\alpha$ for $\alpha = 1, \ldots, 6$ (the index of the Markov step is omitted); at the rightmost step the forward walking is finished (after $m = 4$ steps) and the values of the operators are changed according to the branching procedure.

This reconfiguration plays the same stabilization effect of the conventional branching scheme (Ceperley and Alder, 1980; Reynolds et al., 1982; Foulkes et al., 2001), but with the advantage that the number of walkers is kept fixed along the simulation. Indeed, one of the problems of the traditional branching scheme is to control the walker population and the fluctuations of the weights, which may induce some bias in the simulation.

Within the reconfiguration of Eqs. (8.72) and (8.73), some kind of correlation among the walkers is introduced, however, we can rigorously prove that the statistical average of Eq. (8.71) over the probability $\mathcal{P}_n(\mathbf{x}, \mathbf{w})$ is equal to the one performed with the new probability $\mathcal{P}'_n(\mathbf{x}', \mathbf{w}')$, namely we have that:

$$\Psi_G(x)\Psi'_n(x) = \Psi_G(x)\Psi_n(x). \tag{8.74}$$

In fact, the branching process is described by:

$$\mathcal{P}'_n(\mathbf{x}', \mathbf{w}') = \int d\mathbf{w} \sum_{\mathbf{x}} T_{\mathrm{B}}(\mathbf{x}', \mathbf{w}'|\mathbf{x}, \mathbf{w})\, \mathcal{P}_n(\mathbf{x}, \mathbf{w}), \tag{8.75}$$

where the transition probability $T_{\mathrm{B}}(\mathbf{x}', \mathbf{w}'|\mathbf{x}, \mathbf{w})$ is given by:

$$T_{\mathrm{B}}(\mathbf{x}', \mathbf{w}'|\mathbf{x}, \mathbf{w}) = \prod_{\beta} \left(\frac{\sum_\alpha w_\alpha \delta_{x'_\beta, x_\alpha}}{\sum_\alpha w_\alpha} \right) \delta\left(w'_\beta - \frac{\sum_\alpha w_\alpha}{N_{\mathrm{w}}} \right). \tag{8.76}$$

Therefore, we have that the new wave function, after the branching process, is given by:

$$\Psi_G(x)\Psi'_n(x) = \int d\mathbf{w}' \sum_{\mathbf{x}'} \left(\frac{\sum_\alpha w'_\alpha \delta_{x,x'_\alpha}}{N_w} \right) \mathcal{P}'_n(\mathbf{x}',\mathbf{w}')$$

$$= \int d\mathbf{w}' \sum_{\mathbf{x}'} \left(\frac{\sum_\alpha w'_\alpha \delta_{x,x'_\alpha}}{N_w} \right) \int d\mathbf{w} \sum_{\mathbf{x}} T_B(\mathbf{x}',\mathbf{w}'|\mathbf{x},\mathbf{w}) \mathcal{P}_n(\mathbf{x},\mathbf{w}).$$
(8.77)

Here, it is convenient to single out each term of the sum, i.e., $w'_\gamma \delta_{x,x'_\gamma}$ and sum (integrate) over all the other variables x'_α (w'_α) with $\alpha \neq \gamma$. The contribution of this term (indicated by $\left[\Psi_G(x)\Psi'_n(x)\right]_\gamma$) is given by:

$$\left[\Psi_G(x)\Psi'_n(x)\right]_\gamma = \int dw'_\gamma \sum_{x'_\gamma} \left(\frac{w'_\gamma \delta_{x,x'_\gamma}}{N_w} \right) \int d\mathbf{w} \sum_{\mathbf{x}} \left(\frac{\sum_\alpha w_\alpha \delta_{x'_\gamma,x_\alpha}}{\sum_\alpha w_\alpha} \right)$$

$$\times \delta\left(w'_\gamma - \frac{\sum_\alpha w_\alpha}{N_w} \right) \mathcal{P}_n(\mathbf{x},\mathbf{w}).$$
(8.78)

Then, the factor $\sum_\alpha w_\alpha$, appearing in the previous equation, cancels out after integrating over w'_γ and summing over x'_γ, due to the delta-function that imposes $w'_\gamma = 1/N_w \sum_\alpha w_\alpha$:

$$\left[\Psi_G(x)\Psi'_n(x)\right]_\gamma = \frac{1}{N_w} \int d\mathbf{w} \sum_{\mathbf{x}} \left(\frac{\sum_\alpha w_\alpha \delta_{x,x_\alpha}}{N_w} \right) \mathcal{P}_n(\mathbf{x},\mathbf{w}),$$
(8.79)

namely, $\left[\Psi_G(x)\Psi'_n(x)\right]_\gamma = \Psi_G(x)\Psi_n(x)/N_w$. By summing over γ we have proven the statement of Eq. (8.74).

In practice, the reconfiguration scheme of Eq. (8.75) is performed every n_{bra} power iterations. We would like to emphasize that taking a large n_{bra} implies a rather inefficient branching process, since the walkers will have considerably different weights and only few of them will survive the reconfiguration (here the inefficiency comes from the fact that we loose time in evolving walkers that will not contribute to the final averages). By contrast, a too small n_{bra} is also not optimal, since in this case all the walkers will be unaffected and we spend time in performing a reconfiguration that does not change the distribution $\mathcal{P}_n(\mathbf{x},\mathbf{w})$.

It is useful, after each reconfiguration, to store the average weight $\overline{w}_n$ and put $w'_{\alpha,n} = 1$ for all walkers $\alpha = 1,\ldots,N_w$ (instead of taking $w'_{\alpha,n} = \overline{w}$). The value of the accumulated weight G^p_n can be recovered by following the evolution of the weights in the p reconfiguration processes and reads:

$$G^p_n = \prod_{j=1}^{p} \overline{w}_{n-j},$$
(8.80)

which corresponds to the application of $p \times n_{bra}$ iterations of the power method.

The continuous-time approach of section 8.4 can be also applied; in this case the branching scheme can be done after an evolution of length τ_{bra}, i.e., after a propagation given by $\exp(-\tau_{\text{bra}}\mathcal{H})$.

Once the branching process is done, it is no longer possible to obtain an explicit form of the equilibrium distribution of the marginal probability:

$$\Pi_n(x) = \int d\mathbf{w} \sum_{\mathbf{x}} \left(\frac{\sum_\alpha \delta_{x,x_\alpha}}{N_{\text{w}}} \right) P_n(\mathbf{x}, \mathbf{w}). \tag{8.81}$$

Indeed, by repeating the same steps as before, we can show that the new marginal probability after the reconfiguration is given by:

$$\Pi'_n(x) = \int d\mathbf{w} \sum_{\mathbf{x}} \left(\frac{\sum_\alpha w_\alpha}{N_{\text{w}}} \right)^{-1} \left(\frac{\sum_\alpha w_\alpha \delta_{x,x_\alpha}}{N_{\text{w}}} \right) P_n(\mathbf{x}, \mathbf{w}), \tag{8.82}$$

which depends upon the number of walkers N_{w} and recovers the wave function of Eq. (8.77) when increasing N_{w}, i.e., when the term in the first parenthesis approaches a constant value (by the central limit theorem, see section 2.7).

8.5.1 Ground-State Energy and Correlation Functions

As in the case of the single-walker technique, the ground-state energy can be computed by using a single simulation, with configurations that are equilibrated according to $\Pi_{\text{eq}}(x)$ (not explicitly known) and considering the accumulated weight of Eq. (8.80):

$$E_0 \approx \frac{\sum_n G_n^p e_L(x_n)}{\sum_n G_n^p}; \tag{8.83}$$

where n denotes the n-th reconfiguration process; the local energy can be evaluated either immediately after the reconfiguration process, when all the walkers have the same weight:

$$e_L(x_n) \equiv \frac{1}{N_{\text{w}}} \sum_\alpha e_L(x_{\alpha,n}), \tag{8.84}$$

or, for a better statistical error, just before the reconfiguration, taking properly into account the weight of each single walker:

$$e_L(x_n) \equiv \frac{\sum_\alpha w_{\alpha,n} e_L(x_{\alpha,n})}{\sum_\alpha w_{\alpha,n}}. \tag{8.85}$$

An example of the convergence for two values of N_{w} is reported in Fig. 8.4, for the continuous-time algorithm of section 8.4. Here, we show the case of the

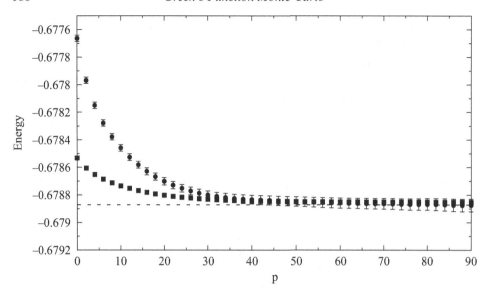

Figure 8.4 Energy per site for the Heisenberg model on the 6×6 cluster (square lattice) as a function of the projection index p, see Eq. (8.83). Two cases with $N_w = 10$ (circles) and $N_w = 50$ (squares) are shown. The continuous-time approach with $\tau = 0.1$ is used. The dashed line indicates the exact result obtained by Lanczos diagonalization.

ground-state energy of the Heisenberg model on the square lattice for the 6×6 cluster (where the exact solution is available by using Lanczos diagonalizations). Notice that the starting point with $p = 0$ is different for $N_w = 10$ and 50, the latter case being closer to the exact result.

Let us finish this part by discussing how to compute correlation functions. For local operators $\mathcal{O}$ that satisfy the condition of Eq. (8.47), we can perform the calculation by using a single Markov chain for both the numerator and the denominator of Eq. (8.48), as discussed in section 8.3 for the single-walker technique. However, in order to control the forward walking technique, the set of measured values $O_{\alpha,n}$, for $\alpha = 1, \ldots, N_w$ at the iteration n, must be modified after each reconfiguration process occurring in the forward direction. Indeed, within the Markov process with the reconfiguration of Eq. (8.76), the denominator can be easily computed by constructing the accumulated weight G_n^p; instead, for the numerator, the branching scheme must be corrected by the fact that the application of the operator at the iteration n would change the reconfigurations at the next iterations, since it modifies the weights. In practice, after each branching it is important to bookkeep only the values of the observables that correspond to the walkers that survive the reconfiguration. This can be done by storing (after each branching) the value $\mathcal{O}_{\alpha,n}$ of the operator for each walker and a table $\beta = j_n(\alpha)$ that gives the old walker β from which the new one α has been generated. Then, the contribution after m reconfiguration steps is given by the recursive application of the function $j_n(\alpha)$:

$$\mathcal{O}'_{\alpha,n+m} = \mathcal{O}_{j_{n+1}(j_{n+2}(...(j_{n+m}(\alpha)))),n}. \tag{8.86}$$

It must be stressed that the forward walking technique may be unstable when few walkers are taken and a large propagation is considered: in fact, only few values of the observable $\mathcal{O}_{\alpha,n}$ survive the forward walk and eventually, for $m \to \infty$, only one value will remain, see Fig. 8.3. Therefore, this approach gives reliable results only for small enough propagation, before some instability appears.

Examples of the convergence as a function of the reconfiguration steps are shown in Fig. 8.5 for $N_w = 50$, by using the continuous-time algorithm of section 8.4. Here, we show the case of the spin-spin correlations in real space for the Heisenberg model on the square lattice (again for the 6×6 cluster, where the exact solution is possible):

$$S(r) = \frac{1}{L} \sum_i S^z_{i+r} S^z_i, \tag{8.87}$$

where L is the number of sites and S^z_i is the z-component of the spin operator at the site i.

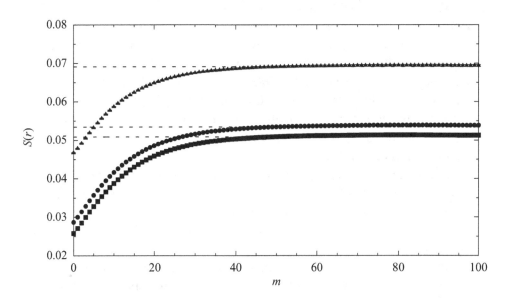

Figure 8.5 Spin-spin correlations $S(r)$ of Eq. (8.87) for the Heisenberg model on the 6×6 cluster (square lattice) as a function of the reconfiguration step m, see Eq. (8.86). The three independent distances along the diagonal of the cluster are considered (with increasing distance from top to bottom). Here, $N_w = 50$ and the continuous-time approach with $\tau = 0.1$ is used. The dashed lines indicate the exact results obtained by Lanczos diagonalization.

8.6 Practical Implementation

We finally give a brief summary of the important steps in a practical implementation of the Green's function Monte Carlo algorithm.

1. **Initialization** at the beginning of the calculation.
   - Generate N_w random configurations $|x_\alpha\rangle$ and initialize the weights $w_\alpha = 1$. As in the variational Monte Carlo method, configurations can be stored into a set of vectors $\mathtt{iconf}(L, N_w)$, whose elements give the local state on the site $i = 1, \ldots, L$ for each walker $\alpha = 1, \ldots, N_w$, and $\mathtt{kel}(2L, N_w)$, whose non-zero elements give, for each site, the position of the creation operators in the string defining the sampled configuration $|x\rangle$ of Eq. (5.44).
   - Verify that the initial configurations are not singular, i.e., $\langle x_\alpha | \Psi_J \rangle \neq 0$, similarly to what has been discussed in the variational Monte Carlo method.
   - As for the variational Monte Carlo approach, compute the table of Eq. (5.33) and all the Green's functions that are necessary to compute $\tilde{\mathcal{G}}$ with importance sampling and perform the time evolution. For the determinant case, only the static Green's function of Eq. (5.80) is necessary, while for the Pfaffian case both the standard Green's function (5.129) and the anomalous ones (5.130) and (5.131) are needed.
2. **GFMC projection (with the continuous-time approach).**
   - For each walker α independently, select the new configuration $|x'_\alpha\rangle$ among the ones that are connected to $|x_\alpha\rangle$ through the stochastic matrix $t_{x'x}$ of Eq. (8.62).
   - For each walker α, update the weight factor w_α by using Eq. (8.66).
   - Update the table for the Jastrow factor (5.34) and all the Green's functions, i.e., Eq. (5.79) for determinants or Eqs. (5.125) and (5.128) for Pfaffians.
   - After a time propagation of length τ_{bra}, perform a reconfiguration of the walkers, according to Eqs. (8.72) and (8.73), see also Table 8.1.
3. **Computation of observables.**
   Observables can be computed every time a branching reconfiguration is performed (either just before or after it). For the calculation of the ground-state energy, it is recommended to write the average weight $\overline{w}$ of Eq. (8.72) and the local energy of Eq. (8.84) (before the branching) or Eq. (8.85) (after the branching) on the hard disk and then perform the calculations of the accumulated weight of Eq. (8.45), with the smallest value of p. For the calculation of local observables, it is recommended to write all the configurations $\mathtt{iconf}(L, N_w)$ and the table $\beta = j_n(\alpha)$ (obtained at each n branching procedure), in order to perform the reshuffling of Eq. (8.86) that is needed within the forward-walker technique.

9

Reptation Quantum Monte Carlo

9.1 A Simple Path Integral Technique

In Chapter 8, we described the Green's function Monte Carlo (GFMC) in which the ground-state wave function is filtered out by using the power method. There, the projection technique is statistically implemented by introducing walkers that are labelled by the electron configuration $|x\rangle$ and the weight w. Then, we have seen that it is possible to solve the problem of the exponential growth of the weights by considering many walkers and iteratively propagate them with approximately the same weight. Although this kind of Monte Carlo approach is largely used, it may become inefficient when the number of walkers, which are necessary to carry out the simulation, becomes too large. This usually happens when the wave function used for the importance sampling is not particularly accurate. In addition, an exceedingly large number of walkers may be necessary for a large number of forward projections. For these reasons, the reptation quantum Monte Carlo (RQMC) algorithm has been introduced, as a simple and efficient way to perform a projection method within a path-integral representation (Baroni and Moroni, 1998; Moroni and Baroni, 1999). The main advantage of the RQMC is that it requires only one walker without any weight, as in the variational Monte Carlo method, thus allowing us to compute correlation functions, even for large projection times. However, in this case, the autocorrelation time may become very large.

The aim of the RQMC approach is to define a Markov process that allows us to sample:

$$\mathcal{Z} = \langle \Psi_{\text{var}} | (\Lambda - \mathcal{H})^N | \Psi_{\text{var}} \rangle, \tag{9.1}$$

where $|\Psi_{\text{var}}\rangle$ is a given initial state (for example the best variational *Ansatz*) from which the projection with N steps is performed (as usual Λ is a diagonal operator with $\Lambda_{x,x} = \lambda$, where λ is a sufficiently large number to guarantee the convergence to the ground state). Notice that, in contrast to the GFMC method, which is defined by Eq. (8.2), here the projection is done for a *fixed* number N of steps (and then the overlap with the initial state is computed). Within RQMC, $\mathcal{Z}$ plays the role of a pseudo-partition function that defines *classical* averages. Indeed, a natural representation of $\mathcal{Z}$ is given by using a discrete path integral representation (where the integral over different paths is substituted by sums over configurations); then, physical quantities will be written as classical averages by using this pseudo-partition function.

As for the variational and the GFMC approaches (see Chapters 5 and 8), we fix a complete basis set $\{|x\rangle\}$ in the many-body Hilbert space:

$$\sum_x |x\rangle\langle x| = \mathbb{I}, \tag{9.2}$$

in which the states are taken to be *orthogonal* and *normalized*. Furthermore, we assume that the matrix elements of the Hamiltonian in this basis $\mathcal{H}_{x'x} = \langle x'|\mathcal{H}|x\rangle$ can be computed efficiently for each $|x\rangle$ and $|x'\rangle$. For a local Hamiltonian, the number of non-zero entries $\mathcal{H}_{x'x}$ is extremely small compared to their total number; indeed, for a given $|x\rangle$, the number of non-zero matrix elements is of the order of the system size L, and not of the full Hilbert space. Therefore, for a given state $|x\rangle$, all entries $\mathcal{H}_{x'x}$ can be computed with a reasonable computational effort.

By inserting $N + 1$ completeness relations for the chosen orthonormal basis, the pseudo-partition function $\mathcal{Z}$ is written as a path integral over N time slices:

$$\mathcal{Z} = \sum_{x_0,\ldots,x_N} \mathcal{W}(x_N,\ldots,x_0); \tag{9.3}$$

here, the weight $\mathcal{W}(x_N,\ldots,x_0)$ is defined by:

$$\mathcal{W}(x_N,\ldots,x_0) = \mathcal{G}_{x_N,x_{N-1}} \ldots \mathcal{G}_{x_2,x_1}\mathcal{G}_{x_1,x_0} \Psi^2_{\text{var}}(x_0), \tag{9.4}$$

where $\Psi_{\text{var}}(x) = \langle x|\Psi_{\text{var}}\rangle$ and, in analogy with the GFMC approach, we have defined the Green's function $\mathcal{G}_{x'x}$ with importance sampling:

$$\mathcal{G}_{x'x} = \langle x'|(\Lambda - \mathcal{H})|x\rangle \frac{\Psi_{\text{var}}(x')}{\Psi_{\text{var}}(x)}. \tag{9.5}$$

Thereafter, in order to simplify the notation, we do not put a tilde over the various quantities with importance sampling.

Similarly to the GFMC approach, in order to have a non-negative weight $\mathcal{W}(x_N,\ldots,x_0)$, it is necessary to have that *all* matrix elements of the Green's function $\mathcal{G}_{x'x}$ are non-negative. Whenever some couples (x',x) give $\mathcal{G}_{x'x} < 0$, we face the *sign problem*, which will be discussed in Chapter 10.

9.1.1 Ground-State Energy and Correlation Functions

The ground-state energy and correlation functions of local operators can be written in terms of the pseudo-partition function $\mathcal{Z}$. For example, the ground-state energy can be estimated (for large enough N) by:

$$E_0 \approx \frac{\langle \Psi_{\text{var}} | \mathcal{H} \mathcal{G}^N | \Psi_{\text{var}} \rangle}{\langle \Psi_{\text{var}} | \mathcal{G}^N | \Psi_{\text{var}} \rangle} = \frac{1}{\mathcal{Z}} \sum_{x_0, \ldots, x_N} e_L(x_N) \mathcal{W}(x_N, \ldots, x_0), \qquad (9.6)$$

where, as for the GFMC technique, the local energy is defined by:

$$e_L(x) = \frac{\langle \Psi_{\text{var}} | \mathcal{H} | x \rangle}{\langle \Psi_{\text{var}} | x \rangle}. \qquad (9.7)$$

A slightly better estimation of the ground-state energy can be obtained by noticing that $\mathcal{H}$ commutes with $\mathcal{G}$ in Eq. (9.6), so that:

$$E_0 \approx \frac{1}{\mathcal{Z}} \sum_{x_0, \ldots, x_N} \left[\frac{e_L(x_0) + e_L(x_N)}{2} \right] \mathcal{W}(x_N, \ldots, x_0). \qquad (9.8)$$

Similarly, for local operators that are diagonal in the basis set, i.e., $\mathcal{O}|x\rangle = \mathcal{O}(x)|x\rangle$, where $\mathcal{O}(x)$ is the eigenvalue corresponding to the configuration $|x\rangle$, we have that:

$$\frac{\langle \Psi_{\text{var}} | \mathcal{G}^M \mathcal{O} \mathcal{G}^{N-M} | \Psi_{\text{var}} \rangle}{\langle \Psi_{\text{var}} | \mathcal{G}^N | \Psi_{\text{var}} \rangle} = \frac{1}{\mathcal{Z}} \sum_{x_0, \ldots, x_N} \mathcal{O}(x_{N-M}) \mathcal{W}(x_N, \ldots, x_0). \qquad (9.9)$$

Therefore, quantum expectation values are written in terms of a *classical* probability $\mathcal{W}(x_N, \ldots, x_0)$ that depends upon the *reptile* (or *polymer*) defined by all the configurations from x_0 to x_N along the projection:

$$\mathcal{R} \equiv (x_N, \ldots, x_0). \qquad (9.10)$$

A picture of the reptile is reported in Fig. 9.1.

9.2 A Simple Way to Sample Configurations

In order to evaluate the ground-state energy and correlation functions, we can consider a Markov process with the Metropolis algorithm to sample the classical weight $\mathcal{W}(\mathcal{R})$. We define two basic moves $\mathcal{R} \rightarrow \mathcal{R}'$ that shift the reptile on the right or left, adopting the convention that the right (left) move is denoted by $d = +1$ ($d = -1$), see Fig. 9.1. In the simplest RQMC sampling, the variable d is chosen randomly at each step with equal probability for left or right moves. We denote by $T(\mathcal{R}', d' | \mathcal{R}, d)$ the transition probability that defines the proposed move within the

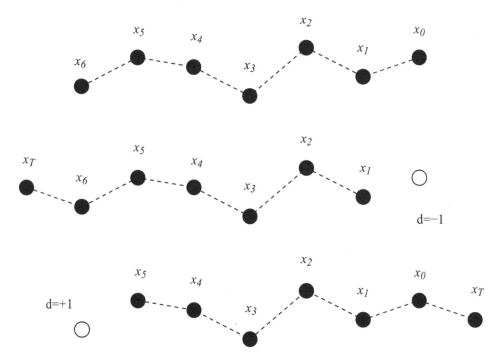

Figure 9.1 Pictorial representation of the reptile $\mathcal{R}$ defined by all the configurations from x_0 to x_N (here $N = 6$). The new reptile $\mathcal{R}'$ is generated from the old one performing left/right moves where a new head/tail is added and the old tail/head is discarded.

Metropolis algorithm. Since the new direction d' is chosen randomly with equal probability for left and right moves, regardless the actual value of d, we have:

$$T(\mathcal{R}', d'|\mathcal{R}, d) = \frac{1}{2} t^{(d')}(\mathcal{R}'|\mathcal{R}), \tag{9.11}$$

where $t^{(d')}(\mathcal{R}'|\mathcal{R})$ defines a change of the reptile $\mathcal{R} \to \mathcal{R}'$ for a given direction d'. The acceptance probability $a^{(d')}(\mathcal{R}'|\mathcal{R})$ can be obtained by exploiting the detailed balance condition:

$$t^{(d')}(\mathcal{R}'|\mathcal{R}) a^{(d')}(\mathcal{R}'|\mathcal{R}) \mathcal{W}(\mathcal{R}) = t^{(d)}(\mathcal{R}|\mathcal{R}') a^{(d)}(\mathcal{R}|\mathcal{R}') \mathcal{W}(\mathcal{R}'); \tag{9.12}$$

in this equation $d = -d'$ because the reverse move is possible only by changing the direction of the propagation d. Then, within the Metropolis algorithm, the acceptance probability is given by:

$$a^{(d')}(\mathcal{R}'|\mathcal{R}) = \mathrm{Min}\left[1, r^{(d')}(\mathcal{R}', \mathcal{R})\right], \tag{9.13}$$

where

$$r^{(d')}(\mathcal{R}',\mathcal{R}) = \frac{W(\mathcal{R}')t^{(-d')}(\mathcal{R}|\mathcal{R}')}{W(\mathcal{R})t^{(d')}(\mathcal{R}'|\mathcal{R})}. \tag{9.14}$$

Within this approach, the probability distribution of the reptile evolves along the Markov iteration n according to:

$$\mathcal{P}_{n+1}(\mathcal{R}',d') = \sum_{\mathcal{R},d} K(\mathcal{R}',d'|\mathcal{R},d)\mathcal{P}_n(\mathcal{R},d), \tag{9.15}$$

where the full transition probability is given by:

$$K(\mathcal{R}',d'|\mathcal{R},d) = T(\mathcal{R}',d'|\mathcal{R},d)a^{(d')}(\mathcal{R}'|\mathcal{R}). \tag{9.16}$$

Notice that the full transition probability satisfies the detailed balance:

$$K(\mathcal{R}',d'|\mathcal{R},d)W(\mathcal{R}) = K(\mathcal{R},d|\mathcal{R}',d')W(\mathcal{R}'), \tag{9.17}$$

which can be easily verified by using Eqs. (9.11) and (9.12). After a thermalization time, e.g., for large n, $\mathcal{P}_n(\mathcal{R},d)$ converges to the desired equilibrium probability $\mathcal{P}_{\text{eq}}(\mathcal{R}) \equiv W(\mathcal{R})/(2\mathcal{Z})$, see section 3.8.

In the following, we decompose the Green's function $\mathcal{G}$ as already done within the GFMC approach, i.e., $\mathcal{G}_{x',x} = p_{x',x}b_x$, such that the classical weight becomes:

$$W(\mathcal{R}) = p_{x_N,x_{N-1}} \cdots p_{x_2,x_1}p_{x_1,x_0}b_{x_{N-1}} \cdots b_{x_1}b_{x_0}\Psi^2_{\text{var}}(x_0), \tag{9.18}$$

and discuss in detail the two right/left moves. Let us start with $d = +1$ (right move). In this case:

$$\mathcal{R}' = (x'_N,\ldots,x'_0) \equiv (x_{N-1},\ldots,x_0,x_T), \tag{9.19}$$

where x_T is the trial move at the rightmost side of the reptile. The transition probability of this process is:

$$t^{(+1)}(\mathcal{R}'|\mathcal{R}) = p_{x_T,x_0}. \tag{9.20}$$

The corresponding weight on the new reptile $\mathcal{R}'$ is given by:

$$W(\mathcal{R}') = p_{x_{N-1},x_{N-2}} \cdots p_{x_1,x_0}p_{x_0,x_T}b_{x_{N-2}} \cdots b_{x_0}b_{x_T}\Psi^2_{\text{var}}(x_T). \tag{9.21}$$

Thus in the ratio $W(\mathcal{R}')/W(\mathcal{R})$, which is required to implement the Metropolis algorithm, almost all factors cancel out, yielding:

$$\frac{W(\mathcal{R}')}{W(\mathcal{R})} = \frac{b_{x_T}p_{x_0,x_T}\Psi^2_{\text{var}}(x_T)}{b_{x_{N-1}}p_{x_N,x_{N-1}}\Psi^2_{\text{var}}(x_0)} = \frac{b_{x_0}p_{x_T,x_0}}{b_{x_{N-1}}p_{x_N,x_{N-1}}}, \tag{9.22}$$

where we used that $\mathcal{G}_{x',x} = p_{x',x}b_x$ is the Green's function with importance sampling that satisfies $\mathcal{G}_{y,x} = \mathcal{G}_{x,y}\Psi_{\mathrm{var}}^2(y)/\Psi_{\mathrm{var}}^2(x)$. By contrast, for $d = -1$ (left move), we have:

$$\mathcal{R}' = (x'_N, \ldots, x'_0) \equiv (x_T, x_N, \ldots, x_1), \tag{9.23}$$

where now x_T is the trial move at the leftmost side of the reptile. The transition probability for this process is:

$$t^{(-1)}(\mathcal{R}'|\mathcal{R}) = p_{x_T,x_N}. \tag{9.24}$$

The corresponding weight on the new reptile $\mathcal{R}'$ is given by:

$$\mathcal{W}(\mathcal{R}') = p_{x_T,x_N}p_{x_N,x_{N-1}} \cdots p_{x_2,x_1}b_{x_N}b_{x_{N-1}} \cdots b_{x_1}\Psi_{\mathrm{var}}^2(x_1). \tag{9.25}$$

Thus the ratio $\mathcal{W}(\mathcal{R}')/\mathcal{W}(\mathcal{R})$, entering the Metropolis algorithm, acquires a simple form:

$$\frac{\mathcal{W}(\mathcal{R}')}{\mathcal{W}(\mathcal{R})} = \frac{b_{x_N}p_{x_T,x_N}\,\Psi_{\mathrm{var}}^2(x_1)}{b_{x_0}p_{x_1,x_0}\,\Psi_{\mathrm{var}}^2(x_0)} = \frac{b_{x_N}p_{x_T,x_N}}{b_{x_1}p_{x_0,x_1}}, \tag{9.26}$$

where, as before, we used the fact that $\mathcal{G}_{x',x} = p_{x',x}b_x$ is the Green's function with importance sampling, which satisfies $\mathcal{G}_{y,x} = \mathcal{G}_{x,y}\Psi_{\mathrm{var}}^2(y)/\Psi_{\mathrm{var}}^2(x)$.

We are now in the position to simplify the term appearing in the Metropolis algorithm of Eq. (9.14). For the right move, the opposite one that brings back $\mathcal{R}' \to \mathcal{R}$ is a left move with $x_T = x_N$:

$$t^{(-1)}(\mathcal{R}|\mathcal{R}') = p_{x_N,x_{N-1}}, \tag{9.27}$$

where we have used the fact that, after the first right move, the leftmost configuration of the $\mathcal{R}'$ reptile is x_{N-1}. Therefore, we obtain:

$$r^{(+1)}(\mathcal{R}',\mathcal{R}) = \frac{b_{x_0}p_{x_T,x_0}}{b_{x_{N-1}}p_{x_N,x_{N-1}}}\frac{p_{x_N,x_{N-1}}}{p_{x_T,x_0}} = \frac{b_{x_0}}{b_{x_{N-1}}}. \tag{9.28}$$

Analogously, for the left move, we have that the opposite move that brings back $\mathcal{R}' \to \mathcal{R}$ is a right move with $x_T = x_0$:

$$t^{(+1)}(\mathcal{R}|\mathcal{R}') = p_{x_0,x_1}, \tag{9.29}$$

where we have used the fact that the rightmost configuration of the $\mathcal{R}'$ reptile is x_1. Then, we get:

$$r^{(-1)}(\mathcal{R}',\mathcal{R}) = \frac{b_{x_N}p_{x_T,x_N}}{b_{x_1}p_{x_0,x_1}}\frac{p_{x_0,x_1}}{p_{x_T,x_N}} = \frac{b_{x_N}}{b_{x_1}}. \tag{9.30}$$

In summary, Eqs. (9.28) and (9.30) completely define the rules for accepting or rejecting the new proposed reptile $\mathcal{R}'$ within the standard Metropolis algorithm (9.13).

9.3 The Bounce Algorithm

The main problem with the previous updating procedure is related to the presence of long correlation times. Indeed, choosing a different direction at every step makes the reptile shaking its head and tail (i.e., few configurations at the leftmost and rightmost positions) without changing its body (i.e., all the inner configurations). This is a particularly serious problem whenever a large number of time slices N is considered. The bounce algorithm has been proposed to achieve a small correlation time by performing many steps along one direction (right or left) before changing it (Pierleoni and Ceperley, 2005). In practice, the variable d is no longer randomly sampled but changes sign only when the move is rejected in Eq. (9.13). In this way, the detailed balance is no longer satisfied. However, the general theory of Markov chains (Meyer, 2000) guarantees the existence of a unique equilibrium distribution, provided ergodicity is satisfied. In the following, we will show that the bounce algorithm admits $\mathcal{W}(\mathcal{R})$ as an equilibrium distribution but we will not discuss the question about the convergence to it.

The full transition probability of the bounce algorithm is given by:

$$K_B(\mathcal{R}', d'|\mathcal{R}, d) = t^{(d')}(\mathcal{R}'|\mathcal{R}) a^{(d')}(\mathcal{R}'|\mathcal{R}) \delta_{d',d} + \delta(\mathcal{R}' - \mathcal{R}) B^{(d)}(\mathcal{R}) \delta_{d',-d}, \quad (9.31)$$

where $B^{(d)}(\mathcal{R})$ can be determined by the normalization condition:

$$\sum_{\mathcal{R}',d'} K_B(\mathcal{R}', d'|\mathcal{R}, d) = 1, \quad (9.32)$$

which leads to:

$$B^{(d)}(\mathcal{R}) = 1 - \sum_{\mathcal{R}'} t^{(d)}(\mathcal{R}'|\mathcal{R}) a^{(d)}(\mathcal{R}'|\mathcal{R}). \quad (9.33)$$

The transition probability $K_B(\mathcal{R}', d'|\mathcal{R}, d)$ (that does not satisfy the detailed balance) determines both the new reptile $\mathcal{R}'$ and the new value of the direction d', e.g., $d' = d$ if the move $\mathcal{R} \to \mathcal{R}'$ is accepted (first term) or $d' = -d$ if the move is rejected (second term).

Then, the Master equation is given by:

$$P_{n+1}(\mathcal{R}', d') = \sum_{\mathcal{R}, d} K_B(\mathcal{R}', d'|\mathcal{R}, d) P_n(\mathcal{R}, d), \quad (9.34)$$

which admits as a stationary solution $P_{eq}(\mathcal{R}, d) \equiv \mathcal{W}(\mathcal{R})/(2\mathcal{Z})$. Indeed, by using the explicit form of the transition probability of Eq. (9.31) with $P_n(\mathcal{R}, d) = \mathcal{W}(\mathcal{R})/(2\mathcal{Z})$, we get:

$$P_{n+1}(\mathcal{R}', d') = \sum_{\mathcal{R}} t^{(d')}(\mathcal{R}'|\mathcal{R}) a^{(d')}(\mathcal{R}'|\mathcal{R}) \frac{\mathcal{W}(\mathcal{R})}{2\mathcal{Z}} + \frac{\mathcal{W}(\mathcal{R}')}{2\mathcal{Z}} B^{(-d')}(\mathcal{R}'); \quad (9.35)$$

then, by using Eq. (9.12) with $d = -d'$, we have:

$$P_{n+1}(\mathcal{R}', d') = \frac{\mathcal{W}(\mathcal{R}')}{2\mathcal{Z}} \sum_{\mathcal{R}} t^{(-d')}(\mathcal{R}|\mathcal{R}') a^{(-d')}(\mathcal{R}|\mathcal{R}') + \frac{\mathcal{W}(\mathcal{R}')}{2\mathcal{Z}} B^{(-d')}(\mathcal{R}'). \quad (9.36)$$

Finally, by using Eq. (9.33), we obtain:

$$P_{n+1}(\mathcal{R}', d') = \frac{\mathcal{W}(\mathcal{R}')}{2\mathcal{Z}} \left[1 - B^{(-d')}(\mathcal{R}') \right] + \frac{\mathcal{W}(\mathcal{R}')}{2\mathcal{Z}} B^{(-d')}(\mathcal{R}') = \frac{\mathcal{W}(\mathcal{R}')}{2\mathcal{Z}}, \quad (9.37)$$

which proves the fact that $\mathcal{W}(\mathcal{R})/(2\mathcal{Z})$ is a stationary solution of the stochastic process determined by $K_B(\mathcal{R}', d'|\mathcal{R}, d)$.

9.4 The Continuous-Time Limit

As for the GFMC approach, it is possible to consider the limit of $\lambda \to \infty$ in the Green's function $\mathcal{G} = (\Lambda - \mathcal{H})$, e.g., by considering the imaginary time evolution of the exact propagator $\exp(-\beta\mathcal{H})$ applied statistically (see section 8.4). This can be done *without* any Trotter error (Trotter, 1959; Suzuki, 1976a,b). Within the RQMC method, we can take $\beta = N\tau$ and consider the path-integral representation with N time slices, each of them consisting into a projection with $\exp(-\tau\mathcal{H})$. Then, the previous updating schemes can be easily performed by propagating either the leftmost (rightmost) configuration $|x_N\rangle$ ($|x_0\rangle$) by using the transition probability of Eq. (8.62) for a time τ where $|x_T\rangle$ is eventually reached. Finally, the Metropolis algorithm is used to accept or reject the move (i.e., the entire propagation by τ); in this case, $r^{(d)}(\mathcal{R}'|\mathcal{R})$ is still given by Eq. (9.28) or (9.30), but the weights at numerator and denominator are the ones accumulated in the corresponding time slices, similarly to Eq. (8.66).

An example of the convergence with the number of time slices N is reported in Fig. 9.2. Here, we show the ground-state energy of the Heisenberg model on the square lattice for the 6×6 cluster (where the exact solution can be obtained by using Lanczos diagonalization).

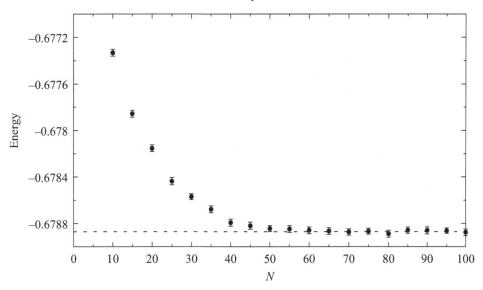

Figure 9.2 Energy per site for the Heisenberg model on the 6×6 cluster (square lattice) as a function of the number of the time slices N (the continuous-time approach with $\tau = 0.1$ is considered). The dashed line indicates the exact result obtained by Lanczos diagonalization.

9.5 Practical Implementation

We finally give a brief summary of the important steps in a practical implementation of the reptation quantum Monte Carlo algorithm.

1. **Initialization** at the beginning of the calculation.
   - Generate the same random configuration $|x_i\rangle$ for all the time slices. These configurations can be stored into a set of vectors $\texttt{iconf}(L, N + 1)$, whose elements give the local state on each site $i = 1, \ldots, L$ and time slice, and $\texttt{kel}(2L, N + 1)$, whose non-zero elements give, for each site, the position of the creation operators in the string defining the sampled configuration $|x\rangle$ of Eq. (5.44).
   - Verify that the initial configurations are not singular, i.e., $\langle x_i | \Psi_J \rangle \neq 0$, similarly to what has been discussed in the variational Monte Carlo method.
   - As for the variational Monte Carlo approach, compute the table of Eq. (5.33) to perform the fast update of the Jastrow factor and all the Green's functions that are necessary to perform the fast update. For the determinant case, only the static Green's function of Eq. (5.80) is necessary, while for the Pfaffian case both the standard Green's function (5.129) and the anomalous ones (5.130) and (5.131) are needed.

2. **RQMC projection (with the continuous-time approach)**
   - Within the given direction $d = \pm 1$, propagate the leftmost (rightmost) configuration $|x_N\rangle$ ($|x_0\rangle$) by using the transition probability of Eq. (8.62) for an imaginary-time τ, until the trial configuration $|x'_T\rangle$ is reached. Accumulate the corresponding weight for the Metropolis algorithm, similarly to Eq. (8.66) to compute $r^{(d)}$. Accept or reject the trial move. If the trial move is accepted then keep the same d for the next one, otherwise change $d \rightarrow -d$.
   - Whenever the trial move is accepted, update the configuration, the table for the Jastrow factor (5.34) and all the Green's functions, i.e., Eq. (5.79) for determinants or Eqs. (5.125) and (5.128) for Pfaffians.
3. **Computation of observables**.

   Observables (e.g., the ground-state energy) can be computed every $O(L)$ steps.

10

Fixed-Node Approximation

10.1 The Sign Problem

In the previous Chapters, we have discussed two approaches that can be used to sample the exact ground-state wave function of a correlated many-body problem. Both of them are based upon the projection technique, in which the ground state is filtered out from an initial trial state; for example, within the Green's function Monte Carlo technique, we have that:

$$\lim_{n \to \infty} (\Lambda - \mathcal{H})^n |\Psi_0\rangle \approx (\lambda - E_0)^n |\Upsilon_0\rangle, \tag{10.1}$$

where Λ is a diagonal operator, with $\Lambda_{x,x} = \lambda$ being a sufficiently large constant, and E_0 is the ground-state energy. Here, the only assumption is that the initial state $|\Psi_0\rangle$ must have a finite overlap with the ground state $|\Upsilon_0\rangle$. Within the stochastic implementation of the power method, it is necessary that the so-called Green's function $\mathcal{G}_{x',x}$ defined in Eq. (8.5) is *non negative* for *all* the elements of the basis set $\{|x\rangle\}$. In some cases, negative terms can be changed into positive ones by including importance sampling; this approach leads to the modified Green's function given in Eq. (8.31). However, in the generic case, the importance sampling cannot remove all the negative elements and we have to face the so-called *sign problem*. In turn, the signs of the Green's function induce the sign structure of the ground-state wave function:

$$|\Upsilon_0\rangle = \sum_x \Upsilon_0(x)|x\rangle, \tag{10.2}$$

where $\Upsilon_0(x) = \langle x|\Upsilon_0\rangle$ can be either positive or negative and its sign is usually *a priori* unknown. Of course, when $\mathcal{G}_{x,x'} \geq 0$, corresponding to a Hamiltonian with all non-positive off-diagonal elements $\mathcal{H}_{x,x'}$, the ground-state wave function has $\Upsilon_0(x) \geq 0$; the strict positivity of the amplitudes can be demonstrated by using the Perron-Frobenius theorem (Meyer, 2000).

Here, we would like to discuss the main issues related to the presence of negative elements in the Green's function and one possible remedy to it, namely the *fixed-node* approximation. In the following, we will limit ourself to *real* wave functions (corresponding to Hamiltonians with real matrix elements), a case that is already tremendously complicated. First of all, we emphasize that the sign problem is *basis dependent*: while the Green's function may have negative entries in a given basis $\{|x\rangle\}$, it may have all positive entries in another basis $\{|y\rangle\}$. Notice that the diagonal elements are not relevant, since they can be made always positive by a suitable shift of $\Lambda_{x,x} = \lambda$. Therefore, the straightforward solution of the sign problem is to choose a clever basis set, such that $\mathcal{G}_{x,x'} \geq 0$ for all the matrix elements. Unfortunately, this possibility can be obtained only in very limited cases. Indeed, for computational reasons of efficiency, the most commonly used basis sets are local ones, which are defined on each individual site or, at most, involve few neighboring sites. In the general case, local basis sets lead to a severe sign problem. Instead, the sign problem would be absent in the basis set that diagonalizes the Hamiltonian; unfortunately, this basis cannot be determined in a reasonable computing time.

In principle the stochastic implementation of the projection scheme can be generalized to the cases where the Green's function has negative entries, by associating the sign to the weight of the walker. In particular, we can consider the following decomposition of the Green's function (here, to simplify the notation, we do not put a tilde over the various quantities, assuming that importance sampling is considered):

$$\mathcal{G}_{x'x} = |\mathcal{G}_{x'x}|s_{x'x} = p_{x'x}b_x s_{x'x}, \tag{10.3}$$

where, similarly to Eq. (8.7), we have defined a transition probability $p_{x'x}$, a weight b_x, and also a sign $s_{x'x}$ as:

$$b_x = \sum_{x'} |\mathcal{G}_{x'x}|, \tag{10.4}$$

$$p_{x'x} = \frac{|\mathcal{G}_{x'x}|}{b_x}, \tag{10.5}$$

$$s_{x'x} = \mathrm{sign}(\mathcal{G}_{x'x}). \tag{10.6}$$

Then, a Markov process can be implemented, generalizing the one used within the Green's function Monte Carlo, see Eqs. (8.10) and (8.11):

1) generate $x_{n+1} = x'$ with probability $p_{x'x_n}$,   (10.7)

2) update the weight with $w_{n+1} = w_n b_x$,   (10.8)

3) attach the sign to the weight $w_{n+1} = w_{n+1}s_{x_{n+1},x_n}$.   (10.9)

This approach can be generalized to N_{w} walkers, as discussed in section 8.5; the branching procedure can be performed according to the absolute value of the weights $|w_{\alpha,n}|$, with $\alpha = 1,\ldots,N_w$. In this way, walkers will be distributed according to the distribution probability $\mathcal{P}_n^F(x,w)$, where now the important difference with what has been discussed in Chapter 8 is that $\mathcal{P}_n^F(x,w) \neq 0$ also for $w < 0$. Then, in analogy with Eq. (8.16), the fermionic wave function is given by:

$$\Psi_n^F(x) \equiv \int dw\, w\, \mathcal{P}_n^F(x,w). \tag{10.10}$$

The fermionic properties are encoded in the presence of non-trivial signs, i.e., within a small asymmetry of the distribution function:

$$\mathcal{P}_n^F(x,w) \neq \mathcal{P}_n^F(x,-w). \tag{10.11}$$

Instead, by disregarding the signs, i.e., by eliminating the third step of Eq. (10.9), the Markov process converges to the "bosonic" ground state given by the lowest-energy state of the Hamiltonian $\mathcal{H}_{x,x'}^B = -|\mathcal{H}_{x,x'}|$ for $x \neq x'$ and $\mathcal{H}_{x,x}^B = \mathcal{H}_{x,x}$. In this case, the walkers will have a non-vanishing probability only for $w > 0$ and the ground-state wave function will be given by:

$$\Psi_n^B(x) \equiv \int dw\, |w|\, \mathcal{P}_n^F(x,w). \tag{10.12}$$

The crucial point is that the fermionic character can be hardly detected in Monte Carlo simulations. Indeed, let us consider a generic observable $\mathcal{O}$ and take the mixed average [see Eq. (8.53)] with $|\Psi_G\rangle = \sum_x |x\rangle$:

$$\frac{\langle \Psi_G|\mathcal{O}|\Psi_n^F\rangle}{\langle \Psi_G|\Psi_n^F\rangle} = \frac{\sum_x \int dw\, w\mathcal{P}_n^F(x,w)\mathcal{O}(x)}{\sum_x \int dw\, w\mathcal{P}_n^F(x,w)}, \tag{10.13}$$

where $\mathcal{O}(x) = \sum_{x'} \mathcal{O}_{x',x}$. Then, by using the reweighting technique described in section 3.2 and denoting the sign of the weight w by $s(w)$, we get:

$$\frac{\sum_x \int dw\, w\mathcal{P}_n^F(x,w)\mathcal{O}(x)}{\sum_x \int dw\, w\mathcal{P}_n^F(x,w)} = \frac{\dfrac{\sum_x \int dw\, |w|s(w)\mathcal{P}_n^F(x,w)\mathcal{O}(x)}{\sum_x \int dw\, |w|\mathcal{P}_n^F(x,w)}}{\dfrac{\sum_x \int dw\, |w|s(w)\mathcal{P}_n^F(x,w)}{\sum_x \int dw\, |w|\mathcal{P}_n^F(x,w)}}. \tag{10.14}$$

Therefore, the previous fermionic quantity can be obtained by sampling over the "bosonic" probability $|w|\mathcal{P}_n^F(x,w)$:

$$\frac{\sum_x \int dw\, w\mathcal{P}_n^F(x,w)\mathcal{O}(x)}{\sum_x \int dw\, w\mathcal{P}_n^F(x,w)} \approx \frac{\langle\langle s\,\mathcal{O}\rangle\rangle_B}{\langle\langle s\rangle\rangle_B}, \tag{10.15}$$

where $\langle\langle\ldots\rangle\rangle_B$ denotes the statistical average over the probability distribution $|w|\mathcal{P}_n^F(x,w)$. In the general case, almost half of the walker's population will have a positive sign, while the other half population will have a negative sign; this fact leads to an almost exact cancellation of both numerator and denominator of Eq. (10.15). Indeed, from Eq. (10.1), we have that:

$$\Psi_n^F(x) \approx (\lambda - E_0^F)^n \Upsilon_0^F(x) \approx e^{-nE_0^F} \Upsilon_0^F(x), \qquad (10.16)$$

$$\Psi_n^B(x) \approx (\lambda - E_0^B)^n \Upsilon_0^B(x) \approx e^{-nE_0^B} \Upsilon_0^B(x), \qquad (10.17)$$

where E_0^F and E_0^B are the ground-state energies of the "fermionic" (i.e., described by $\mathcal{H}$) and "bosonic" (i.e., described by $-|\mathcal{H}|$) systems. Since the "bosonic" ground-state energy is usually much lower than the "fermionic" one, the average sign goes to zero exponentially with the number of particles N_p (both E_0^F and E_0^B being proportional to N_p):

$$\langle\langle s \rangle\rangle_B = \frac{\sum_x \int dw \ |w| s(w) \mathcal{P}_n^F(x, w)}{\sum_x \int dw \ |w| \mathcal{P}_n^F(x, w)} \approx \exp\left[n\left(E_0^B - E_0^F\right)\right] \approx 0. \qquad (10.18)$$

Similarly, also $\langle\langle s \ \mathcal{O} \rangle\rangle_B \approx 0$. In particular, in any numerical evaluation of these quantities that is based upon Monte Carlo sampling, the errorbars will become much larger than the value of the observable making prohibitive its estimation (Ceperley and Alder, 1980; Loh et al., 1990).

Historically, the sign problem first appeared in fermionic models on the continuum, for this reason it is also usually called "fermionic sign problem," which is related to the fact that the fermionic wave function is anti-symmetric when interchanging two particles. Therefore, a fermionic wave function cannot be interpreted as a probability density.

On the continuum, a particularly important concept is given by the *nodal surface*, namely the location of the points where $\Upsilon_0(x) = 0$, which separates the regions with opposite signs. In a d-dimensional system with N_p particles, the dimension of the whole phase space is dN_p. Obviously, the dimension of the nodal surface is $dN_p - 1$; instead, the locus of the points where a fermionic wave function vanishes because of the Pauli principle (i.e., whenever the coordinates of two particles coincide) generates a set of surfaces of dimension $dN_p - d$ (denoted as "Pauli surface"). Therefore, the location of the nodes due to the anti-symmetry does not exhaust the nodal surface, except for $d = 1$, but builds its "skeleton." In order to exemplify this concept, we show in Fig. 10.1 the case with $N_p = 49$ free spinless fermions in a two-dimensional box with side $L = 10$ and periodic-boundary conditions; here, we fixed the coordinates of all particles except one and computed the signs of the many-body wave function (a Slater determinant constructed by filling the lowest-energy one-body states, i.e., plane waves) when displacing the last particle in the box (Ceperley, 1991). Remarkably, in one spatial dimension, the Pauli surface alone completely determines the nodal surface, apart from further accidental nodes that may be caused by degeneracies or singular interactions.

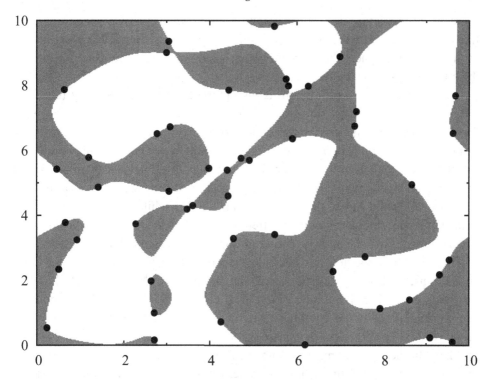

Figure 10.1 The nodal and Pauli surfaces are shown for a system of 49 free spinless fermions in a two-dimensional box of side $L = 10$ with periodic boundary conditions. The coordinates of 48 particles are fixed (randomly) in the box and denoted by black dots. The last particle is displaced in the box and the signs of the many-body wave function are computed. The Pauli surface (or better, its projection in the subspace with the 48 particles frozen in the box) coincides with the black dots. The many-body wave function is positive (negative) in grey (white) regions; therefore, the nodal surface is located at the border between these two regions. Notice that the black dots belongs to the full nodal surface, but do not exhaust it.

The non-trivial aspect is that the nodal surface changes in the presence of interaction and may become very complicated, while the Pauli surface does not change, since it is determined only by symmetry. On the continuum, a straightforward implementation of the projection technique is given by the so-called *diffusion Monte Carlo*, in which the Schrödinger equation (in imaginary time) is solved by using a random sample of the wave function (e.g., walkers) that undergoes diffusion and branching processes, generated by the kinetic and potential terms of the Hamiltonian (Kalos et al., 1974; Ceperley and Alder, 1980; Foulkes et al., 2001). However, without any further constraint, walkers may diffuse in the whole phase space (corresponding to a "bosonic" state) and the "fermionic" character is hidden in the extremely small disproportion between walkers in positive and negative regions; this fact leads to insurmountable difficulties when performing simulations with a large number of electrons. In principle, we could solve the problem by restricting the simulations to regions where the wave function has a fixed sign and then use the symmetrization procedure to generate the wave function in the rest of the space; the problem is that we do not know the exact location of the boundaries of this region.

On the lattice, the sign problem is also present but it is not strictly related to the presence of fermions, since, also in bosonic systems, the ground-state wave function may have non-trivial signs, due to a "frustrating" kinetic energy with positive hoppings on generic lattices. The main difference between the continuum and the lattice is that, in the former one, the imaginary-time propagation with the free-electron kinetic term can be associated to a pure diffusion process (Kalos et al., 1974; Ceperley and Alder, 1980; Foulkes et al., 2001), while this is not always true in the latter one. Another important difference is that on the lattice the nodal surface is not defined, since the allowed configurations span a *discrete* Hilbert space.

The *fixed-node* method has been developed to deal with stable Monte Carlo simulations, for both continuum (Anderson, 1975, 1976; Moskowitz et al., 1982; Reynolds et al., 1982) and lattice (ten Haaf et al., 1995) models. This approach can be generalized to the complex case, by the *fixed-phase* approximation (Ortiz et al., 1993). The fixed-node approximation is a simple way to avoid the sign problem. In the continuum, it consists in forbidding electron moves that change the sign of the guiding function. In practice, we consider a diffusion process determined by the kinetic energy with the additional boundary condition that prevents node crossings: walkers that cross the node in a given diffusion step $|x\rangle \rightarrow |x'\rangle$ (i.e., having $\Psi_G(x)\Psi_G(x') < 0$, where $\Psi_G(x)$ denotes the guiding function used within the importance sampling procedure to fix the nodal surface) are killed. Here, only walkers close to the nodal surface are involved in this process. The antisymmetry of the wave function is replaced by boundary conditions and the approximation is equivalent to solving the Schrödinger equation inside one nodal pocket. Then, the full solution is uniquely determined by the shape of the nodal surface (and the potential). The fixed-node method gives a variational bound to the exact ground-state energy, since it corresponds to the best possible wave function with the boundary conditions determined by the guiding function. On the lattice, any configuration can be directly connected (through the Hamiltonian) to other configurations with opposite signs and, therefore, the fixed-node approximation is not limited to walkers that are "close" to the nodes.

10.2 A Simple Example on the Continuum

In order to have a first idea on how the sign problem arises in the projection technique, we consider a very simple example of a single particle moving in a one-dimensional box with hard walls at its boundaries. The Hamiltonian is given by:

$$\mathcal{H} = -\frac{1}{2}\frac{d^2}{dx^2} + V(x), \tag{10.19}$$

where $0 \leq x \leq L$ and the potential $V(x)$ is even under reflection symmetry $\mathcal{R}$ with respect to the center of the box $x_c = L/2$. All wave functions must vanish at $x = 0$ and $x = L$ and can be classified according to their parity. The actual ground state has no nodes and can be called "bosonic": it is positive for all configurations. Instead, the first excited state has one node at x_c and can be called "fermionic": here, the wave function has both positive and negative regions. In this sense, the symmetry properties associated to a permutation of two particles in a many-body state are translated into the transformations under reflection symmetry $\mathcal{R}$. The ground state is even under reflection, while the first excited state is odd, implying that regions with positive and negative signs must exist.

Let us define two operators P_+ and P_- that project over the subspace of even (symmetric) and odd (anti-symmetric) wave functions:

$$P_\pm \Psi(x) = \frac{1}{2}\left[\Psi(x) \pm \Psi(L-x)\right]. \tag{10.20}$$

In the following, we will define an algorithm that will allow to sample the first excited state $\Upsilon_1(x)$ with *odd* reflection symmetry around the center, i.e., $\Upsilon_1(L-x) = -\Upsilon_1(x)$. On the continuum, it is necessary to use an exponential form of the projector, given the presence of an unbounded spectrum:

$$\lim_{\tau \to \infty} e^{-\tau\mathcal{H}}|\Psi_0\rangle \propto |\Upsilon_0\rangle. \tag{10.21}$$

In this case, the Green's function (without importance sampling) for an imaginary time-evolution τ is given by (Anderson, 1976; Ceperley and Alder, 1980; Reynolds et al., 1982):

$$\mathcal{G}(x',x) = \langle x'|e^{-\tau\mathcal{H}}|x\rangle, \tag{10.22}$$

which can be evaluated by using a Trotter approximation for a small evolution $\Delta\tau$ (Trotter, 1959; Suzuki, 1976a,b):

$$\mathcal{G}(x',x) \approx \frac{1}{\sqrt{2\pi\Delta\tau}}\exp\left[-\frac{(x'-x)^2}{2\Delta\tau}\right] \times \exp\left[-\Delta\tau V(x)\right], \tag{10.23}$$

where the first term can be related to a diffusion process, while the second one is the responsible for the branching scheme (Foulkes et al., 2001). An algorithm implementing the above iteration is given by Eqs. (10.7) and (10.8), where:

$$p_{x'x} = \frac{1}{\sqrt{2\pi\Delta\tau}}\exp\left[-\frac{(x'-x)^2}{2\Delta\tau}\right], \tag{10.24}$$

$$b_x = \exp\left[-\Delta\tau V(x)\right]. \tag{10.25}$$

The former one can be simulated by considering a simple Langevin process for the configuration of the walker (see Chapter 4):

$$x_{n+1} = x_n + \sqrt{\Delta\tau}\,\eta_n, \tag{10.26}$$

where η_n is a random number with a Gaussian distribution (with zero mean and unit variance). The latter one determines the evolution of the weight of the walker.

Let us now consider the simplest case with $V(x) = 0$. Then, the exact ground-state ("bosonic") and the first-excited ("fermionic") wave functions are given by:

$$\Upsilon_0(x) = \sqrt{\frac{2}{L}}\sin\left(\frac{\pi x}{L}\right), \tag{10.27}$$

$$\Upsilon_1(x) = \sqrt{\frac{2}{L}}\sin\left(\frac{2\pi x}{L}\right). \tag{10.28}$$

with energies $E_0 = \pi^2/(2L^2)$ and $E_1 = (2\pi)^2/(2L^2)$. By performing the Markov process of Eqs. (10.24) and (10.25) without any other constraint, the walker will diffuse into the whole region $0 \le x \le L$ and equilibrate to the "bosonic" ground state. In order to obtain the "fermionic" wave function, we must consider a restricted space of wave functions $\Psi_{fn}(x)$ that vanish outside a given interval $0 \le x \le L_{fn}$. Then, the full anti-symmetrized state is obtained by acting with the projector P_- on such a wave function, as shown in Fig. 10.2:

$$\Psi_{as}(x) = P_-\Psi_{fn}(x). \tag{10.29}$$

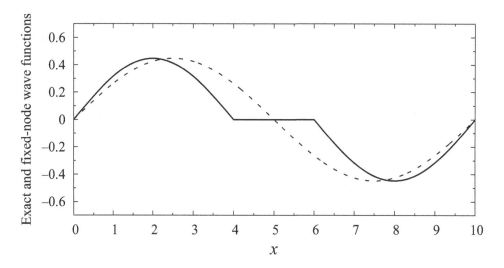

Figure 10.2 Wave functions for one particle in a box with $L = 10$ and $V(x) = 0$: the exact first-excited state (dashed line) with the node at $x = 5$ is compared with the fixed-node approximation (solid line) of it when the node is fixed at $x = 4$.

In practice, we do not allow the walker to cross the barrier introduced with L_{fn}. We can easily see that the extended wave function $\Psi_{as}(x)$ has the same variational energy than $\Psi_{fn}(x)$, which is restricted in the nodal pocket. Indeed, we have that:

$$
E_{fn} = \frac{\int_0^L dx\, \Psi_{as}(x) \left(-\frac{1}{2}\frac{d^2}{dx^2}\right)\Psi_{as}(x)}{\int_0^L dx\, \Psi_{as}^2(x)} = \frac{\int_0^{L_{fn}} dx\, \Psi_{fn}(x) \left(-\frac{1}{2}\frac{d^2}{dx^2}\right)\Psi_{fn}(x)}{\int_0^{L_{fn}} dx\, \Psi_{fn}^2(x)}.
$$
(10.30)

It is important to emphasize that this equality holds despite the fact that the wave function has a discontinuous first derivative at the wrong nodal point (see Fig. 10.2), leading to:

$$
\frac{d^2 \Psi_{fn}(x)}{dx^2}\bigg|_{x=L_{fn}} = -\delta(x - L_{fn})\Psi_{fn}'(L_{fn}),
$$
(10.31)

where $\Psi_{fn}'(L_{fn}) = \lim_{x \to L_{fn}^-} d\Psi_{fn}(x)/dx$. In fact, the singular contribution coming from the delta-function does not play any role in the integrals, since the wave function vanishes at the nodal point $x = L_{fn}$, and similarly for $x = L - L_{fn}$.

Given the equality of Eq. (10.30), it is clear that the lowest possible energy can be obtained by optimizing the wave function just in the nodal pocket, which becomes a "bosonic" (i.e., node-less) problem that is suitable for the diffusion Monte Carlo approach. Notice that, in this simple case, we can also provide the analytical form of the best wave function, which is given by:

$$
\Psi_{fn}(x) = \sqrt{\frac{2}{L}} \sin\left(\frac{\pi x}{L_{fn}}\right).
$$
(10.32)

The corresponding energy is $E_{fn} = \pi^2/(2L_{fn}) \geq E_1$.

In general, we would like to emphasize the most important properties of the fixed-node approximation:

- The method is variational, as the projected wave function $\Psi_{as}(x)$ has exactly the energy that is obtained with a "bosonic" ground-state calculation within a single nodal region.
- The fixed-node energy is very sensitive to the accuracy of the nodal surface of the wave function. In the previous example, the systematic error is linear in $\epsilon = L/2 - L_{fn}$. Therefore, it is important to obtain a good description of the nodal surface (e.g., by a suitable optimization of the variational wave function).
- Although the energy corresponding to the projected wave function $\Psi_{as}(x)$ and the one defined in the nodal pocket $\Psi_{fn}(x)$ coincide, the same does not hold for the variance. For example, it is simple to show that the variance of the fixed-node ground state (10.32) is zero in the nodal pocket, but when the physical wave

function is considered in the whole space $0 \leq x \leq L$, an infinite contribution to the variance comes from the delta-functions of Eq. (10.31). In this case, the infinite terms involving the square of delta-functions cancel only when the position of the node is exact.

In the previous example, the sign problem mainly arises because the Schrödinger equation (in imaginary time) is interpreted as a diffusion equation, whose steady state at large times is "bosonic," namely it is positive everywhere. This is the absolute ground state of any Hamiltonian with positive definite kinetic term and diagonal potentials. Instead, "fermionic" states can be seen as excited states of the same Hamiltonian, whenever the statistics of the particles is not specified *a priori* (in the spirit of the first-quantization formalism). The difficulty in this kind of calculation is embodied in the fact that we must extract the "fermionic" character of the observables from a small signal and a large noise, see Eq. (10.18). The previous example for the single particle in the box enlightens this issue.

10.3 A Simple Example on the Lattice

On the lattice, we are used to work in the second-quantization formalism, where the statistics of the particles is fully specified. Therefore, in this case, the diffusion process does not generically converge to the "bosonic" ground state, as in the continuum. However, whenever the kinetic terms have some negative signs, a sign problem is present. The simplest case is given by the lattice version of the previous example. Let us consider a single spinless particle on a one-dimensional lattice with L sites and open-boundary conditions, described by:

$$\mathcal{H} = -t_1 \sum_{i=1}^{L-1} c_i^\dagger c_{i+1} + \text{h.c.} - t_2 \sum_{i=1}^{L-2} c_i^\dagger c_{i+2} + \text{h.c.}, \qquad (10.33)$$

where $c_i^\dagger$ (c_i) creates (destroys) the particle on the site i; t_1 and t_2 are the hopping amplitudes for nearest- and next-nearest-neighbor hoppings. The energy scale can be set by taking $t_1 = 1$. In the following, we will work in the local basis, where the states of the Hilbert space are specified by the position of the particle, i.e., $|i\rangle = c_i^\dagger |0\rangle$.

For $t_2 \geq 0$, the Hamiltonian has non-negative matrix elements and, therefore, can be sampled without sign problem by using the Green's function Monte Carlo. In this case a fully positive guiding function can be taken. Instead, for $t_2 < 0$, the Hamiltonian has both positive and negative elements and the sign problem is present. In particular, this happens even when the ground state is positive for all configurations (i.e., for small values of $|t_2|$). Therefore, on the lattice, the sign problem may exist even when the ground state is positive definite. Moreover, we emphasize that the sign problem on the lattice is not related to configurations that are close to the nodal surface but involves essentially all the configurations in the Hilbert space. For example, for the Hamiltonian of Eq. (10.33), all configurations have "frustrating" matrix elements with $\mathcal{H}_{x,x'} > 0$ when $t_2 < 0$. Finally, the sign problem is not related to the statistics of the particles (in this example, there is just one particle). Therefore, also bosonic models may have a sign problem, e.g., in presence of hopping terms that have the wrong sign.

On the lattice, the fixed-node approximation can be reformulated in terms of an effective Hamiltonian, which has all non-positive off-diagonal elements, such that a stable projection technique is possible (either by using the Green's function or reptation Monte Carlo approaches). Moreover, the calculations are kept under control, since a variational bound exists for the fixed-node energy.

10.4 The Fixed-Node Approximation on the Lattice

In this section, we present the fixed-node approximation on the lattice, as introduced by ten Haaf et al. (1995) and developed by Sorella (2002). This technique allows a refinement of a given guiding (variational) wave function $\Psi_G(x)$ by improving its amplitudes in a stable and controllable way (i.e., by lowering the energy with a reasonable computational effort). Before discussing the fixed-node approximation, it is useful to introduce a scheme that allows us to obtain the guiding function $\Psi_G(x)$ as the ground state of an effective Hamiltonian. For a given state, there are infinite "parent" Hamiltonians; here, we want to focus on the ones that are as close as possible to the original model and, most importantly, can be treated within stable Monte Carlo approaches. i.e., they have non-positive off-diagonal elements. Therefore, we consider the following family of effective (i.e., "variational") Hamiltonians:

$$\mathcal{H}_{x,x'}^{\mathrm{var},\gamma} = \begin{cases} \mathcal{H}_{x,x} + (1+\gamma)V_{\mathrm{sf}}(x) - e_L(x) & \text{for } x' = x, \\ \mathcal{H}_{x,x'} & \text{for } x' \neq x, \ s_{x,x'} < 0, \\ -\gamma \mathcal{H}_{x,x'} & \text{for } x' \neq x, \ s_{x,x'} > 0, \end{cases} \tag{10.34}$$

where $\gamma \geq 0$ is a real parameter, $e_L(x)$ is the local energy of the guiding function:

$$e_L(x) = \sum_{x'} \mathcal{H}_{x,x'} \frac{\Psi_G(x')}{\Psi_G(x)}, \tag{10.35}$$

and the off-diagonal elements of the effective Hamiltonian are modified with respect to the original ones, according to their signs (including the ones of the guiding function):

$$s_{x,x'} = \Psi_G(x)\mathcal{H}_{x,x'}\Psi_G(x'). \tag{10.36}$$

Finally, $\mathcal{V}_{\text{sf}}(x)$ is a local potential that includes all the off-diagonal matrix elements generating sign problem, i.e., the ones with $s_{x,x'} > 0$:

$$\mathcal{V}_{\text{sf}}(x) = \sum_{x':s_{x,x'}>0} \mathcal{H}_{x,x'} \frac{\Psi_G(x')}{\Psi_G(x)}. \tag{10.37}$$

Within this construction, the guiding function is an eigenstate of $\mathcal{H}^{\text{var},\gamma}$ with zero eigenvalue, as can be verified by direct inspection; most importantly, for $\gamma \geq 0$, it is the unique ground state. Indeed, the unitary transformation $|x\rangle \rightarrow \text{Sign}\Psi_G(x)|x\rangle$ changes all the off-diagonal matrix elements into non-positive ones. Therefore, by applying the Perron-Frobenius theorem (Meyer, 2000) (or, equivalently, by replicating the steps of section 3.8), we can show that $\Psi_G(x)$ is the non-degenerate ground state of the effective Hamiltonian (10.34). In other words, once including the importance sampling with $\Psi_G(x)$, the effective Hamiltonian has no sign problem, since $\mathcal{G}^{\text{var},\gamma}_{x',x} = -\Psi_G(x')\mathcal{H}^{\text{var},\gamma}_{x',x}/\Psi_G(x)$ is non-negative for $x' \neq x$. Hence, the standard Green's function Monte Carlo technique can be used to sample configurations corresponding to $\Psi_G^2(x)$. Then, an estimation of the variational energy E_G can be obtained by averaging the local energy $e_L(x)$. Notice that the diagonal term of the effective Hamiltonian can be arbitrarily large, due to the presence of the sign-flip potential (10.37); therefore, the continuous-time approach of section 8.4 is needed to have stable simulations.

We are now in the position to improve the variational description of the ground state of $\mathcal{H}$ and define a fixed-node scheme that is built from a slightly different effective Hamiltonian. Indeed, there are many configurations with a local energy that is below its average value and these configurations are likely to be energetically favorable in the true ground state. From this intuitive argument, we can expect to gain some energy by adding a diagonal potential term, which favors configurations that have a local energy below its average value, thus increasing the corresponding amplitudes. More precisely, from the variational Hamiltonian $\mathcal{H}^{\text{var},\gamma}$, it is possible to define the fixed-node approximation on the lattice by adding a term that is linear in the local energy, with the purpose to improve the amplitudes of the variational *Ansatz*:

$$\mathcal{H}^{\text{fn},\gamma}_{x,x'} = \mathcal{H}^{\text{var},\gamma}_{x,x'} + \delta_{x,x'}e_L(x). \tag{10.38}$$

Again, for $\gamma \geq 0$, this Hamiltonian has non-positive off-diagonal elements (once importance sampling is included) and its ground state can be sampled by using projection techniques. The ground state of $\mathcal{H}^{\text{fn},\gamma}$ has the same signs of $\Psi_G(x)$, thus justifying the name of "fixed-node" approximation, and has a lower energy than the variational one. Indeed, we can easily verify that:

$$\mathcal{H}^{\text{fn},\gamma}|\Psi_G\rangle = \mathcal{H}|\Psi_G\rangle, \tag{10.39}$$

which immediately implies that the expectation value of the fixed-node Hamiltonian over the guiding function is equal to the variational energy E_G:

$$\frac{\langle \Psi_G | \mathcal{H}^{\mathrm{fn},\gamma} | \Psi_G \rangle}{\langle \Psi_G | \Psi_G \rangle} = \frac{\langle \Psi_G | \mathcal{H} | \Psi_G \rangle}{\langle \Psi_G | \Psi_G \rangle} = E_G. \tag{10.40}$$

Therefore, the ground-state energy $E^{\mathrm{fn},\gamma}$ of the fixed-node Hamiltonian is certainly lower than (or at most equal to) the variational energy E_G. In order to prove that $E^{\mathrm{fn},\gamma}$ gives an upper bound of the true ground-state energy E_0 of $\mathcal{H}$, we re-write the fixed-node Hamiltonian in a compact way:

$$\mathcal{H}^{\mathrm{fn},\gamma} = \mathcal{H} + (1+\gamma)\mathcal{O}^{\mathrm{fn}}, \tag{10.41}$$

where $\mathcal{O}^{\mathrm{fn}}$ is defined in terms of the guiding function:

$$\mathcal{O}^{\mathrm{fn}}_{x,x'} = \begin{cases} \sum_{y: s_{x,y} > 0} \frac{s_{x,y}}{\Psi_G^2(x)} & \text{for } x' = x, \\ -\mathcal{H}_{x,x'} & \text{for } x' \neq x \text{ and } s_{x,x'} > 0. \end{cases} \tag{10.42}$$

This operator is semi-positive definite. In fact, for any (real) state $|\Phi\rangle$, we have that:

$$\langle \Phi | \mathcal{O}^{\mathrm{fn}} | \Phi \rangle = \sum_{x,x': s_{x,x'} > 0} s_{x,x'} \frac{\Phi^2(x)}{\Psi_G^2(x)} - \sum_{x,x': s_{x,x'} > 0} s_{x,x'} \frac{\Phi(x)\Phi(x')}{\Psi_G(x)\Psi_G(x')}, \tag{10.43}$$

where, the first (second) term comes from the diagonal (off-diagonal) matrix elements the operator $\mathcal{O}^{\mathrm{fn}}$, respectively. By using the fact that $s_{x,x'}$ is symmetric, we can interchange the dummy variables x and x' in the first term and obtain:

$$\langle \Phi | \mathcal{O}^{\mathrm{fn}} | \Phi \rangle = \frac{1}{2} \sum_{x,x': s_{x,x'} > 0} s_{x,x'} \left[\frac{\Phi(x)}{\Psi_G(x)} - \frac{\Phi(x')}{\Psi_G(x')} \right]^2 \geq 0. \tag{10.44}$$

Therefore, denoting the ground state of the fixed-node Hamiltonian by $|\Upsilon^{\mathrm{fn},\gamma}\rangle$, we have that:

$$E_0 \leq \frac{\langle \Upsilon^{\mathrm{fn},\gamma} | \mathcal{H} | \Upsilon^{\mathrm{fn},\gamma} \rangle}{\langle \Upsilon^{\mathrm{fn},\gamma} | \Upsilon^{\mathrm{fn},\gamma} \rangle} = \frac{\langle \Upsilon^{\mathrm{fn},\gamma} | [\mathcal{H}^{\mathrm{fn},\gamma} - (1+\gamma)\mathcal{O}^{\mathrm{fn}}] | \Upsilon^{\mathrm{fn},\gamma} \rangle}{\langle \Upsilon^{\mathrm{fn},\gamma} | \Upsilon^{\mathrm{fn},\gamma} \rangle} \leq E^{\mathrm{fn},\gamma}, \tag{10.45}$$

where we have used that $\mathcal{H}^{\mathrm{fn},\gamma} | \Upsilon^{\mathrm{fn},\gamma} \rangle = E^{\mathrm{fn},\gamma} | \Upsilon^{\mathrm{fn},\gamma} \rangle$ and the fact that the operator $\mathcal{O}^{\mathrm{fn}}$ is semi-positive definite. This result concludes the proof that $E^{\mathrm{fn},\gamma}$ improves the variational estimate: the fixed-node energy is lower than (or at most equal to) the variational estimate E_G and higher than (or at most equal to) the exact one:

$$E_0 \leq E^{\mathrm{fn},\gamma} \leq E_G. \tag{10.46}$$

We would like to emphasize that the minimum energy (that can be computed without sign problem) is obtained for $\gamma = 0$. Indeed, by exploiting the Hellmann-Feynman theorem and Eq. (10.41), we have that:

$$\frac{dE^{\text{fn},\gamma}}{d\gamma} = \frac{\langle \Upsilon^{\text{fn},\gamma} | \mathcal{O}^{\text{fn}} | \Upsilon^{\text{fn},\gamma} \rangle}{\langle \Upsilon^{\text{fn},\gamma} | \Upsilon^{\text{fn},\gamma} \rangle} \geq 0, \tag{10.47}$$

which implies that $E^{\text{fn},\gamma}$ is a monotonically increasing function of γ.

Moreover, it is worth mentioning that, within the projection technique, we get $E^{\text{fn},\gamma}$ and not the expectation value of the original Hamiltonian over the fixed-node state, which is given by:

$$E^{\text{fn},\gamma}_{\text{ev}} = \frac{\langle \Upsilon^{\text{fn},\gamma} | \mathcal{H} | \Upsilon^{\text{fn},\gamma} \rangle}{\langle \Upsilon^{\text{fn},\gamma} | \Upsilon^{\text{fn},\gamma} \rangle}. \tag{10.48}$$

This quantity can be estimated by using Eq. (10.47):

$$E^{\text{fn},\gamma}_{\text{ev}} = E^{\text{fn},\gamma} - (1+\gamma)\frac{dE^{\text{fn},\gamma}}{d\gamma}. \tag{10.49}$$

Notice that, by performing several calculations at different values of $\gamma \geq 0$, we could in principle extrapolate $E^{\text{fn},\gamma}$ to $\gamma = -1$ and obtain the exact ground-state energy. However, in the general case, such extrapolation is not easy to perform.

An important remark is that, on the lattice case, $E^{\text{fn},\gamma}$ not necessarily gives the lowest energy that is compatible with the nodes of the guiding function. In fact, for the ground state of the fixed-node Hamiltonian $|\Upsilon^{\text{fn},\gamma}\rangle$, the error is given by:

$$\Delta E = \langle \Upsilon^{\text{fn},\gamma} | (\mathcal{H}^{\text{fn},\gamma} - \mathcal{H}) | \Upsilon^{\text{fn},\gamma} \rangle = (1+\gamma)\langle \Upsilon^{\text{fn},\gamma} | \mathcal{O}^{\text{fn}} | \Upsilon^{\text{fn},\gamma} \rangle, \tag{10.50}$$

which is vanishing only if each individual term in the summation (10.44) vanishes. Therefore, having a guiding function with the correct signs is not sufficient to obtain the exact ground state. Instead, it is necessary that for *all* sign-flip pairs with $s_{x,x'} > 0$:

$$\frac{\Psi_G(x')}{\Psi_G(x)} = \frac{\Upsilon^{\text{fn},\gamma}(x')}{\Upsilon^{\text{fn},\gamma}(x)}. \tag{10.51}$$

Indeed, if this relation is verified, $|\Upsilon^{\text{fn},\gamma}\rangle$ is a simultaneous eigenstate of $\mathcal{O}^{\text{fn}}$ (with vanishing eigenvalue) and $\mathcal{H}^{\text{fn},\gamma}$ and, therefore, it must be also an exact eigenstate of the original Hamiltonian, as $\mathcal{H} = \mathcal{H}^{\text{fn},\gamma} - (1+\gamma)\mathcal{O}^{\text{fn}}$. This is an important difference with respect to the fixed-node approach on the continuum, where it is only the sign of the guiding function that matters (if the nodes are correctly placed, the exact result is obtained). By contrast, on the lattice, the sign and the relative amplitudes of the wave function in configurations that are connected by a sign flip must be correct. For example, in Fig. 10.3, we report the results for the model (10.33) on 30 sites, with $t_2 = -0.2$ and -0.4. In the former case, the true ground state is positive everywhere, but the fixed-node approach is not exact when taking a positive guiding

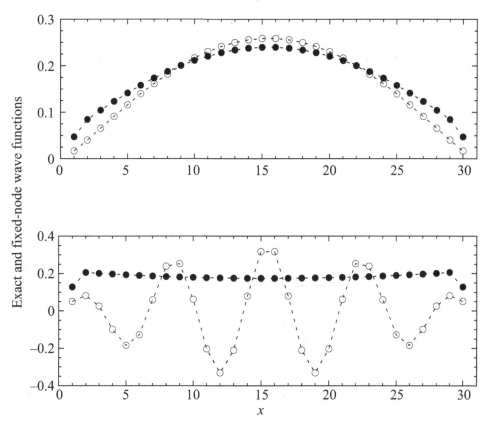

Figure 10.3 Comparison between the exact (empty circles) and the fixed-node (full circles) wave functions for the ground state of the Hamiltonian of Eq. (10.33) for the cases with $t_2 = -0.2$ (upper panel) and $t_2 = -0.4$ (lower panel); the number of sites is $L = 30$. The guiding function is taken to be constant, e.g., $\Psi_G(x) = 1/\sqrt{L}$.

function, e.g., $\Psi_G(x) = 1/\sqrt{L}$; indeed, we get $E^{\text{fn},0}/t_1 = -1.59196$, to be compared with $E_0/t_1 = -1.59775$. The situation is much worse when also the true ground state is not positive definite, i.e., for $t_2 = -0.4$; here, still using the same guiding function as before, we get $E^{\text{fn},0}/t_1 = -1.20192$, while the true ground-state energy is $E_0/t_1 = -1.41588$.

10.5 Practical Implementation

The fixed-node approximation can be applied to the Green's function and Reptation Monte Carlo techniques. The practical implementation follows the ones described in Chapters 8 and 9 with the only difference that all the new configurations are selected by considering the fixed-node Hamiltonian defined by Eqs. (10.34) and (10.38) instead of the original one.

11

Auxiliary Field Quantum Monte Carlo

11.1 Introduction

One of the most important advantages of the Hartree-Fock approach (or more generally any mean-field theory) is that a complete and efficient numerical solution is possible in terms of simple linear algebra operations, possibly including diagonalization of small matrices. However, as discussed in Chapter 1, these techniques, which are based upon independent-electron wave functions, are usually not accurate to describe correlated systems, as for example the Hubbard model. In this respect, projection methods that filter out the ground state by the iterative application of a suitable operator have been introduced; two examples have been discussed in Chapters 8 and 9. Here, we would like to introduce an alternative projection technique, which is particularly suited for the fermionic Hubbard model (Sorella et al., 1989), the so-called auxiliary-field quantum Monte Carlo (AFQMC). The main task of this approach is to reduce the many-body problem to the solution of several mean-field-like calculations, which are affordable by using a stochastic method in the space of Slater determinants. In this way, statistical errors can be controlled, provided the sign problem does not appear during the mentioned transformation; remarkably, this is indeed possible in many interesting cases. Moreover, here, the sign problem is not related to the bosonic-fermionic instability, which has been discussed in section 10.1 since explicitly antisymmetric Slater determinants are sampled within AFQMC) and, therefore, it is often milder than the conventional case, e.g., within the Green's function Monte Carlo, which is defined in configuration space (Fahy and Hamann, 1991).

The general idea is to filter out the ground-state wave function by applying an imaginary-time τ propagation (see section 1.7):

$$\lim_{\tau \to \infty} e^{-\tau \mathcal{H}} |\Psi_0\rangle \propto |\Upsilon_0\rangle, \qquad (11.1)$$

as before, the necessary requirement is to choose an initial state $|\Psi_0\rangle$ such that $\langle \Upsilon_0 | \Psi_0 \rangle \neq 0$. First of all, as done in section 5.6, we contract the spin index σ and the lattice site i into a single index I running from 1 to $2L$:

$$c_{i,\uparrow} \equiv d_i, \tag{11.2}$$

$$c_{i,\downarrow} \equiv d_{i+L}. \tag{11.3}$$

Then, we will specialize to the case in which the Hamiltonian contains kinetic and potential terms:

$$\mathcal{H} = \mathcal{K} + \mathcal{V}, \tag{11.4}$$

where

$$\mathcal{K} = \sum_{I,J} K_{I,J} d_I^\dagger d_J, \tag{11.5}$$

$$\mathcal{V} = U \sum_i n_i n_{i+L}; \tag{11.6}$$

here, we have denoted the densities on site $i = 1, \ldots, L$, for electrons with spin up and down, by:

$$n_i = d_i^\dagger d_i \equiv n_{i,\uparrow} = c_{i,\uparrow}^\dagger c_{i,\uparrow}, \tag{11.7}$$

$$n_{i+L} = d_{i+L}^\dagger d_{i+L} \equiv n_{i,\downarrow} = c_{i,\downarrow}^\dagger c_{i,\downarrow}. \tag{11.8}$$

The kinetic energy is written in terms of a $2L \times 2L$ matrix $\mathbf{K}$; in the simplest case, $K_{i,j} = K_{i+L,j+L} = -t$ if i and j are nearest neighbors and 0 otherwise. In general cases with conserved number of particles, it can be taken as a generic Hermitian matrix. Moreover, extensions to an arbitrary bilinear form in the fermion fields, namely without assuming that the kinetic energy commutes with the number of particles, are also possible within this framework.

In the following, we consider the initial state as a non-interacting wave function with N_e electrons:

$$|\Psi_0\rangle = \prod_{\alpha=1}^{N_e} \left(\sum_I \psi_{I,\alpha} d_I^\dagger \right) |0\rangle, \tag{11.9}$$

which is described by the $2L \times N_e$ matrix ψ. This is the general case where the orbitals defined by $\{\psi_{I,\alpha}\}$ do not have a definite spin value (which includes BCS-like states, once a particle-hole transformation has been done on spin-down electrons). By contrast, whenever there is no mixing between spin-up and spin-down components, the matrix ψ can be written in terms of two blocks with dimensions $L \times N_e^\uparrow$ and $L \times N_e^\downarrow$.

In the absence of the interaction term $\mathcal{V}$, the projection technique of Eq. (11.1) can be implemented in a simple and efficient way. Indeed, the one-body propagator

$\exp(-\tau \mathcal{K})$ transforms an uncorrelated wave function into another state that is still uncorrelated:

$$e^{-\tau \mathcal{K}}|\Psi_0\rangle = \prod_{\alpha=1}^{N_e} \left(\sum_I \psi_{I,\alpha}(\tau) d_I^\dagger \right) |0\rangle, \qquad (11.10)$$

where the evolved orbitals are given by:

$$\psi_{I,\alpha}(\tau) = \sum_\beta e^{-\tau \epsilon_\beta} \left(\sum_J \psi_{J,\alpha} U_{J,\beta}^* \right) U_{I,\beta}; \qquad (11.11)$$

here $\{\epsilon_\beta\}$ are the eigenvalues of the kinetic term $\mathcal{K}$ and the matrix $\mathbf{U}$ contains its eigenvectors, i.e., $\phi_\beta^\dagger = \sum_I U_{I,\beta} d_I^\dagger$. Therefore, the imaginary-time evolution of a Slater determinant can be conveniently written as a matrix-matrix multiplication of the original $2L \times N_e$ matrix $\boldsymbol{\psi}$ and the $2L \times 2L$ propagation matrix $\mathbf{P}(\tau)$:

$$\boldsymbol{\psi}(\tau) = \mathbf{P}(\tau)\boldsymbol{\psi}, \qquad (11.12)$$

where

$$P_{I,J}(\tau) = \sum_\beta e^{-\tau \epsilon_\beta} U_{I,\beta} U_{J,\beta}^*. \qquad (11.13)$$

   In summary, we have shown that the imaginary-time propagation with only the kinetic operator transforms a Slater determinant into another Slater determinant; in addition, this kind of propagation is achieved by a simple (and efficient) matrix-matrix multiplication. When the interaction term $\mathcal{V}$ is non-zero, the time evolution is no longer so simple, since the application of a many-body propagator $\exp(-\tau \mathcal{H})$ to a Slater determinant does not lead to a single Slater determinant. The AFQMC is based upon the Trotter approximation (Trotter, 1959; Suzuki, 1976a,b), which allows us to express the full evolution $\exp(-\tau \mathcal{H})$ in terms of a sequence of distinct propagations involving either the kinetic or the potential term. Then, the latter one is handled by using the so-called Hubbard-Stratonovich transformation (Hubbard, 1959; Stratonovich, 1957), which replaces a many-body term with a superposition of one-body propagations depending on classical auxiliary variables, which have to be integrated out by means of a stochastic technique.

11.2 Trotter Approximation

Let us consider a small step $\Delta \tau$ in the imaginary-time evolution of Eq. (11.1); then, we can define an effective Hamiltonian $\overline{\mathcal{H}}$ that is close to the original one:

$$\exp\left(-\Delta \tau \overline{\mathcal{H}}\right) = \exp\left(-\frac{\Delta \tau}{2} \mathcal{K}\right) \exp\left(-\Delta \tau \mathcal{V}\right) \exp\left(-\frac{\Delta \tau}{2} \mathcal{K}\right). \qquad (11.14)$$

Indeed, for small values of $\Delta\tau$, a straightforward expansion of both sides of the previous equation allows us to evaluate $\overline{\mathcal{H}}$ as:

$$\overline{\mathcal{H}} = \mathcal{H} + \Delta\tau^2 \mathcal{O}_{\text{Tr}} + \mathcal{O}(\Delta\tau^3), \tag{11.15}$$

where the Hermitian operator $\mathcal{O}_{\text{Tr}}$ is given by:

$$\mathcal{O}_{\text{Tr}} = \frac{1}{12}[[\mathcal{K},\mathcal{V}],\mathcal{V}] + \frac{1}{24}[[\mathcal{K},\mathcal{V}],\mathcal{K}]. \tag{11.16}$$

We would like to emphasize that the operator $\mathcal{O}_{\text{Tr}}$ is extensive (i.e., it scales with the number of the sites L). Indeed, given the actual form of the operators $\mathcal{K}$ and $\mathcal{V}$ (which are sums of local operators), the commutator $[\mathcal{K},\mathcal{V}]$ is extensive; more generally, the commutator of two extensive operators is extensive. Therefore, all the properties that we compute with the effective Hamiltonian $\overline{\mathcal{H}}$ at fixed $\Delta\tau$ are related to the ones of the original Hamiltonian $\mathcal{H}$ by a small error in $\Delta\tau^2$. Thus, the error associated to the Trotter decomposition of Eq. (11.14) is perfectly under control, since it can be reduced systematically by decreasing $\Delta\tau$. Moreover, for a given accuracy on thermodynamic quantities (e.g., the energy per site), we do not need to decrease the Trotter time $\Delta\tau$ when increasing the system size, namely, the error associated to the Trotter approximation is size consistent.

11.3 Hubbard-Stratonovich Transformation

The Hubbard-Stratonovich transformation allows us to replace a propagator containing in its exponent two-body interaction terms in the density (or spin), with much simpler one-body propagators. This is obtained at the cost of introducing auxiliary *classical* fields (variables) on each site i, hereafter denoted by $\{\sigma_i\}$. Since $\exp(-\Delta\tau\mathcal{V})$ can be factorized in L terms, each one acting on a single site i, we can focus on every site independently. Let us discuss the decouplings that are usually employed for the repulsive Hubbard model. The simplest example is obtained by using *continuous* auxiliary fields:

$$\exp\left[\frac{g}{2}(n_{i,\uparrow} - n_{i,\downarrow})^2\right] = \int_{-\infty}^{+\infty} \frac{d\sigma_i}{\sqrt{2\pi}} \exp\left[-\frac{1}{2}\sigma_i^2 + \lambda\sigma_i(n_{i,\uparrow} - n_{i,\downarrow})\right], \tag{11.17}$$

with $\lambda^2 = g$. Indeed, in the basis where $(n_{i,\uparrow} - n_{i,\downarrow})$ is diagonal, the r.h.s. of the above equation becomes a standard Gaussian integral, where $(n_{i,\uparrow} - n_{i,\downarrow})$ can be considered just as a number. This transformation can be simplified and extended to *discrete*, i.e., Ising, auxiliary fields $\sigma_i = \pm 1$ by means of the following relation (Hirsch, 1985):

$$\exp\left[\frac{g}{2}(n_{i,\uparrow} - n_{i,\downarrow})^2\right] = \frac{1}{2}\sum_{\sigma_i=\pm 1} \exp\left[\lambda\sigma_i(n_{i,\uparrow} - n_{i,\downarrow})\right], \tag{11.18}$$

where now $\cosh \lambda = \exp(g/2)$. This expression can be easily proved by expanding $\exp(\lambda n_{i,\sigma})$ and noting that all powers k different from zero contribute with $\lambda^k/k!$ (since $n_{i,\sigma}^k = n_{i,\sigma}$ for $k \geq 1$, as a consequence of the Pauli principle). This fact leads to:

$$\exp(\lambda n_{i,\sigma}) = 1 + \left(e^\lambda - 1\right) n_{i,\sigma}. \tag{11.19}$$

Therefore, we have:

$$\exp\left[\frac{g}{2}(n_{i,\uparrow} - n_{i,\downarrow})^2\right] = 1 + \left[\exp\left(\frac{g}{2}\right) - 1\right](n_{i,\uparrow} + n_{i,\downarrow})$$
$$+ 2\left[1 - \exp\left(\frac{g}{2}\right)\right] n_{i,\uparrow} n_{i,\downarrow}, \tag{11.20}$$

$$\frac{1}{2} \sum_{\sigma_i = \pm 1} \exp\left[\lambda \sigma_i(n_{i,\uparrow} - n_{i,\downarrow})\right] = 1 + (\cosh \lambda - 1)(n_{i,\uparrow} + n_{i,\downarrow})$$
$$+ 2(1 - \cosh \lambda) n_{i,\uparrow} n_{i,\downarrow}, \tag{11.21}$$

which imply that Eq. (11.18) is indeed verified with $\cosh \lambda = \exp(g/2)$.

The Hubbard-Stratonovich transformation is very useful since it turns a complicated two-body operator, containing in the exponent the square of $(n_{i,\uparrow} - n_{i,\downarrow})$, into a superposition of simple one-body terms, which depend upon an auxiliary field. In particular, by using that $n_{i,\sigma}^2 = n_{i,\sigma}$, the many-body propagator can be recast in a form that is suitable for the Hubbard-Stratonovich transformation:

$$\exp(-\Delta\tau\mathcal{V}) = \exp\left(-\frac{\Delta\tau}{2} U N_e\right) \prod_i \exp\left[\frac{\Delta\tau}{2} U (n_{i,\uparrow} - n_{i,\downarrow})^2\right]. \tag{11.22}$$

In this way, by introducing an independent Ising field $\sigma_i = \pm 1$ for each site, we can write:

$$\exp(-\Delta\tau\mathcal{V}) \propto \sum_{\sigma_i = \pm 1} \exp\left[\lambda \sum_i \sigma_i(n_{i,\uparrow} - n_{i,\downarrow})\right], \tag{11.23}$$

where we have used Eq. (11.18) for $g = U\Delta\tau$, which implies that:

$$\cosh \lambda = \exp\left(\frac{U\Delta\tau}{2}\right). \tag{11.24}$$

In Eq. (11.23), we have omitted an irrelevant constant $\exp(-UN_e\Delta\tau/2)/2^L$, since we assume that the total number of particles N_e is fixed. Therefore, we are now in the position to apply a many-body operator $\exp(-\tau\mathcal{V})$ to the non-interacting state of Eq. (11.9). In fact, by defining:

$$h_\sigma(\lambda) = \lambda \sum_i \sigma_i(n_{i,\uparrow} - n_{i,\downarrow}), \tag{11.25}$$

such that:

$$\exp(-\Delta\tau V) \propto \sum_{\sigma_i = \pm 1} \exp\left[h_\sigma(\lambda)\right], \tag{11.26}$$

we obtain:

$$e^{h_\sigma(\lambda)}|\Psi_0\rangle = \prod_{\alpha=1}^{N_e}\left(\sum_I \psi_{I,\alpha}(\lambda)d_I^\dagger\right)|0\rangle, \tag{11.27}$$

where the evolved orbitals are given by:

$$\psi_{I,\alpha}(\lambda) = \begin{cases} \exp(\lambda\sigma_I)\psi_{I,\alpha} & \text{if } I \le L, \\ \exp(-\lambda\sigma_{I-L})\psi_{I,\alpha} & \text{if } I > L. \end{cases} \tag{11.28}$$

In this way, not only the kinetic term, but also the propagation with the potential energy can be written in terms of a matrix-matrix multiplication, through a diagonal $2L \times 2L$ matrix $\mathbf{V}(\lambda)$:

$$\boldsymbol{\psi}(\lambda) = \mathbf{V}(\lambda)\boldsymbol{\psi}, \tag{11.29}$$

where

$$V_{I,J}(\lambda) = \delta_{I,J} \begin{cases} \exp(\lambda\sigma_I) & \text{if } I \le L, \\ \exp(-\lambda\sigma_{I-L}) & \text{if } I > L. \end{cases} \tag{11.30}$$

Finally, we would like to remark that there are several other ways for performing the Hubbard-Stratonovich decoupling. For example, the two-body operator can be written as:

$$V = \frac{U}{2}\sum_i n_i^2 - \frac{UN_e}{2}. \tag{11.31}$$

Then, the propagator $\exp(-\Delta\tau V)$ contains squared one-body operators with a *negative* prefactor; therefore, we can use the transformation of Eq. (11.17) with an *imaginary* constant $\lambda \to i\lambda$. This introduces a *phase problem* in quantum Monte Carlo calculations that can be afforded by using some approximation, e.g., the so-called constrained-path quantum Monte Carlo (Zhang et al., 1995). Remarkably, the phase problem is absent at half filling $N_e = L$. To our knowledge, no systematic studies have been done to identify the most efficient Hubbard-Stratonovich decoupling for a given model, namely the one that minimizes the statistical errors for a given computational time.

11.4 The Path-Integral Representation

Now, we are in the position to compute all the properties of the ground state of the effective Hamiltonian $\overline{\mathcal{H}}$, which can be taken arbitrarily close to the exact

Hamiltonian, i.e., with an error vanishing as $\Delta\tau^2$. To this purpose we consider the pseudo-partition function:

$$\mathcal{Z} = \langle\Psi_0|e^{-\tau\overline{\mathcal{H}}}|\Psi_0\rangle = \langle\Psi_0|\left[\exp(-\Delta\tau\overline{\mathcal{H}})\right]^{2T}|\Psi_0\rangle, \tag{11.32}$$

where $2T$ is the number of Trotter slices and $\tau = 2T\Delta\tau$. For $T \to \infty$ the many-body state

$$|\Psi_T\rangle = \left[\exp(-\Delta\tau\overline{\mathcal{H}})\right]^T|\Psi_0\rangle \tag{11.33}$$

converges to the exact ground state $|\Upsilon_0\rangle$ of $\overline{\mathcal{H}}$. Thus, any operator or correlation function $\mathcal{O}$ can be evaluated as:

$$\frac{\langle\Upsilon_0|\mathcal{O}|\Upsilon_0\rangle}{\langle\Upsilon_0|\Upsilon_0\rangle} = \lim_{T\to\infty}\frac{\langle\Psi_T|\mathcal{O}|\Psi_T\rangle}{\mathcal{Z}}. \tag{11.34}$$

Then, a path-integral representation can be obtained by replacing in both the numerator and the denominator of Eq. (11.34) the auxiliary-field transformation of Eq. (11.23) for *each* Trotter slice:

$$\mathcal{Z} = \sum_{\sigma_i^q=\pm1}\langle\Psi_0|U_\sigma(2T,0)|\Psi_0\rangle, \tag{11.35}$$

where we have introduced the compact definition of the one-body propagator $U_\sigma(2T,0)$ defined by the auxiliary-field transformation (11.23), which implies a corresponding one-body operator $h_\sigma^q(\lambda) = \lambda\sum_i\sigma_i^q(n_{i,\uparrow} - n_{i,\downarrow})$ for each Trotter slice $1 \leq q \leq 2T$; here, the field $\sigma_i^q = \pm1$ acquires also a discrete time index q, besides the one associated to the lattice site i. Then, we have that:

$$U_\sigma(2T,0) = \exp\left(-\frac{\Delta\tau}{2}\mathcal{K}\right)\exp\left[h_\sigma^{2T}(\lambda)\right]\exp\left(-\Delta\tau\mathcal{K}\right)\exp\left[h_\sigma^{2T-1}(\lambda)\right]\cdots$$

$$\cdots\exp\left[h_\sigma^2(\lambda)\right]\exp\left(-\Delta\tau\mathcal{K}\right)\exp\left[h_\sigma^1(\lambda)\right]\left(-\frac{\Delta\tau}{2}\mathcal{K}\right). \tag{11.36}$$

Since $U_\sigma(2T,0)$ is a one-body propagator, when applied to a Slater determinant, it gives back another Slater determinant. Therefore, the quantity $\langle\Psi_0|U_\sigma(2T,0)|\Psi_0\rangle$ can be numerically evaluated for any *fixed* configuration of the Ising fields $\{\sigma_i^q\}$. By using the same fields also in the numerator of Eq. (11.34), we obtain:

$$\frac{\langle\Psi_T|\mathcal{O}|\Psi_T\rangle}{\mathcal{Z}} = \frac{\sum_{\sigma_i^q=\pm1}\langle\Psi_0|U_\sigma(2T,T)\mathcal{O}U_\sigma(T,0)|\Psi_0\rangle}{\sum_{\sigma_i^q=\pm1}\langle\Psi_0|U_\sigma(2T,0)|\Psi_0\rangle}. \tag{11.37}$$

Then, a Monte Carlo scheme can be devised to evaluate the r.h.s. of this equation through the computation of the ratio of the average of two random variables. In particular, we consider the positive weight:

$$\mathcal{W}(\sigma) = |\langle\Psi_0|U_\sigma(2T,0)|\Psi_0\rangle|, \tag{11.38}$$

such that, the expectation value of any operator or correlation function $\mathcal{O}$ is given by:

$$\frac{\langle \Psi_T | \mathcal{O} | \Psi_T \rangle}{\mathcal{Z}} \approx \frac{\langle\langle \mathcal{O}(\sigma) S(\sigma) \rangle\rangle}{\langle\langle S(\sigma) \rangle\rangle}, \tag{11.39}$$

where $\langle\langle \dots \rangle\rangle$ indicates the statistical average over the distribution defined by the weight (11.38), $S(\sigma)$ is the sign of $\langle \Psi_0 | U_\sigma(2T, 0) | \Psi_0 \rangle$, and

$$\mathcal{O}(\sigma) = \frac{\langle \Psi_0 | U_\sigma(2T, T) \mathcal{O} U_\sigma(T, 0) | \Psi_0 \rangle}{\langle \Psi_0 | U_\sigma(2T, T) U_\sigma(T, 0) | \Psi_0 \rangle}. \tag{11.40}$$

Notice that within the AFQMC technique, the sign (or phase) problem appears whenever $\langle \Psi_0 | U_\sigma(2T, 0) | \Psi_0 \rangle$ has not a definite sign for all the field configurations $\{\sigma_i^q\}$.

All quantities (i.e., random variables and weights) can be computed in polynomial time once the fields σ_i^q are given for each site and time slice q. The forward-propagated state $|\bar{R}_T\rangle = U_\sigma(T, 0) | \Psi_0 \rangle$ can be written in the same form as in Eq. (11.9) by means of a $2L \times N_e$ matrix:

$$\bar{\mathbf{R}}_T = \mathbf{P}\left(\frac{\Delta\tau}{2}\right) \mathbf{R}_T, \tag{11.41}$$

$$\mathbf{R}_T = \mathbf{V}_T(\lambda) \mathbf{P}(\Delta\tau) \dots \mathbf{P}(\Delta\tau) \mathbf{V}_1(\lambda) \mathbf{P}\left(\frac{\Delta\tau}{2}\right) \boldsymbol{\psi}, \tag{11.42}$$

where $\mathbf{P}(\tau)$ and $\mathbf{V}_q$ are $2L \times 2L$ matrices defined in Eqs. (11.13) and (11.30), for each time slice q, i.e., with fields $\{\sigma_i^q\}$. Notice that, the matrix $\mathbf{R}_T$ can be computed iteratively over the time slices for increasing values of q, by applying matrix-matrix multiplications with $O(8TL^2N_e)$ total operations:

$$\mathbf{R}_{q+1} = \mathbf{V}_{q+1}(\lambda) \mathbf{P}(\Delta\tau) \mathbf{R}_q, \tag{11.43}$$

with the initial condition:

$$\mathbf{R}_1 = \mathbf{V}_1(\lambda) \mathbf{P}\left(\frac{\Delta\tau}{2}\right) \boldsymbol{\psi}. \tag{11.44}$$

The backward-propagated state $\langle \bar{L}_T | = \langle \Psi_0 | U_\sigma(2T, T)$ can be computed in an analogous way:

$$\bar{\mathbf{L}}_T = \mathbf{P}\left(-\frac{\Delta\tau}{2}\right) \mathbf{L}_T, \tag{11.45}$$

$$\mathbf{L}_T = \mathbf{P}(\Delta\tau) \mathbf{V}_{T+1}^\dagger(\lambda) \dots \mathbf{P}(\Delta\tau) \mathbf{V}_{2T}^\dagger(\lambda) \mathbf{P}\left(\frac{\Delta\tau}{2}\right) \boldsymbol{\psi}, \tag{11.46}$$

which can be again computed iteratively:

$$\mathbf{L}_{q-1} = \mathbf{P}(\Delta\tau)\mathbf{V}_q^\dagger\mathbf{L}_q, \tag{11.47}$$

with the initial condition:

$$\mathbf{L}_{2T} = \mathbf{P}\left(\frac{\Delta\tau}{2}\right)\psi. \tag{11.48}$$

Then, the overlap between the two wave functions is just a conventional determinant of an $N_e \times N_e$ matrix:

$$(\overline{\mathbf{L}}_T)^\dagger\overline{\mathbf{R}}_T = (\mathbf{L}_T)^\dagger\mathbf{R}_T = \mathbf{M}, \tag{11.49}$$

which is computable in $O(N_e^3)$ operations by standard numerical libraries. Finally, the quantum expectation value $\mathcal{O}(\sigma)$ of Eq. (11.40) can be also calculated in polynomial time by using the Wick theorem, as shown in section 11.5.

11.4.1 Stable Imaginary-Time Propagation

The imaginary-time evolution of Eqs. (11.42) and (11.46) is defined for an infinite precision arithmetic. Instead, actual computations are affected by truncation errors, which lead to an instability problem when applying the one-body operator to a Slater determinant for large imaginary time τ. Indeed, let us consider the kinetic part of the propagator, for which the evolved orbitals are given by Eq. (11.11). Then, the factor $\exp(-\tau\epsilon_\beta)$ will induce an exponential instability of the algorithm, since all the orbitals will be dominated by the ones corresponding to the lowest eigenvalues; more precisely, in double-precision floating point $1 + 10^{-15} = 1$, and, therefore, whenever $\exp[-\tau(\epsilon_{N_e} - \epsilon_1)] < 10^{-15}$, there will be only $N_e - 1$ linearly independent orbitals within numerical accuracy; this fact leads to a vanishing Slater determinant. In order to overcome this problem, we must take $\Delta\tau$ such that:

$$\Delta\tau(\epsilon_{N_e} - \epsilon_1) \ll \ln 10^{15}. \tag{11.50}$$

Notice that this condition is size consistent, because in a lattice model the single-particle bandwidth is finite in the thermodynamic limit (e.g., it is $4dt$ in the d-dimensional case with nearest-neighbor hopping t). Then, after each short-time evolution, the propagated orbitals can be orthogonalized without loosing information of the Slater determinant. Indeed, by using any linear transformation generated by the matrix $\mathbf{A}$:

$$\psi'_{I,\alpha} = \sum_\beta \psi_{I,\beta}A_{\beta,\alpha}, \tag{11.51}$$

the Slater determinant is not changed, apart from an overall constant $\det\mathbf{A}$, which can be saved and updated during the total propagation. In this context, the

Gram-Schmidt orthogonalization is the simplest algorithm for this purpose; however, a more efficient one is obtained by applying the Cholesky decomposition. Here, the $N_e \times N_e$ overlap matrix is considered and decomposed:

$$\mathbf{M} = \boldsymbol{\psi}^\dagger \boldsymbol{\psi} = \mathbf{L}^\dagger \mathbf{L}, \tag{11.52}$$

where $\boldsymbol{\psi}$ is now a generic matrix and $\mathbf{L}$ is a lower-triangular matrix. Then, the new orbitals given by:

$$\boldsymbol{\psi}' = \boldsymbol{\psi} \mathbf{L}^{-1} \tag{11.53}$$

will be orthogonal to each other. Indeed, we have that:

$$(\boldsymbol{\psi}')^\dagger \boldsymbol{\psi}' = (\mathbf{L}^\dagger)^{-1} \boldsymbol{\psi}^\dagger \boldsymbol{\psi} \mathbf{L}^{-1} = (\mathbf{L}^\dagger)^{-1} \mathbf{L}^\dagger \mathbf{L} \mathbf{L}^{-1} = 1. \tag{11.54}$$

These orbitals will describe the same determinant apart for an overall constant equal to det $\mathbf{L}$, which is simple to compute, since it is equal to the product of the diagonal elements of $\mathbf{L}$. In the ground-state technique described here, the normalization of the Slater determinant is an irrelevant constant that does not appear in any physical quantity and, therefore, can be disregarded. After the orthogonalization, another stable propagation can be done without facing any instability problems.

11.5 Sequential Updates

In this section, we describe a simple Markov process that can be used to sample the weight of Eq. (11.38). In particular, for each time slice q, a single Ising variable on the site k is changed. In order to speed up the update, it is important to propose the new values of the auxiliary fields for all sites in a given time slice before moving to the next one. If we flip the field $\sigma_k^q \to -\sigma_k^q$ at the site k (a *sequential* loop over the sites is implied) only the right wave function $|R_q\rangle$ will change in a simple way:

$$|R_q'\rangle = e^{-2\lambda \sigma_k^q (n_{k,\uparrow} - n_{k,\downarrow})} |R_q\rangle = (1 + \lambda_\downarrow n_{k,\downarrow})(1 + \lambda_\uparrow n_{k,\uparrow})|R_q\rangle, \tag{11.55}$$

where $\lambda_\uparrow = \exp(-2\lambda\sigma_k^q) - 1$ and $\lambda_\downarrow = \exp(2\lambda\sigma_k^q) - 1$. For a proposed flip of the Ising variable, the change of the weight $\mathcal{W}(\sigma) = |\langle L_q|R_q\rangle|$ and the corresponding sign (or phase) $\mathcal{S}(\sigma)$ can be updated by means of the ratio:

$$\mathcal{R}_k^q = \frac{\langle L_q|R_q'\rangle}{\langle L_q|R_q\rangle} = \frac{\langle L_q|(1 + \lambda_\downarrow n_{k,\downarrow})(1 + \lambda_\uparrow n_{k,\uparrow})|R_q\rangle}{\langle L_q|R_q\rangle}, \tag{11.56}$$

which represents a many-body correlation function computed between two different Slater determinants $\langle L_q|$ and $|R_q\rangle$. Finally, the acceptance probability is obtained by the standard Metropolis algorithm, described in section 3.9; a slightly better scheme, with smaller correlation times and faster equilibration, is given by the so-called heat-bath algorithm (Krauth, 2006), where the proposed configuration σ_k' is accepted with probability:

$$A(\sigma_k'|\sigma_k) = \frac{|\mathcal{R}_k^q|}{1 + |\mathcal{R}_k^q|}. \tag{11.57}$$

The ratio $\mathcal{R}_k^q$ can be computed in terms of the equal-time Green's function defined by a $2L \times 2L$ matrix:

$$(G_q)_{J,I} = \frac{\langle L_q | d_I^\dagger d_J | R_q \rangle}{\langle L_q | R_q \rangle}. \tag{11.58}$$

In the most general case, $(G_q)_{J,I}$ can be non-zero even for the off-diagonal spin components with $I \leq L$ and $J > L$ or $I > L$ and $J \leq L$, which are present whenever BCS-like wave functions are considered (these elements are related to the anomalous averages, once a particle-hole transformation on the spin-down electrons is performed). Instead, whenever the Slater determinants are factorized into spin-up and spin-down components, $(G_q)_{J,I}$ has a block diagonal form. Having defined the above Green's function, all possible many-body correlation functions between two different determinants can be computed by using the Wick theorem (for the case when *bra* and *ket* states are different).

Let us now compute the Green's function of Eq. (11.58). First of all, the overlap between two Slater determinants is given by the determinant of the corresponding overlap matrix, see Eq. (11.49); for example, the denominator of the r.h.s. of Eq. (11.58) is given by:

$$\langle L_q | R_q \rangle = \det(\mathbf{M}) = \det\left[(\mathbf{L}_q)^\dagger \mathbf{R}_q \right]. \tag{11.59}$$

The numerator can be evaluated in an analogous way, by noticing that it can be written in terms of the overlap between two Slater determinants with $N_e + 1$ electrons:

$$\langle L_q | d_I^\dagger d_J | R_q \rangle = \delta_{I,J} \langle L_q | R_q \rangle - \langle \Psi_J | \Psi_I \rangle, \tag{11.60}$$

with

$$|\Psi_I\rangle = d_I^\dagger |R_q\rangle, \tag{11.61}$$

$$|\Psi_J\rangle = d_J^\dagger |L_q\rangle. \tag{11.62}$$

The new matrix $\mathbf{M}'$ corresponding to the overlap $\langle \Psi_J | \Psi_I \rangle$ can be computed realizing that the extra $N_e + 1$ orbital enter in a particularly simple way:

$$\mathbf{M}' = \begin{pmatrix} \delta_{I,J} & (R_q)_{J,\beta} \\ (L_q)_{I,\alpha}^* & M_{\alpha,\beta} \end{pmatrix} \tag{11.63}$$

where we have highlighted the contribution of the extra row/column, while all the other matrix elements remain exactly equal to the original $N_e \times N_e$ matrix $\mathbf{M}$. In order to compute the determinant of the above matrix, we use the property that the determinant is unchanged if we add to the left-most column an appropriate linear combination of the other columns. This can be done in a way to cancel all its values but the diagonal one: the coefficients x_β of the linear combination are given by:

$$\sum_\beta M_{\alpha,\beta} x_\beta = -(L_q)^*_{I,\alpha}, \tag{11.64}$$

which leads to:

$$x_\beta = -\sum_\alpha M^{-1}_{\beta,\alpha} (L_q)^*_{I,\alpha}. \tag{11.65}$$

Therefore, we obtain that:

$$\det \mathbf{M}' = \begin{vmatrix} \delta_{I,J} - \sum_{\alpha,\beta} (R_q)_{J,\beta} M^{-1}_{\beta,\alpha} (L_q)^*_{I,\alpha} & (R_q)_{J,\beta} \\ 0 & M_{\alpha,\beta} \end{vmatrix}$$

$$= \det \mathbf{M} \times \left[\delta_{I,J} - \sum_{\alpha,\beta} (R_q)_{J,\beta} M^{-1}_{\beta,\alpha} (L_q)^*_{I,\alpha} \right]. \tag{11.66}$$

By combining this result with Eq. (11.60), we obtain that:

$$(G_q)_{J,I} = \sum_{\alpha,\beta} (R_q)_{J,\beta} M^{-1}_{\beta,\alpha} (L_q)^*_{I,\alpha}, \tag{11.67}$$

which can be cast in a compact form:

$$\mathbf{G}_q = \mathbf{R}_q \mathbf{M}^{-1} \mathbf{L}^\dagger_q. \tag{11.68}$$

This quantity can be evaluated by scratch at the beginning of the calculation by using standard linear algebra operations and then updated whenever a Monte Carlo move is accepted (see below). For stability reasons, it is necessary to perform a re-computation by scratch of the Green's function at selected time slices, e.g., for $q = m \times p$, where m is an integer and p is a fixed small integer.

By performing a similar derivation, we can get the expressions containing four fermion operators:

$$\frac{\langle L_q | d^\dagger_I d_J d^\dagger_K d_N | R_q \rangle}{\langle L_q | R_q \rangle} = (G_q)_{J,I}(G_q)_{N,K} + (G_q)_{N,I}[\delta_{K,J} - (G_q)_{J,K}], \tag{11.69}$$

which allows us to compute the ratio in Eq. (11.56):

$$\mathcal{R}^q_k = [1 + \lambda_\uparrow (G_q)_{k,k}][1 + \lambda_\downarrow (G_q)_{k+L,k+L}] - \lambda_\uparrow \lambda_\downarrow (G_q)_{k+L,k}(G_q)_{k,k+L}; \tag{11.70}$$

here the last term is only present if the determinants do not factorize into spin-up and spin-down blocks.

Once the trial move $|R_q\rangle \rightarrow |R'_q\rangle$ is accepted, the Green's function must be updated:

$$(G_q)'_{J,I} = \frac{\langle L_q | d_I^\dagger d_J | R'_q \rangle}{\langle L_q | R'_q \rangle}. \tag{11.71}$$

As we will show in the following, this can be done without recomputing it by scratch, but using a simple application of the Wick theorem. Notice that, if the Slater determinants are factorized for spin-up and spin-down electrons, the last term in Eq. (11.70) vanishes and also the updating of the Green's function simplifies. Without assuming a factorized Slater determinant, we obtain the final expression:

$$(G_q)'_{J,I} = (G_q)_{J,I} + a_J^\uparrow (G_q)_{k,I} + a_J^\downarrow (G_q)_{k+L,I}, \tag{11.72}$$

where

$$a_J^\uparrow = \frac{\lambda_\uparrow}{\mathcal{R}_k^q} \left\{ -\overline{(G_q)}_{J,k} + \lambda_\downarrow \left[\overline{(G_q)}_{J,k+L} (G_q)_{k+L,k} - \overline{(G_q)}_{J,k} (G_q)_{k+L,k+L} \right] \right\}, \tag{11.73}$$

$$a_J^\downarrow = \frac{\lambda_\downarrow}{\mathcal{R}_k^q} \left\{ -\overline{(G_q)}_{J,k+L} + \lambda_\uparrow \left[\overline{(G_q)}_{J,k} (G_q)_{k,k+L} - \overline{(G_q)}_{J,k+L} (G_q)_{k,k} \right] \right\}, \tag{11.74}$$

in which $\overline{(G_q)}_{J,I} = (G_q)_{J,I} - \delta_{J,I}$. We would like to mention that whenever the Green's function factorizes into spin up and down components, a much simpler updating, involving each spin part independently, is possible:

$$(G_q)'_{j,i} = (G_q)_{j,i} + a_j^\uparrow (G_q)_{k,i}, \tag{11.75}$$

$$(G_q)'_{j+L,i+L} = (G_q)_{j+L,i+L} + a_{j+L}^\downarrow (G_q)_{k+L,i+L}, \tag{11.76}$$

where now:

$$a_j^\uparrow = -\frac{\lambda_\uparrow}{1 + \lambda_\uparrow (G_q)_{k,k}} \overline{(G_q)}_{j,k}, \tag{11.77}$$

$$a_{j+L}^\downarrow = -\frac{\lambda_\downarrow}{1 + \lambda_\downarrow (G_q)_{k+L,k+L}} \overline{(G_q)}_{j+L,k+L}. \tag{11.78}$$

In summary, the updating scheme for a given time slice q can be done with $O(L^3)$ operations. In order to go from one time slice to the neighboring one, we need to propagate the left Slater determinant $\langle L_q|$ with the matrix-matrix multiplications of Eq. (11.47), which requires again at most $O(L^3)$ operations. Instead, the right Slater determinant $|R_q\rangle$ can be stored at the beginning and saved in memory (otherwise, it can be backward propagated by using Eq. (11.43), i.e., $\mathbf{R}_{q-1} = \mathbf{P}(-\Delta\tau)[\mathbf{V}_q(\lambda)]^{-1}\mathbf{R}_q$). Also the Green's function can be backward propagated from time slice q to time slice $q - 1$ with analogous matrix-matrix operations. Indeed, from Eq. (11.68) and the propagation of the left and right matrices, we have:

$$\mathbf{G}_{q-1} = \mathbf{P}(-\Delta\tau)[\mathbf{V}_q(\lambda)]^{-1}\mathbf{G}_q\mathbf{V}_q(\lambda)\mathbf{P}(\Delta\tau). \tag{11.79}$$

Thus, a sweep over all the time slices costs at most $O(TL^3)$ operations. Therefore, this algorithm is particularly efficient for ground-state properties since the $T \to \infty$ limit can be reached with a cost that is only linear with the number of time slices.

11.5.1 Delayed Updates

The basic operation in the updating procedure Eq. (11.72) is the rank-1 update of a generic $2L \times 2L$ matrix:

$$(G_q)'_{J,I} = (G_q)_{J,I} + a_J^{\sigma} b_I^{\sigma}, \tag{11.80}$$

where a_J^{σ} is defined in Eqs. (11.73) and (11.74), while $b_I^{\uparrow} = (G_q)_{k,I}$ and $b_I^{\downarrow} = (G_q)_{k+L,I}$. As already emphasized in section 5.6, this kind of operation can be computationally inefficient when, for large sizes, the matrix $\mathbf{G}_q$ is not completely contained in the cache of the processor. A way to overcome this drawback is to delay the update of the matrix, without loosing its information. This can be obtained by storing a set of left and right vectors $\mathbf{a}^{(p)}$ and $\mathbf{b}^{(p)}$ with $p = 1, \ldots, 2m$ (including the "spin" index σ), as well as the "initial" matrix, denoted by $\mathbf{G}_q^0$, from which the delayed updates start. Then, the matrix $\mathbf{G}_q$, after m updates is given by:

$$(G_q)_{J,I} = (G_q)_{J,I}^0 + \sum_{p=1}^{2m} a_J^{(p)} b_I^{(p)}. \tag{11.81}$$

Every time we accept a new configuration, new vectors $\mathbf{a}^{(m+1)}$, $\mathbf{a}^{(m+2)}$, $\mathbf{b}^{(m+1)}$ $\mathbf{b}^{(m+2)}$ are computed in few operations in term of $\mathbf{G}_q^0$ and the previous vectors with $p = 1, \ldots, 2m$. Notice that, once the matrix $\mathbf{G}_q$ is written in the form of Eq. (11.81), the number of operations required to evaluate the factors in the sum is $O(8mL)$, which is negligible compared to the full update for $m \ll L$.

By performing this kind of delayed update, we can find an optimal $m_{\max}$, for which we can evaluate the full matrix $\mathbf{G}_q$ by a standard matrix multiplication:

$$\mathbf{G}_q = \mathbf{G}_q^0 + \mathbf{A}\mathbf{B}^T, \tag{11.82}$$

where $\mathbf{A}$ and $\mathbf{B}$ are $2L \times 2m_{\max}$ matrices, which are made of the $p = 1, \ldots, 2m_{\max}$ vectors $\mathbf{a}^{(p)}$ and $\mathbf{b}^{(p)}$, respectively. After that, we can continue with a new delayed update with a new $\mathbf{G}_q^0 = \mathbf{G}_q$, by initializing to zero the integer m. The advantage of this updating procedure is that after a cycle of $2m_{\max}$ steps the bulk of the computation is given by the matrix-matrix multiplication in Eq. (11.82), which is much more efficient (and is not limited by the dimension of the cache memory) than the $2m_{\max}$ rank-1 updates. For large number of electrons, the delayed update procedure allows us to improve the speed of the Monte Carlo code by about an order of magnitude.

11.6 Ground-State Energy and Correlation Functions

The ground-state energy or any other observable, including correlation functions, can be easily computed by using the Wick theorem that has been discussed in section 11.5. Indeed, let us consider a generic n-body operator:

$$\mathcal{O} = \sum_{I_1,\ldots,I_n} \sum_{J_1,\ldots,J_n} \mathcal{O}_{I_1,\ldots,I_n;J_1,\ldots,J_n} d_{I_1}^{\dagger} d_{J_1} \ldots d_{I_n}^{\dagger} d_{J_n}, \tag{11.83}$$

where $d_I^{\dagger}$ (d_I) creates (destroys) an electron on the site $I = 1,\ldots,2L$. Then, the correlation function of Eq. (11.34) is evaluated by computing $\mathcal{O}(\sigma)$ given by Eq. (11.40), which can be done by using the fact that both $|\bar{R}_T\rangle = U_\sigma(T,0)|\Psi_0\rangle$ and $\langle \bar{L}_T| = \langle \Psi_0|U_\sigma(2T,T)$ are Slater determinants and, therefore, the Wick theorem can be applied.

Finally, it is worth mentioning that the error on the ground-state energy, which is obtained for $T \to \infty$, vanishes as $O(\Delta\tau^4)$. Indeed, given the symmetric Trotter approximation of Eq. (11.14), the effective Hamiltonian $\overline{\mathcal{H}}$ differs from the original one $\mathcal{H}$ by a perturbation that vanishes as $O(\Delta\tau^2)$, see Eq. (11.15). The ground state of $\overline{\mathcal{H}}$, which is obtained by the projection technique of Eq. (11.33), can be related to the exact ground state of $\mathcal{H}$ by using standard perturbation theory, i.e., $|\overline{\Upsilon}_0\rangle = |\Upsilon_0\rangle + \Delta\tau^2|\Upsilon'\rangle + O(\Delta\tau^4)$ (where $|\Upsilon'\rangle$ is orthogonal to $|\Upsilon_0\rangle$). Thus, when computing the expectation value of $\mathcal{H}$ over $|\overline{\Upsilon}_0\rangle$ the leading correction $O(\Delta\tau^2)$ vanishes and we obtain that the leading error is $O(\Delta\tau^4)$.

11.7 Simple Cases without Sign Problem

Let us discuss few examples in which the AFQMC technique does not suffer from the sign problem. First of all, we consider the case of the attractive Hubbard model, i.e., the case with $U < 0$ (and any real kinetic term). In this case, an auxiliary-field transformation is possible with a real field σ_i that is coupled only to the total density $\sigma_i(n_{i,\uparrow}+n_{i,\downarrow}-1)$. Indeed, the discrete transformation of Eq. (11.18) can be modified into:

$$\exp\left[\frac{g}{2}(n_{i,\uparrow} + n_{i,\downarrow} - 1)^2\right] = \frac{1}{2}\sum_{\sigma_i=\pm 1} \exp\left[\lambda\sigma_i(n_{i,\uparrow} + n_{i,\downarrow} - 1)\right], \tag{11.84}$$

where $\cosh \lambda = \exp(g/2)$ with $g = |U|\Delta\tau$. Then, we take a trial wave function $|\Psi_0\rangle$ with the *same* real orbitals for both spin components, namely $|\Psi_0\rangle = |\Psi_{0,\uparrow}\rangle \otimes |\Psi_{0,\downarrow}\rangle$, where:

$$|\Psi_{0,\sigma}\rangle = \prod_{\alpha=1}^{N_e/2} \sum_i \psi_{i,\alpha} c_{i,\sigma}^{\dagger} |0\rangle. \tag{11.85}$$

In this case, the one-body propagator $U_\sigma(2T, 0)$ factorizes into up and down terms, i.e., $U_\sigma(2T, 0) = U_\sigma^{\uparrow}(2T, 0) \times U_\sigma^{\downarrow}(2T, 0)$, because the total propagation acts in the same way over the spin-up and the spin-down components of the wave function. The consequence is that the pseudo-partition function $\mathcal{Z}$ of Eq. (11.35) becomes:

$$\mathcal{Z} = \sum_{\sigma_i^q = \pm 1} \left(\langle \Psi_{0,\uparrow} | U_\sigma^{\uparrow}(2T, 0) | \Psi_{0,\uparrow} \rangle \right)^2. \tag{11.86}$$

Therefore, the Monte Carlo weight is strictly positive, being the square of a real number. In summary, the AFQMC has no sign problem for attractive (spin-independent) interactions, for any lattice structure, whenever the number of spin-up particles is exactly equal to the number of spin-down ones.

Another example in which there is no sign problem is the case of the repulsive Hubbard model on bi-partite lattices (e.g., with a kinetic term that only couples opposite sublattices A and B) at half-filling, i.e., when $N_e^{\uparrow} + N_e^{\downarrow} = L$. Indeed, we can always map the repulsive model into the attractive one by performing a particle-hole transformation on the spin-down particles:

$$c_{i,\uparrow} \rightarrow f_{i,\uparrow}, \tag{11.87}$$

$$c_{i,\downarrow} \rightarrow s_i f_{i,\downarrow}^{\dagger}, \tag{11.88}$$

where $s_i = 1$ ($s_i = -1$) if the site i belongs to the A (B) sublattice; this sign is necessary in order not to change the kinetic term. Then, the number of spin-down electrons is changed by the particle-hole transformation (see section 5.6):

$$u_{i,\uparrow} = f_{i,\uparrow}^{\dagger} f_{i,\uparrow} = c_{i,\uparrow}^{\dagger} c_{i,\uparrow} = n_{i,\uparrow}, \tag{11.89}$$

$$u_{i,\downarrow} = f_{i,\downarrow}^{\dagger} f_{i,\downarrow} = 1 - c_{i,\downarrow}^{\dagger} c_{i,\downarrow} = 1 - n_{i,\downarrow}, \tag{11.90}$$

which imply that the total number of "new" spin-up electrons is $N_e^{\uparrow}$, while the total number of "new" spin-down electrons is $L - N_e^{\downarrow}$. Therefore, based upon the results for the attractive model, we conclude that there is no sign problem when the number of spin-down electrons after the particle-hole transformation is exactly equal to the number of spin-up electrons, namely $N_e^{\downarrow} + N_e^{\uparrow} = L$, which is exactly the half-filled condition.

The results for the repulsive Hubbard model with $U/t = 4$ are shown in Fig. 11.1 for $N_e = 42$ on 50 sites. The initial wave function $|\Psi_0\rangle$ is given by the

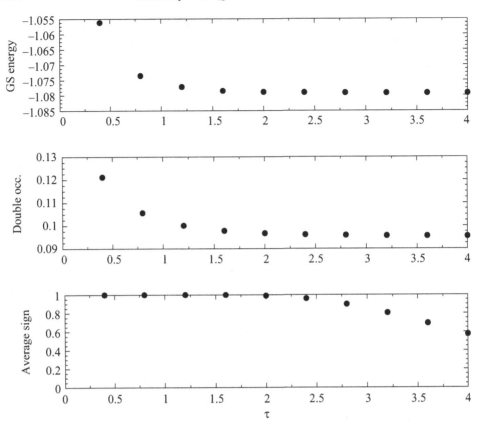

Figure 11.1 Results for the Hubbard model for $N_e = 42$ electrons on 50 sites, with $U/t = 4$; the time discretization in the Trotter approximation is $\Delta\tau = 0.1$ All the quantities are shown as a function of the projection time τ: ground-state energy (upper panel), double occupations $D = 1/L \sum_i n_{i,\uparrow} n_{i,\downarrow}$ (middle panel), and average sign (lower panel).

non-interacting one, which corresponds to a closed-shell configuration (i.e., there is a finite-size gap between the highest occupied and the lowest unoccupied levels). We show the results for the ground-state energy, double occupations $D = 1/L \sum_i n_{i,\uparrow} n_{i,\downarrow}$, and also the average sign $\langle\langle S(\sigma)\rangle\rangle$ along the simulation. We remark that, in this case, a fast convergence to the exact values is obtained as a function of the projection time τ (the discretization time associated to the Trotter decomposition is $\Delta\tau = 0.1$). We would like to emphasize that, given the closed-shell condition, the average sign remains close to 1 up to $\tau \approx 2$ and falls down to 0 very slowly, allowing a precise evaluation of the physical quantities.

Finally, we would like to mention that a sign-free decoupling is possible also when long-range interactions are present (on bi-partite lattices). Indeed, let us take:

$$V_{LR} = \sum_{l,k} U_{l,k}(n_l - 1)(n_k - 1);$$ (11.91)

then, we can consider the following Hubbard-Stratonovich transformation for the many-body propagator:

$$\exp\left(-\Delta\tau V_{LR}\right) = \int d\rho \exp\left[-\frac{1}{2}\sum_{l,k}\rho_l U_{l,k}^{-1}\rho_k + i\sqrt{\Delta\tau}\sum_k \rho_k(n_k - 1)\right],$$ (11.92)

where the only constraint is given by the fact that $\mathbf{U}$ must be a positive-definite matrix. Therefore, whenever the kinetic term acts separately on electrons with up and down spin, the propagator is factorized. Moreover, by performing the particle-hole transformation of Eqs. (11.87) and (11.88), we immediately obtain that the spin-up and spin-down propagators are one the complex conjugated of the other, thus leading to a positive weight:

$$\mathcal{Z} = \sum_{\sigma_i^q = \pm 1} \left|\langle\Psi_{0,\uparrow}|U_\sigma^\uparrow(2T,0)|\Psi_{0,\uparrow}\rangle\right|^2.$$ (11.93)

11.8 Practical Implementation

Here, we would like to sketch the important steps in a practical implementation of the auxiliary-field quantum Monte Carlo algorithm.

1. **Initialization** at the beginning of the calculation.
   - Before starting, we have to decide the total number of Trotter slices $2T$ and the number p of time slices between two consecutive orthogonalizations (with the Cholesky algorithm). In particular, p should be chosen small enough to guarantee a sufficient accuracy of the calculation, which can be systematically tested by decreasing p ($p = 1$ being the most accurate case).
   - Start with randomly generated Ising fields for each site and time slice $\sigma_i^q = \pm 1$. The $2L \times 2L$ matrices $\mathbf{P}(\pm\Delta\tau)$ and $\mathbf{P}(\pm\Delta\tau/2)$ must be computed at the beginning and stored in memory by means of Eq. (11.13). Instead, the diagonal matrices $\mathbf{V}_q(\lambda)$ do not need to be stored, since all the calculations can be done efficiently by using the fact that these matrices are diagonal (a diagonal matrix times a full matrix costs only $4L^2$ operations).
2. **Initial propagation**.
   - Propagate forward $\mathbf{R}_q$ for $q = 1, \ldots, 2T$ by using Eq. (11.43) and, for stability reasons, orthogonalize the propagation each p steps; save, in the computer memory, the associated $2L \times N_e$ matrices corresponding to the wave function $|R_q\rangle$ for $q = k \times p$ (where k is an integer). Arrived at the final interval $q = 2T$,

initialize the back-propagated wave function $\mathbf{L}_{2T}$ of Eq. (11.48) and compute the Green's function $\mathbf{G}_{2T}$ of Eq. (11.68) by scratch.

3. **Markov process** with the Metropolis algorithm.

   - At the time slice q, the sweep on the sites $i = 1, \ldots 2L$ begins. At each step, we can use Eq. (11.70) to compute the ratio between the new and old weights. Then, either the Metropolis algorithm or the heat-bath one can be used to generate a Markov process. Whenever the proposed configuration is accepted, the Green's function is updated according to Eq. (11.72). Delayed updates can be also used for an improved performances with large-size calculations.

   - At the end of the sweep over the lattice sites, we change the backward propagated state from q to $q - 1$, by using Eq. (11.47). Notice that, after p propagations a Cholesky orthogonalization has to be employed for stability reasons. At this stage, it is also necessary to propagate the Green's function according to Eq. (11.79). Only when $\mathbf{L}_q$ is orthogonalized, we compute the Green's function by scratch (by using $\mathbf{R}_q$ that have been saved in the initialization).

   - Go back to the initial propagation and continue the Monte Carlo procedure in order to accumulate statistics. Notice that a minor improvement of the algorithm is to reverse back the direction of the sweep, namely going from $q = 1$ to $q = 2T$, so that the initial propagation can be avoided. This approach typically saves approximately 10% of the total amount of computation.

4. **Computation of observables**.

   Observables (i.e., the total energy) can be computed along the sweeps over the time slices. In particular, for $q \approx T$ (i.e., around the middle interval) we get the expectation value over the state $|\Psi_T\rangle$ of Eq. (11.33). In this case, we should remind that the Green's function $\mathbf{G}_q$ that is used to compute correlation functions has to be propagated by half-time with the kinetic energy:

$$\mathbf{G}_q \rightarrow \mathbf{P}\left(\frac{\Delta\tau}{2}\right) \mathbf{G}_q \mathbf{P}\left(-\frac{\Delta\tau}{2}\right), \tag{11.94}$$

which is particularly important for computing physical quantities with a sufficiently small Trotter error.

Part V
Advanced Topics

12

Realistic Simulations on the Continuum

12.1 Introduction and Motivations

In the previous chapters, we introduced efficient variational and projection Monte Carlo techniques, which provide very accurate descriptions of the ground-state properties of lattice models. The restriction of the electronic coordinates on a given lattice has been proven to be very useful and successful to describe many phenomena, including Anderson localization, Kondo screening, and the existence of Mott insulators. In numerical calculations, the lattice has enormous advantages because (i) configurations can be easily enumerated and most importantly (ii) there is a natural short-distance cutoff that allows us to treat the model without further approximations. Within the computational approach, the continuous limit represents a mathematical abstraction, which cannot be captured on a computer, given the finite-precision arithmetic that is at the basis of any numerical calculations. Therefore, we can imagine that the continuous space is obtained by considering a lattice with spacing a and taking the limit $a \rightarrow 0$. In this case, electrons hop performing discrete jumps of length a and interact through the original Coulomb potential. This approach is called "lattice regularization" and provides the advantage that, for any given lattice spacing a, the system can be directly simulated in the computer. Then, we are able to get the correct physical properties of the continuous Hamiltonian by taking sufficiently small lattice spacings.

However, within the variational approach, whenever the wave function is expressed in terms of few parameters (McMillan, 1965; Ceperley et al., 1977), the lattice regularization introduces an unnecessary approximation, without improving the efficiency of the algorithm. Nevertheless, we can be tempted to use the lattice regularization to consider a systematic parametrization of the wave function on a grid. Unfortunately, this kind of approach is not suitable for *ab-initio* calculations of realistic materials, whenever a full optimization of the electronic orbitals is desirable; indeed, the lattice regularization would introduce too many parameters that cannot be handled at present within stochastic techniques. For example, let us consider liquid Hydrogen with 200 atoms (corresponding to the same number of electrons and 100 molecular orbitals) in a $100 \times 100 \times 100$ grid; this case would imply 10^8 variational parameters, which is impossible to handle by any stochastic optimization method: despite the remarkable progress in recent years, at present this number cannot exceed $\approx 10^4$ (Neuscamman et al., 2012). Therefore, as in lattice models, the variational state can be written as a product of an uncorrelated (e.g., Slater) part, which embodies the correct anti-symmetric fermionic properties, and a correlation (e.g., Jastrow) factor, which includes correlation among electrons. The Slater part of the wave function is usually obtained by using a deterministic mean-field approach, where the problem of having many variational parameters can be tackled by using very cheap and efficient self-consistent methods. Then, the Slater part is not changed in presence of the Jastrow factor, which can be optimized by using a suitable parametrization (Filippi and Umrigar, 1996; Wagner et al., 2009; Needs et al., 2010; Kim et al., 2012).

Here, we want to follow the idea that the Slater determinant given by a mean-field-like approach is not the optimal solution. Indeed, whenever the mean field does not describe correctly the system, this procedure has no hope to capture the correct low-energy behavior. For example, within this kind of approach, we will never be able to describe superconductivity driven by repulsive interactions only. Instead, a qualitative correct description of the ground state can be achieved by including a correlation (e.g., Jastrow) factor, even when the Slater determinant is built from a small basis set but it is optimized in presence of the Jastrow factor. In this case, the optimization procedure can be mapped into an iterative linear problem in a reduced sub-space spanned by the variational parameters. We remark that an optimized wave function with a limited number of parameters corresponds to the ground state of a local Hamiltonian that, for a good *Ansatz*, is "sufficiently close" to the original one. By contrast, a variational approach may be intrinsically biased because it cannot describe properties that have not been explicitly considered in the wave function. Nevertheless, in our opinion, the variational method is tremendously important, even in presence of this limitation, since a black-box tool that solves the electron problem does not exist yet. Indeed, for a given system, we should be able to construct reasonable wave functions and get quantitative results. Then, this method can be validated by explicit and direct comparison with experiments, without tunable parameters, because the variational wave function is determined unambiguously (i.e., by an *ab-initio* procedure) by the minimization of the expectation value of the energy. Moreover, the variational approach will remain useful even when it does not compare favorably with experiments, since this case will falsify the form of the wave function, stimulating the introduction of new physical ingredients in it. In this spirit, we mention successful examples, as the resonating valence-bond state introduced by Anderson (1987) to describe High-temperature superconductivity and the Laughlin (1983) wave function for the fractional quantum Hall effect.

12.2 Variational Wave Function with Localized Orbitals

Here, we discuss the general form of the correlated wave functions that we use for the continuum case. We denote the coordinates of the N_e electrons with $\mathbf{x} = \{\mathbf{r}_1, \ldots \mathbf{r}_{N_e}\}$. In the following, we will consider an equal number of spin-up and spin-down electrons (i.e., $N_e/2$ pairs of electrons with opposite spins). For practical purposes, we can fix the spin of each electron in such a way that the first (last) $N_e/2$ particles have spin-up (spin-down). Within this choice, the coordinates of the spin-up and spin-down electrons are denoted by $\{\mathbf{r}_i^\uparrow\}$ and $\{\mathbf{r}_i^\downarrow\}$ (with $i = 1, \ldots, N_e/2$), respectively. Then, because of the Pauli principle, the orbital part of the variational wave function $\Psi(\mathbf{x})$ (associated to this spin configuration) must be anti-symmetric for all exchanges of particles with the same spin (while there are no symmetry properties when exchanging particles with opposite spins). The variational *Ansatz* for the many-body state is written as the product of two terms:

$$\Psi(\mathbf{x}) = \mathcal{J}(\mathbf{x})F(\mathbf{x}), \tag{12.1}$$

where the (positive) Jastrow factor $\mathcal{J}(\mathbf{x})$ is symmetric under all the permutations among the coordinates of the particles and is given in terms of the two-body pseudo-potential $u(\mathbf{r}, \mathbf{r}')$:

$$\mathcal{J}(\mathbf{x}) = \exp\left[\sum_{i<j} u(\mathbf{r}_i, \mathbf{r}_j)\right], \tag{12.2}$$

and the determinant part $F(\mathbf{x})$, enforcing the anti-symmetry among the particles with the same spin, is defined in terms of the pairing function $f(\mathbf{r}, \mathbf{r}')$ (defined in the configuration space):

$$F(\mathbf{x}) = \det f\left(\mathbf{r}_i^\uparrow, \mathbf{r}_j^\downarrow\right). \tag{12.3}$$

In the following, we will restrict ourselves to the case of a *singlet* wave function, implying that the pairing function is symmetric, i.e., $f(\mathbf{r}, \mathbf{r}') = f(\mathbf{r}', \mathbf{r})$ (which, in most cases, is also taken *real*).

The determinant, which is computed in terms of the $N_e/2 \times N_e/2$ matrix of the pairing function is called *anti-symmetrized geminal power* (AGP). We remark that the uncorrelated part $F(\mathbf{x})$ can be reduced to a Slater determinant for particular choices of the pairing function (see below), but it allows us to obtain much more general quantum states, like for example BCS superconductors. Moreover, it is important to emphasize that, similarly to what happens in lattice models, the Jastrow factor is the simplest term to describe the Coulomb repulsion between electrons, introducing a "renormalization" of the uncorrelated $F(\mathbf{x})$ when two electrons are very close. This effect cannot be taken into account by any one-electron *Ansatz*, since, in such a case, each electron behaves as an independent particle that experiences a mean-field potential determined by the average effect of all the other electrons. By using the above expression, a standard variational Monte Carlo approach is possible, allowing the evaluation of the expectation value of the energy and correlation functions, with a computational time that scales with $O(N_e^3)$ (Ceperley et al., 1977).

Here, we describe in detail how the two functions $u(\mathbf{r}, \mathbf{r})$ and $f(\mathbf{r}, \mathbf{r}')$ (entering in the definition of the Jastrow factor and the determinant part, respectively) can be conveniently expanded by using a set of *atomic orbitals*. For this purpose, we consider an atomic basis $\{\phi_p(\mathbf{r} - \mathbf{R}_I)\}$, where each element has an orbital index p and depends upon a given index I, denoting the ion where the orbital is centered. Here, $p = 1, \ldots, N_{\text{orb}}(I)$ and $I = 1, \ldots N_{\text{ion}}$, $N_{\text{orb}}(I)$ and N_{ion} being the total number of orbitals associated to a given ion I and the number of ions, respectively. In the following, to simplify the notation, we use a single index μ to indicate both the orbital p and the atomic center I, leading to $\{\phi_\mu(\mathbf{r})\}$. Within this compact notation, both the positions of the ions and the orbital indices are integer functions of μ, namely $I = I(\mu)$ and $p = p(\mu)$.

The atomic basis is not necessarily orthonormal. Actually, for practical purposes, it is convenient to choose simple (e.g., Slater or Gaussian, see below) localized orbitals. In this scheme, we define the overlap matrix $\mathbf{S}$ by:

$$S_{\mu,\nu} = \int \mathbf{dr}\ \phi_\mu(\mathbf{r})\phi_\nu(\mathbf{r}). \qquad (12.4)$$

In practice, the matrix $\mathbf{S}$ can be computed exactly for Gaussian orbitals in open systems; nevertheless, we have experienced that a very good approximation can be also obtained by using a finite mesh of lattice points and then evaluating numerically the corresponding integrals. This latter approach is very general and allows us to consider also Slater orbitals (and even more general ones). We also emphasize that, on modern supercomputers, the numerical evaluation of $\mathbf{S}$ is very efficient and usually requires a negligible computational effort.

Then, the pairing function can be generally written as:

$$f(\mathbf{r}, \mathbf{r}') = \sum_{\mu,\nu} f_{\mu,\nu}\phi_\mu(\mathbf{r})\phi_\nu(\mathbf{r}'), \qquad (12.5)$$

where $f_{\mu,\nu}$ defines a real and symmetric matrix. We point out that, whenever the atomic basis is large enough, it is possible to represent any (normalizable) function $f(\mathbf{r}, \mathbf{r}')$ exactly. Analogously, the correlation term $u(\mathbf{r}_i, \mathbf{r}_j)$ can be also expanded in a possibly distinct set of atomic orbitals $\{\chi_\mu(\mathbf{r})\}$, with the same compact notation used before. However, in order to speed up the convergence to the complete basis set limit (or, in other words, to parametrize satisfactorily the Jastrow term within a small basis), it is important to fulfill the so-called cusp conditions that are implied by the divergence of the Coulomb potential at short distances. In this way, the leading singular behaviors when $\mathbf{r} \to \mathbf{r}'$ (electron-electron) or when $\mathbf{r} \to \mathbf{R}_I$ (electron-ion) are satisfied exactly, namely $u(\mathbf{r}, \mathbf{r}') \approx |\mathbf{r} - \mathbf{r}'|/2$ and $u(\mathbf{r}, \mathbf{r}') \approx -Z_I |\mathbf{r} - \mathbf{R}_I|/(N_e - 1)$ (where Z_I indicates the charge number of the atom I), respectively. Therefore, the general form of the correlator factor $u(\mathbf{r}, \mathbf{r}')$ is written as (here, we assume that $N_e > 1$):

$$u(\mathbf{r}, \mathbf{r}') = \left[\frac{u_{ei}(\mathbf{r}) + u_{ei}(\mathbf{r}')}{N_e - 1} \right] + u_{ee}(\mathbf{r}, \mathbf{r}'), \tag{12.6}$$

where the first two terms in the bracket describe the electron-ion interaction and the third one is related to the genuine electron-electron correlations:

$$u_{ei}(\mathbf{r}) = u_{ei}^{cusp}(\mathbf{r}) + \sum_\nu w_\nu \chi_\nu(\mathbf{r}), \tag{12.7}$$

$$u_{ee}(\mathbf{r}, \mathbf{r}') = u_{ee}^{cusp}(\mathbf{r}, \mathbf{r}') + \sum_{\mu,\nu} u_{\mu,\nu} \chi_\mu(\mathbf{r}) \chi_\nu(\mathbf{r}'); \tag{12.8}$$

here, $u_{ei}^{cusp}(\mathbf{r})$ and $u_{ee}^{cusp}(\mathbf{r}, \mathbf{r}')$ are assumed to satisfy the electron-electron and electron-ion cusp conditions, respectively. The simplest expressions that are always bounded and widely used are (Foulkes et al., 2001):

$$u_{ei}^{cusp}(\mathbf{r}) = -\sum_I \frac{Z_I |\mathbf{r} - \mathbf{R}_I|}{1 + \sqrt{2Z_I} b_{ei} |\mathbf{r} - \mathbf{R}_I|}, \tag{12.9}$$

$$u_{ee}^{cusp}(\mathbf{r}, \mathbf{r}') = \frac{|\mathbf{r} - \mathbf{r}'|}{2(1 + b_{ee} |\mathbf{r} - \mathbf{r}'|)}, \tag{12.10}$$

where b_{ei} and b_{ee} are two independent (positive) variational parameters. In addition, $u_{\mu,\nu}$ is a symmetric matrix and w_ν is a vector, which describe the variational freedom of the Jastrow factor in the given (finite) atomic basis. The reason to scale the one-body contribution by $1/(N_e - 1)$ follows from the identity:

$$\exp\left\{ \sum_{i<j} \left[\frac{u_{ei}(\mathbf{r}_i) + u_{ei}(\mathbf{r}'_j)}{N_e - 1} \right] \right\} = \exp\left[\sum_i u_{ei}(\mathbf{r}_i) \right], \tag{12.11}$$

as the r.h.s. of the above equation is the commonly adopted expression for a one-body Jastrow factor in Eq. (12.2). Notice that a general and complete description of both the one-body and two-body terms can be achieved taking a sufficiently large basis set and using appropriate values for $u_{\mu,\nu}$ and w_ν.

In practice, we can adopt the same basis set for both the Jastrow factor and the pairing function. However, usually the dimension of the basis set for the Jastrow term (D_u, including both electron-ion and electron-electron terms) can be taken much smaller than the one used for the pairing function (D_f), since converged results for various chemical properties can be achieved with a small value of D_u. In summary, we are led to optimize a total number of parameters:

$$N_{\text{par}} = \frac{D_f(D_f + 1) + D_u(D_u + 1)}{2} + D_u. \tag{12.12}$$

Fortunately, similarly to what happens in quantum chemistry calculations, a very small basis is enough to describe rather accurately an electronic wave function, even in the presence of electron correlations. Indeed, the largest energy scale of the problem is the strong electron-ion attraction, which represents the main contribution for large values of Z_I and can be very accurately represented with an appropriate atomic basis containing only few elements.

12.2.1 Atomic Orbitals

Here, we would like to discuss the form of the atomic orbitals $\{\phi_\mu(\mathbf{r})\}$, centered at the position $\mathbf{R}_I$ and written in terms of the radial vector $\mathbf{r} - \mathbf{R}_I$ that connects the position $\mathbf{r}$ of the electron to the position $\mathbf{R}_I$ of the ion I. Notice that, in the most general case, there are several orbitals that are defined on a given atom I (which defines the atomic basis on the ion I). The *elementary* objects are determined by a radial part, which is given by a Gaussian or a Slater form, and an angular part, characterized by an angular momentum l and its projection m along the z axis. On the one hand, the Gaussian form is given by:

$$\phi^G_{l,\pm|m|,I}(\mathbf{r};\zeta) \propto |\mathbf{r} - \mathbf{R}_I|^l e^{-\zeta|\mathbf{r}-\mathbf{R}_I|^2} \left[Y_{l,m,I}(\mathbf{\Omega}) \pm Y_{l,-m,I}(\mathbf{\Omega}) \right]; \tag{12.13}$$

on the other hand, the Slater form is:

$$\phi^S_{l,\pm|m|,I}(\mathbf{r};\zeta) \propto |\mathbf{r} - \mathbf{R}_I|^l e^{-\zeta|\mathbf{r}-\mathbf{R}_I|} \left[Y_{l,m,I}(\mathbf{\Omega}) \pm Y_{l,-m,I}(\mathbf{\Omega}) \right]; \tag{12.14}$$

in both cases, $Y_{l,m,I}(\mathbf{\Omega})$ indicates a spherical harmonic centered around $\mathbf{R}_I$ (that depends upon the unit vector $\mathbf{\Omega}$); for each value of the angular momentum l (and its projection along the z axis m), the Slater and Gaussian orbitals depend upon the (variational) parameter ζ. By considering the linear combination of $Y_{l,m,I}(\mathbf{\Omega})$ with its complex conjugate $Y_{l,-m,I}(\mathbf{\Omega})$ (for $m \neq 0$), we get a real atomic basis set; the $m = 0$ case does not require any symmetrization, being $Y_{l,0,l}$ already real. These orbitals can have a simple cartesian representation, which can be efficiently implemented.

The most general atomic orbital $\phi_\mu^{\text{contr}}(\mathbf{r})$ can be written as a linear combination of Gaussian or Slater orbitals (denoted by $X = G$ or S), everyone associated to an angular momentum l and specified by an additional index k. Then, the variational parameter ζ may acquire a dependence on l and k; the contracted orbital is finally given by:

$$\phi_\mu^{\text{contr}}(\mathbf{r}) = \sum_{k,l,m} c_{l,m,I}^k \phi_{l,m,I}^{X_{k,l}}(\mathbf{r}, \zeta_{k,l}). \tag{12.15}$$

In quantum chemistry, this kind of linear combination is called *contraction* and is adopted to reduce the dimension of the atomic basis to describe strongly-localized atomic orbitals (e.g., $1s$), even though it is not very common to hybridize states with different angular momenta. However, in quantum chemistry, the reduction of the atomic basis set is not crucial and the main computational advantage is achieved by using a small number of molecular orbitals. By contrast, in quantum Monte Carlo approaches, it is extremely important to minimize the number of variational parameters and, therefore, it is necessary to reduce the atomic basis dimension D_f as much as possible. Then, it is recommended to reduce D_f by optimizing a number of independent atomic orbitals of the form given in Eq. (12.15), in a large basis of elementary functions. In this way, a small number of contracted orbitals is necessary to reach converged or, at least, accurate results. Thus, this kind of orbitals are generalized hybrid orbitals and appear to be of fundamental importance for describing wave functions in a compact and efficient way (Sorella et al., 2015).

As commonly done in standard electronic structure approaches (Dovesi et al., 2014), both Gaussian and Slater orbitals can be generalized to periodic systems by considering a $L_x \times L_y \times L_z$ supercell. This can be conveniently done by means of vectors $\mathbf{L}_{n_x,n_y,n_z} = (n_x L_x, n_y L_y, n_z L_z)$, labeling all possible periodic images, and by evaluating the sum for generic twisted-boundary conditions (determined by the angles $|\theta_x|$, $|\theta_y|$, and $|\theta_z| \leq \pi$):

$$\phi_{l,m,I}^{G_{\text{PBC}},S_{\text{PBC}}}(\mathbf{r}; \zeta) = \sum_{n_x,n_y,n_z} \phi_{l,m,I}^{G,S}\left(\mathbf{r} + \mathbf{L}_{n_x,n_y,n_z}; \zeta\right) e^{i(n_x\theta_x + n_y\theta_y + n_z\theta_z)}. \tag{12.16}$$

The presence of the Gaussian or exponential factors guarantees that the above series rapidly converges. This new basis satisfies twisted-boundary conditions:

$$\phi_{l,m,I}^{G_{\text{PBC}},S_{\text{PBC}}}(\mathbf{r} + \mathbf{L}_{n_x,n_y,n_z}; \zeta) = e^{-i(n_x\theta_x + n_y\theta_y + n_z\theta_z)} \phi_{l,m,I}^{G_{\text{PBC}},S_{\text{PBC}}}(\mathbf{r}; \zeta), \tag{12.17}$$

which are important to study bulk properties of materials. For example, by using suitably chosen values of the twist (Baldereschi, 1973) or averaging the physical observables over all possible angles (Lin et al., 2001), it is possible to have a smooth and rapid convergence to the thermodynamic limit.

The same procedure can be applied to the orbitals of the Jastrow factor, with $\phi_{l,\pm|m|,I}^{S,G}(\mathbf{r};\zeta) \to \chi_{l,\pm|m|,I}^{S,G}(\mathbf{r};\zeta)$. However, since the Jastrow factor only couples local densities, we limit ourselves to a periodic case with $\theta_x = \theta_y = \theta_z = 0$, which implies $\chi_{l,m,I}^{G_{PBC},S_{PBC}}(\mathbf{r}+\mathbf{L}_{n_x,n_y,n_z};\zeta) = \chi_{l,m,I}^{G_{PBC},S_{PBC}}(\mathbf{r};\zeta)$. Finally, the one- and two-body functions $u_{ei}^{cusp}(\mathbf{r})$ and $u_{ee}^{cusp}(\mathbf{r},\mathbf{r}')$, defined in Eqs. (12.9) and (12.10), are slowing decaying functions and, therefore, it is convenient to express them in terms of a periodic generalization of the Euclidean distance:

$$|\mathbf{r}| \to \sqrt{\left[\frac{L_x}{\pi}\sin\left(\frac{\pi x}{L_x}\right)\right]^2 + \left[\frac{L_y}{\pi}\sin\left(\frac{\pi y}{L_y}\right)\right]^2 + \left[\frac{L_z}{\pi}\sin\left(\frac{\pi z}{L_z}\right)\right]^2}. \quad (12.18)$$

12.2.2 Molecular Orbitals

Now, we consider a different functional form for representing the uncorrelated state described by the pairing function $f(\mathbf{r},\mathbf{r}')$. In this way, the relation between the AGP wave function and the Slater determinant will be very transparent. The (real and symmetric) pairing function can be considered as a linear operator in the continuum space; therefore, it can be diagonalized, by looking for eigenfunctions $\Phi_\alpha(\mathbf{r})$ that satisfy:

$$\int \mathbf{dr}' \, f(\mathbf{r},\mathbf{r}')\Phi_\alpha(\mathbf{r}') = \lambda_\alpha \Phi_\alpha(\mathbf{r}), \quad (12.19)$$

where λ_α are the corresponding eigenvalues. In the finite basis $\{\phi_\mu(\mathbf{r})\}$ of dimension D_f, the eigenfunctions can be written as:

$$\Phi_\alpha(\mathbf{r}) = \sum_\mu P_{\mu,\alpha}\phi_\mu(\mathbf{r}). \quad (12.20)$$

where the elements of the matrix $\mathbf{P}$ are found from the generalized eigenvalue equation that is obtained by substituting this expression, as well as the expansion of the pairing function (12.5), in Eq. (12.19). In this way, we obtain that:

$$\mathbf{fSP} = \mathbf{P\Lambda}, \quad (12.21)$$

where $\mathbf{\Lambda}$ is the diagonal matrix containing the eigenvalues λ_α, $\mathbf{S}$ is the overlap matrix of Eq. (12.4), and $\mathbf{f}$ is the matrix with elements $f_{\mu,\nu}$ of Eq. (12.5). Notice that, as the result of the generalized eigenvalue problem, the matrix $\mathbf{P}$ defines a set of orthonormal states:

$$\int \mathbf{dr} \, \Phi_\alpha(\mathbf{r})\Phi_\beta(\mathbf{r}) = \delta_{\alpha,\beta}, \quad (12.22)$$

which leads to $\mathbf{P}^T\mathbf{SP} = \mathbf{1}$. The one-body wave functions $\{\Phi_\alpha(\mathbf{r})\}$ are generally called *molecular orbitals*. Notice that, depending on the properties of the matrix $\mathbf{P}$, the molecular orbitals may be either *localized* or *extended*.

Finally, from Eq. (12.19), the pairing function can be written in terms of the molecular orbitals:

$$f(\mathbf{r},\mathbf{r}') = \sum_{\alpha=1}^{D_f} \lambda_\alpha \Phi_\alpha(\mathbf{r})\Phi_\alpha(\mathbf{r}'). \tag{12.23}$$

Remarkably, whenever the number of non-zero eigenvalues λ_α is equal to the number of electron pairs $N_e/2$, the determinant $F(\mathbf{x})$ in Eq. (12.3) can be written as the product of three square matrices of dimension $N_e/2 \times N_e/2$:

$$\det f(\mathbf{r}_i^\uparrow,\mathbf{r}_j^\downarrow) = \det(\mathbf{P}_\uparrow\tilde{\mathbf{\Lambda}}\mathbf{P}_\downarrow^T) = \left(\prod_{\alpha=1}^{N_e/2} \lambda_\alpha\right) \times \det\mathbf{P}_\uparrow \times \det\mathbf{P}_\downarrow \tag{12.24}$$

where $\tilde{\mathbf{\Lambda}}$ is the square (diagonal) matrix containing the non-zero eigenvalues only; the matrices $\mathbf{P}_\uparrow$ and $\mathbf{P}_\downarrow$ are given by:

$$(P_\uparrow)_{i,\alpha} = \Phi_\alpha(\mathbf{r}_i^\uparrow), \tag{12.25}$$
$$(P_\downarrow)_{i,\alpha} = \Phi_\alpha(\mathbf{r}_i^\downarrow). \tag{12.26}$$

Therefore, in this special case, we obtain a form that is consistent with the standard expression of a Slater determinant (apart from the irrelevant constant, i.e., the product of non-zero eigenvalues), i.e., the AGP wave function reduces to a Slater determinant with $N_e/2$ molecular orbitals. In other words, the molecular orbitals, which have been defined within the AGP formalism, coincide with the ones that are widely used to define Slater determinants.

The representation of a simple determinant as an AGP state, with a constrained number of molecular orbitals, has important advantages from the numerical point of view. The simplest one is to consider only one determinant of a $N_e/2 \times N_e/2$ matrix to evaluate the wave function by means of Eq. (12.3); by contrast, within the standard representation of the Slater determinant, we would need two determinants of the same dimension. Another advantage comes from the fact that it is possible to exploit symmetries, such as the translation invariance in a crystal, in a simple way. In this case, it is enough to require that the pairing matrix $\mathbf{f}$, corresponding to the same type of orbitals, depends only on the difference between the ion positions, i.e., $f_{\mu,\nu} = f_{\mu',\nu'}$ if $|\mathbf{R}_{I(\mu')} - \mathbf{R}_{I(\nu')}| = |\mathbf{R}_{I(\mu)} - \mathbf{R}_{I(\nu)}|$, which represents a very simple linear constraint.

Finally, we also remark that, even when the AGP state is exactly equivalent to a Slater determinant, the combined optimization of the Jastrow factor and the molecular orbitals may lead to a qualitatively different wave function, namely with chemical and physical properties that are different from the ones obtained within uncorrelated Hartree-Fock or density-functional theory. The fully optimized Jastrow-Slater wave function with $N_e/2$ molecular orbitals provides an accurate description of atoms (Foulkes et al., 2001), with about 90% of the correlation energy (defined as the energy difference between the estimated exact result, obtained by experiments or high-quality calculations, and the best Hartree-Fock value). By releasing the constraint on the molecular orbitals, we reach a larger variational freedom; it is then obvious that the unconstrained AGP wave function can improve the Hartree-Fock approach, especially when also the Jastrow factor is included, leading to an improvement of the accuracy of the chemical bond or even to new qualitative properties. In particular, the AGP state, allowing an explicit pairing between electrons, can describe superconductors and, in presence of a Jastrow factor, also Mott (resonating-valence bond) insulators, as introduced by Anderson (1987) soon after the discovery of high-temperature superconductors.

12.3 Size Consistency of the Variational Wave Functions

A basic notion in quantum chemistry is the so-called size consistency, stating that, whenever we compute the energy of a molecular system in which we clearly distinguish two regions A and B that are spatially well separated, the total energy of the composed system E_{A+B} should be equal to the sum of the energies of the isolated systems E_A and E_B, i.e., $E_{A+B} = E_A + E_B$. The size consistency crucially affects the chemical bond and, therefore, it is considered as one of the most important properties of a wave function that is consistently optimized within its full variational freedom. It is well known that, in general, the AGP wave function is not size consistent, even when the simple Hartree-Fock state is not affected by this problem. Therefore, AGP states have been abandoned in chemistry. Within the AGP approach, size inconsistency originates from projecting over a given number of particles. For example, let us consider the Be_2 molecule, where the two regions A and B contain a single Be atom; then, the AGP wave function for the single atom implies correctly only two electron pairs. However, in the molecule, the constraint on the number of electrons would act only on the total number of four pairs. Thus, the AGP wave function for the $A + B$ system would also allow the case in which there are three or even four pairs close to a given atom and one or zero pairs close to the other one. When the molecule is stretched and the two atoms are at very large distances, this effect will not produce two independent atomic AGP wave functions, as projecting the total number of electrons is not equivalent to projecting the number of electrons in each atom. Therefore, the corresponding energy will be much higher than the one of two independent Be atoms and size consistency will be not verified.

Here, we show that, whenever the AGP wave function is combined with the Jastrow factor, the size consistency is recovered, since the latter one has the variational freedom to allow partial number projections on different space regions (Sorella et al., 2007; Neuscamman, 2012). For this purpose, we can consider a Jastrow factor that projects out all the electronic configurations that do not have exactly N_A and N_B particles in the region A and B, respectively:

$$\mathcal{J}(\mathbf{x}) \propto \exp\left\{-\frac{C}{2}[N_A(\mathbf{x}) - N_A]^2 + A \to B\right\}, \qquad (12.27)$$

where $N_{A,B}(\mathbf{x}) = \sum_i g_{A,B}(\mathbf{r}_i)$, with $g_A(\mathbf{r}) = 1$ [$g_B(\mathbf{r}) = 1$] if $\mathbf{r} \in A$ ($\mathbf{r} \in B$) and zero otherwise, and C is a sufficiently large constant. This form of the Jastrow factor can be obtained by taking a particular form of the pseudo-potential given by:

$$u(\mathbf{r}, \mathbf{r}') = -C\left\{g_A(\mathbf{r})g_A(\mathbf{r}') - \frac{N_A - 1/2}{N_e - 1}\left[g_A(\mathbf{r}) + g_A(\mathbf{r}')\right] + A \to B\right\}. \quad (12.28)$$

Therefore, whenever the basis set of the Jastrow factor is large enough to represent the form given by Eq. (12.28), a full optimization of the global wave function guarantees that the total energy will be below the one of the two fragments (when the two regions A and B are well separated), thus implying size consistency. A *caveat* of the previous rule applies to the case in which the two fragments have non-zero spin: in this case, a wave function that is built from a single determinant is often not size consistent (e.g., the case of two Oxygen atoms, for which each independent atom has $S = 1$, while the O_2 molecule has $S = 0$). In this respect, the restricted Hartree-Fock determinant is not size consistent whenever the whole system has not the maximum spin compatible with the two fragments.

12.4 Optimization of the Variational Wave Functions

A straightforward way to optimize the pairing function is to consider all the matrix elements $f_{\mu,\nu}$ in Eq. (12.5) as independent variational parameters. This would be an *unconstrained* optimization. However, this way of proceeding is not always efficient, since the variational energy may have a very weak dependence on long-range pairings, especially for insulating phases. The main advantage of having a pairing function that is explicitly defined in terms of a localized basis set is the possibility to exploit the locality of the correlations. Whenever the geminal function $f(\mathbf{r}, \mathbf{r}')$ is described by a characteristic length ξ, a useful way to reduce the number of variational parameters is to consider in the optimization only matrix elements $f_{\mu,\nu}$ connecting orbitals at a distance smaller than a certain cutoff R_{cut}. For a given R_{cut}, for each column (or row) of the matrix $f_{\mu,\nu}$, only a fixed number of elements will be non-zero and, therefore, the total number of parameters scales as the total atomic basis dimension, leading to a number of variational parameters $N_{par} \propto D_f \propto N_e$. This approach is much more efficient than an unconstrained minimization where all the matrix elements $f_{\mu,\nu}$ are independently optimized. Whenever a finite correlation length is present in the physical problem, the accuracy should improve exponentially by increasing R_{cut}. For example, if the rank of the geminal matrix is equal to half the number of electrons $N_e/2$, the geminal function corresponds to the density matrix $\rho(\mathbf{r}, \mathbf{r}')$. Then, within insulating phases, we have that:

$$\lim_{|\mathbf{r}-\mathbf{r}'|\to\infty} \rho(\mathbf{r}, \mathbf{r}') \propto \exp\left(-\frac{|\mathbf{r}-\mathbf{r}'|}{\xi}\right), \tag{12.29}$$

which justifies the inclusion of R_{cut} for an efficient minimization of the number of parameters. Unfortunately, this approach becomes quite inefficient when considering metallic phases, where (in three spatial dimensions):

$$\lim_{|\mathbf{r}-\mathbf{r}'|\to\infty} |\rho(\mathbf{r}, \mathbf{r}')| \propto \frac{1}{|\mathbf{r}-\mathbf{r}'|^2}, \tag{12.30}$$

which implies a very slow convergence with R_{cut}. The reason for this slow decay of the density matrix is due to the existence of a Fermi surface that constrains the occupation within partially filled bands.

Here, we define a *constrained* optimization procedure that is able to deal with both insulating and metallic behaviors, even in presence of a finite cutoff R_{cut}. This result can be achieved by fixing the rank of the pairing function to a given number n, namely by taking only n molecular orbitals in the definition of the geminal function:

$$f^c(\mathbf{r}, \mathbf{r}') = \sum_{\alpha=1}^{n} \lambda_\alpha \Phi_\alpha(\mathbf{r}) \Phi_\alpha(\mathbf{r}'), \tag{12.31}$$

where the molecular orbitals $\{\Phi_\alpha(\mathbf{r})\}$ with $\alpha = 1, \ldots, n$ define the "occupied" states. Within this procedure, a Fermi surface can be "reconstructed" by imposing the mentioned constraint of n occupied states, thus allowing us to recover slowly decaying density matrices. We have three main advantages by constraining the optimization of the pairing function with a form given by a finite number n of molecular orbitals. The first one is the dual representation of the pairing function, which allows us to consider the matrix elements $f_{\mu,\nu}$ as variational parameters with the constraint of a fixed rank n or equivalently a fixed number n of optimized molecular orbitals; the second one is that, within an optimization procedure, we can initialize the pairing function (i.e., the matrix $f_{\mu,\nu}$) by taking the molecular orbitals $\Phi_\alpha(\mathbf{r})$ from a mean-field method defined in the same localized basis, such as Hartree-Fock or density-functional theory; the third one is that we can implement locality within a constrained approach, highly reducing the number of variational parameters, without affecting much the accuracy of the calculation, since the long-range part of the pairing function is automatically taken into account by constraining the number of molecular orbitals, namely by forcing the presence of the Fermi surface.

Before discussing the technical details, we would like to mention that on a lattice we have used a different approach to minimize the variational parameters (see Chapters 5 and 6). There, we considered a parametrization given by an effective (BCS) Hamiltonian with finite range (i.e., parameters corresponding only to distances smaller than R_{cut}); the ground state of this effective Hamiltonian implicitly defined the pairing function, which was long range in the case of a metal or a gapless superconductor. In continuous systems, it is difficult to implement an approach that is based upon an auxiliary Hamiltonian and it is much easier to work directly with the pairing function.

For a constrained pairing function with a fixed rank n, by performing an infinitesimal change and keeping the same rank, we obtain that:

$$\delta f^c(\mathbf{r}, \mathbf{r}') = \sum_{\alpha=1}^{n} \delta \lambda_\alpha \Phi_\alpha(\mathbf{r}) \Phi_\alpha(\mathbf{r}') + \lambda_\alpha \left[\delta \Phi_\alpha(\mathbf{r}) \Phi_\alpha(\mathbf{r}') + \Phi_\alpha(\mathbf{r}) \delta \Phi_\alpha(\mathbf{r}') \right]. \quad (12.32)$$

Then, we can easily verify that this kind of modification satisfies the following condition (which is obtained by using the orthonormalization of the original molecular orbitals):

$$\int d\mathbf{r}_1 \int d\mathbf{r}_2 \left[\delta(\mathbf{r} - \mathbf{r}_1) - \Pi(\mathbf{r}, \mathbf{r}_1) \right] \delta f^c(\mathbf{r}_1, \mathbf{r}_2) \left[\delta(\mathbf{r}' - \mathbf{r}_2) - \Pi(\mathbf{r}_2, \mathbf{r}') \right] = 0, \quad (12.33)$$

where the projector onto the occupied states $\Pi(\mathbf{r}, \mathbf{r}')$ is given by:

$$\Pi(\mathbf{r}, \mathbf{r}') = \sum_{\alpha=1}^{n} \Phi_\alpha(\mathbf{r}) \Phi_\alpha(\mathbf{r}'). \quad (12.34)$$

In matrix notations, corresponding to the finite basis *Ansatz* in Eq. (12.5), Eq. (12.33) becomes:

$$(1 - \mathbf{L}) \delta \mathbf{f}^c (1 - \mathbf{R}) = 0 \quad (12.35)$$

where $\mathbf{L}$ and $\mathbf{R}$ are two $D_f \times D_f$ matrices that, since the original basis set $\{\phi_\mu(\mathbf{r})\}$ is not orthogonal, are given by:

$$\mathbf{R} = (\mathbf{SP})\mathbf{P}^T, \quad (12.36)$$

$$\mathbf{L} = (\mathbf{R})^T = \mathbf{P}(\mathbf{SP})^T; \quad (12.37)$$

here, the $D_f \times n$ matrix $\mathbf{P}$ is defined in Eq. (12.20) and the $D_f \times D_f$ matrix $\mathbf{S}$ is the overlap matrix defining the basis, see Eq. (12.4). In practice, since the dimension

n is usually much smaller than D_f, it is convenient to work with only two matrices **P** and **SP**; indeed, all linear operations with **L** and **R** become much faster when performed in terms of the rectangular matrices **P** and **SP**.

At this point, we are able to implement the constraint in a simple way. Suppose that we make an unconstrained variation of the pairing function $\delta\mathbf{f}$, then, from this change, we can remove all the contributions that will affect the rank of the matrix (at linear order in the variation) by considering:

$$\delta\mathbf{f}^c = \delta\mathbf{f} - (\mathbf{1} - \mathbf{L})\delta\mathbf{f}(\mathbf{1} - \mathbf{R}). \tag{12.38}$$

This equation represents a linear mapping between unconstrained and constrained variations. It is important to emphasize that this mapping can be considered a simple parametrization of a variation of the geminal function that is constrained to a given number n of molecular orbitals. Moreover, even if **f** is restricted to a finite range by introducing R_{cut}, the application of Eq. (12.38) may induce a long-range tail in the geminal function. This approach is useful to represent gapless phases, like metals and nodal superconductors and highlights the reason why the convergence with R_{cut} can be very fast, even in case of metallic behavior.

When applying the optimization techniques described in Chapter 6, we have to compute wave function derivatives with respect to arbitrary variations of the geminal matrix **f**. Let us denote the unconstrained derivatives by:

$$D_{\mu,\nu} = \frac{\partial \ln |F(\mathbf{x})|}{\partial f_{\mu,\nu}}. \tag{12.39}$$

The total change with respect to the constrained parameters is given by $\text{Tr}[\mathbf{D}(\delta\mathbf{f}^c)^T]$, where $\delta\mathbf{f}^c$ is given by Eq. (12.38). Therefore, we can define the constrained derivatives, i.e., the variations with respect to the parameters that are subject to the constraint:

$$D^c_{\mu,\nu} = \left[\mathbf{D} - (\mathbf{1} - \mathbf{L}^T)\mathbf{D}(\mathbf{1} - \mathbf{R}^T)\right]_{\mu,\nu}, \tag{12.40}$$

which can be obtained by standard and efficient linear algebra operations.

In summary, the scheme for a constrained optimization of the pairing function is given by:

1. Compute the derivative matrix **D** of Eq. (12.39), without taking into account any constraint.
2. Apply the projection of Eq. (12.40) with the current molecular orbitals. By exploiting the fact that all matrices involved are written in terms of small rectangular matrices, a very convenient computation, scaling with $O(n\, D_f^2)$ can be achieved.

3. After having accumulated enough statistics, apply the steepest-descent method (or more sophisticated schemes, as described in Chapter 6) to obtain $\delta f_{\mu,\nu} \propto D^c_{\mu,\nu}$.

4. Obtain the constrained $\delta \mathbf{f}^c$ by using Eq. (12.38).

5. Change the parameters according to:

$$f_{\mu,\nu} \to f_{\mu,\nu} + \delta f^c_{\mu,\nu}, \qquad (12.41)$$

by considering only the ones allowed by the cutoff R_{cut}.

6. Finally, diagonalize $\mathbf{f}$ and keep only the n eigenvectors corresponding to the eigenvalues with the largest absolute values. The new molecular orbitals are then defined after this diagonalization and the corresponding projection matrices $\mathbf{L}$ and $\mathbf{R}$ are re-computed (together with a possible re-computation of the overlap matrix $\mathbf{S}$ whenever also the parameters of the contracted orbitals are also optimized).

7. Repeat all the previous steps until convergence is achieved.

The above algorithm is very stable and general, since it works also when the quasi-particle energies are degenerate at the Fermi energy, which is typical for metallic behavior. Indeed, within this approach the eigenvalues have nothing to do with quasi-particle energies and represent just average occupations of molecular orbitals. In Table 12.1, we report the case of 128 Hydrogens (i.e., 128 electrons) in a body-centered cubic crystal at $r_s = 1.32$.

To achieve a stable convergence to the minimum, it is important that, at each step, when we define the new matrix $\mathbf{f}$, the non-zero eigenvalues considered are substantially different from zero because if they are too close to zero there may be some problem to set the rank of the matrix within numerical precision. In particular, when $n = N_e/2$ we can, at each iteration of the optimization, reset the largest n eigenvalues $\lambda_\alpha = 1$ (since, as we have discussed, for the Slater determinant the eigenvalues λ_α define only an irrelevant normalization constant) and continue with a very stable optimization.

12.4.1 Beryllium and Fluorine Molecules

Here, we show two examples of the optimization scheme that has been described in the previous section for the F_2 and Be_2 molecules (Marchi et al., 2009). Let us start with the former case, which is relatively simple to treat. Here, the determinant is constructed from a $5s5p2d$ basis of Gaussian orbitals on each ion (leading to $D_f = 60$). The Hartree-Fock approach (with $n = 9$ molecular orbitals), supplemented by the Jastrow factor, gives a considerable error in the binding energy ($E_{\text{binding}} \approx 39mH$, to be compared with the estimated exact value of about $62mH$), while the equilibrium distance is well approximated ($R_{\text{eq}} \approx 2.67a.u.$, which is

Table 12.1. *Optimized energy (a.u.) for 128 Hydrogens in a body-centered cubic crystal at $r_s = 1.32$ as a function of R_{cut}. Twisted boundary conditions with $\theta_x = \theta_y = 2\pi/6$ and $\theta_z = \pi$, see Eq. (12.17) are used. The atomic basis is obtained by contracting the two hybrid orbitals of Eq. (12.15) expanded in a 3s1p basis, containing three s-wave (with different exponents $\zeta_{k,l}$) and one p-wave Gaussian orbitals. In order to appreciate the remarkable accuracy obtained with small R_{cut}, we also report the correlated energy difference (computed with the reweighting technique) with respect to the case with $R_{cut} = 0$. The case with a Slater determinant, taken from a density-functional theory (DFT) calculation, in presence of an optimized Jastrow factor (with two s-wave Gaussian orbitals per atom) is also reported. Notice that already the case with $R_{cut} = 0$ gives an improvement of about 90% with respect to the latter state.*

R_{cut}	# parameters AGP	Total Energy	Corr. diff.
DFT	0	−64.945(2)	0.064(2)
0	11	−65.009(1)	0.0
2.5	43	−65.010(2)	−0.0018(1)
3	67	−65.017(2)	−0.0064(1)
4	115	−65.014(1)	−0.0072(2)
5	243	−65.015(1)	−0.0072(2)
∞	519	−65.015(2)	−0.0071(2)

compatible with the estimated exact value), see Fig. 12.1. The overall energy dispersion as a function of the relative distance R between the two Fluorine atoms shows up an unphysical maximum at $R \approx 4.5 a.u.$. Moreover, at large distances, this wave function, which describes a total singlet, cannot recover the size-consistent result of two Fluorine atoms with $S = 1/2$.

A considerable improvement is achieved by increasing the number of molecular orbitals n. Indeed, already with $n = 10$ the AGP wave function in presence of the Jastrow factor gives a very accurate result for both the binding energy and the equilibrium distance (i.e., $E_{binding} \approx 61mH$), see Fig. 12.1. Moreover, also size consistency is recovered in this case. By further increasing the number of molecular orbitals, we do not substantially improve the results of $n = 10$. We stress the fact that, by using a single determinant, it is possible to reach an accuracy that is comparable to (if not even better than) the one obtained with quantum-chemistry methods using several thousand determinants.

An interesting example for a diatomic molecule, which represents a real challenge for most of the numerical techniques, is given by the Beryllium dimer.

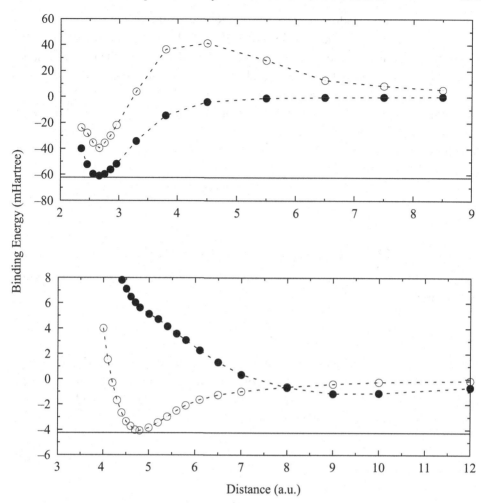

Figure 12.1 Results for the binding energies of the F_2 (upper panel) and Be_2 (lower panel) molecules. The AGP (Hartree-Fock) determinant, supplemented with the Jastrow factor is denoted by filled (empty) circles. The estimated exact results are shown by solid lines. The Hartree-Fock state for the F_2 (Be_2) molecule has $n = 9$ ($n = 4$) molecular orbitals which are constructed from a $5s5p2d$ basis set of Gaussian orbitals ($5s2p1d$ basis set of both Gaussian and Slater types) on each ion. The AGP state for the F_2 (Be_2) molecule has $n = 10$ ($n = 32$) with the same atomic orbitals as the Hartree-Fock wave function. The parameters $\zeta_{k,l}$ are optimized to improve the accuracy of the finite basis set.

Here, the difficulty comes from the small binding energy, for which also the weak Van der Waals forces play an important role. Indeed, the simple Hartree-Fock approach is not able to bind the molecule, while other self-consistent approaches, based upon density-functional theory, widely overestimate the binding energy. We consider a determinant with a $5s2p1d$ basis of both Gaussian and Slater types on each ion (leading to $D_f = 32$). Remarkably, by considering $n = 4$ molecular orbitals (i.e., the case in which the AGP function corresponds to the Hartree-Fock state), we are able to get excellent results for both the binding energy (e.g., $E_{binding} \approx 4.1mH$, to be compared with the estimated exact value of about $4.2mH$) and the equilibrium position ($R_{eq} \approx 4.75a.u.$), see Fig. 12.1. However, by including a larger number of molecular orbitals, the accuracy on the total energy improves, and the overall energy dispersion worsen, giving a much larger bond length and a much smaller binding energy. This outcome is due to the fact that, in this case, the correlated AGP state is much more effective for the single atom (where it reaches 99.9% of the correlation energy) than for the description of the chemical bond.

12.5 Lattice-Regularized Diffusion Monte Carlo

In this section, we will consider an electronic Hamiltonian (written in atomic units) of the form:

$$\mathcal{H} = -\frac{1}{2}\nabla^2 + V(\mathbf{x}) + V_{ii}, \tag{12.42}$$

where $\nabla^2 = \sum_{i=1}^{N_e} \nabla_i^2$ is the Laplacian operator that defines the total kinetic energy, ∇_i^2 being the corresponding Laplacian operator acting on the single electron with coordinates $\mathbf{r}_i$, and $V(\mathbf{x}) = V_{ee}(\mathbf{x}) + V_{ei}(\mathbf{x})$ is the standard many-body potential, which includes the electron-electron interaction:

$$V_{ee}(\mathbf{x}) = \sum_{i<j} \frac{1}{|\mathbf{r}_i - \mathbf{r}_j|}, \tag{12.43}$$

and the electron-ion interaction:

$$V_{ei}(\mathbf{x}) = -\sum_{i} v_{ei}(\mathbf{r}_i), \tag{12.44}$$

where

$$v_{ei}(\mathbf{r}_i) = \sum_{I} \frac{Z_I}{|\mathbf{r}_i - \mathbf{R}_I|} \tag{12.45}$$

represents its contribution for the electron at position $\mathbf{r}_i$; finally, the ion-ion term V_{ii}, which does not depend upon the electron coordinates, is given by:

$$V_{ii} = \sum_{I<J} \frac{Z_I Z_J}{|\mathbf{R}_I - \mathbf{R}_J|}; \tag{12.46}$$

here, the ion coordinates $\mathbf{R}_I$ with $I = 1, \ldots, N_{\text{ion}}$ are just given parameters within the Born-Oppenheimer approximation and, therefore, V_{ii} represents a constant shift of the electronic energy.

In continuous models, the energy spectrum of $\mathcal{H}$ has arbitrary large and positive eigenvalues (e.g., plane waves with large momenta); therefore, in order to filter out the ground-state $|\Upsilon_0\rangle$ from a given trial function $|\Psi_0\rangle$, it is not possible to apply the power method of section 1.7 and it is convenient to introduce an exponential operator:

$$\lim_{\tau \to \infty} e^{-\mathcal{H}\tau}|\Psi_0\rangle \propto |\Upsilon_0\rangle. \tag{12.47}$$

However, this projector operator is difficult to evaluate exactly for large values of τ. The conventional method is to work in the limit of short time $\Delta\tau$, where the Trotter decomposition gives a small error that vanishes for $\Delta\tau \to 0$ (Trotter, 1959; Suzuki, 1976a,b), and apply the approximate form several times. Within this approach, there are several approximations that are often not rigorously justified, especially for fermions. Indeed, few aspects of the diffusion Monte Carlo method and the related fixed-node approximation (Kalos et al., 1974; Ceperley and Alder, 1980; Foulkes et al., 2001) are still debated (as for example, how the error vanishes for $\Delta\tau \to 0$ or the possibility to rigorously define a variational upper-bound property when the so-called pseudo-potentials are used to remove the core electrons).

Here, we take an alternative point of view that relies on a lattice regularization. We start from the approximation of the Laplacian ∇^2 (which defines the kinetic energy of each electron) by using a finite mesh with lattice spacing a:

$$\begin{aligned} \nabla_a^2 g(x, y, z) = \frac{1}{a^2} &\left[g(x + a, y, z) + g(x - a, y, z) - 2g(x, y, z) \right] \\ + \frac{1}{a^2} &\left[g(x, y + a, z) + g(x, y - a, z) - 2g(x, y, z) \right] \\ + \frac{1}{a^2} &\left[g(x, y, z + a) + g(x, y, z - a) - 2g(x, y, z) \right], \end{aligned} \tag{12.48}$$

where $g(x, y, z)$ is a generic function defined in the three-dimensional space (x, y, z). For $a \to 0$, the linear operator ∇_a^2 converges to the exact Laplacian with an error that scales as a^2 for a sufficiently regular function g. This fact can be easily found by performing a Taylor expansion of the function $g(x \pm a, y, z)$ (and analogously for the y and z components) and noticing that all the odd derivatives vanish, leaving a residual term that is $O(a^2)$.

In the following, we can define $\nabla_{i,a}^2$ as the Laplacian acting on the single electron coordinate $\mathbf{r}_i$, according to Eq. (12.48). A lattice regularization of the Hamiltonian $\mathcal{H}$ corresponds to defining an Hermitian operator $\mathcal{H}_a$, such that $\mathcal{H}_a \to \mathcal{H}$ for $a \to 0$; then, the properties of $\mathcal{H}_a$ can be computed numerically with stable simulations, recovering the ones of the continuous limit for $a \to 0$. The above requirement is not at all obvious; indeed, the presence of $V_{ei}(\mathbf{x})$, which is an infinitely attractive potential, may lead to the presence of unbounded negative energies. This fact leads us to define a regularization for the electron-ion interaction, whenever the electronic coordinates are too close to the positions of the ions:

$$v_{ei}^a(\mathbf{r}_i) = \sum_I \frac{Z_I}{\text{Max}\,(|\mathbf{r}_i - \mathbf{R}_I|, a)}, \qquad (12.49)$$

which is clearly cutting the Coulomb singularity at small distances. Therefore, the electron-ion potential is modified as:

$$V_{ei}^a(\mathbf{x}) = -\sum_i v_{ei}^a(\mathbf{r}_i). \qquad (12.50)$$

It is easy to verify that the regularization of Eq. (12.49) introduces errors in the energy of the same order as those due to the discretization of the Laplacian, as the region where the approximation holds (i.e., $|\mathbf{r} - \mathbf{R}_I| < a$) has a volume that is proportional to a^3 and, in this region, the difference between the original electron-ion potential and the regularized one is on average $O(1/a)$. We also notice that the positive divergence in the electron-electron term $V_{ee}(\mathbf{x})$ does not require a particular attention since an infinite repulsive potential does not affect the existence of a finite ground-state energy.

A natural definition for the lattice regularization would correspond to a simple cubic lattice $\mathbf{r}_{l,m,n} = (la, ma, na)$, where l, m, and n are arbitrary integers. However, this scheme will produce uncontrollable errors as a function of a, since, for not too simple systems, the minimum distance between ion positions and lattice points behaves erratically. In order to improve the convergence for $a \to 0$ and define an efficient computational scheme, we notice that the operator ∇_a^2 is not necessarily defined on a *regular* lattice. For example, let us consider electrons in a periodic super-cell $L \times L \times L$. Then, whenever the lattice spacing a is not commensurate with the box side L (i.e., L/a is an irrational number, or very close to it, as in any computer implementations) ∇_a^2 will produce electron "hoppings" that will ergodically fill the whole continuous space of the periodic super-cell. In general, also with open systems, it is possible to produce the same effect. By using two incommensurate lattice spacings in the definition of the Laplacian, e.g., $\nabla_a^2 \to p\nabla_a^2 + (1-p)\nabla_{a'}^2$ with $0 < p < 1$ and a'/a fixed to an irrational number, it is possible to define $\mathcal{H}_a$ acting in the same continuous space of the original Hamiltonian, with the remarkable advantage to improve the smoothness of the extrapolation for $a \to 0$ (Casula et al., 2005). In the following, we continue to denote this approach as "lattice regularization," even if the Hamiltonian $\mathcal{H}_a$ acts in the continuum. This is because we have taken the most important advantages of the lattice: the possibility to enumerate off-diagonal matrix elements of $\mathcal{H}_a$ (see below) and the introduction of a natural cutoff, which is given by the lattice spacing a.

In summary, the lattice regularization of the original Hamiltonian gives:

$$\mathcal{H}_a = -\frac{1}{2}\nabla_a^2 + V^a(\mathbf{x}) + V_{\text{ii}}, \tag{12.51}$$

where the potential term is given by:

$$V^a(\mathbf{x}) = V_{\text{ee}}(\mathbf{x}) + V_{\text{ei}}^a(\mathbf{x}). \tag{12.52}$$

We stress that $\mathcal{H}_a = \mathcal{H} + O(a^2)$ and the fact that both $\mathcal{H}$ and $\mathcal{H}_a$ act on the *same* Hilbert space, as we have explained before. The most important advantage of this approach is that we can apply to the regularized Hamiltonian $\mathcal{H}_a$ the same techniques that we have discussed for lattice models. Indeed, all these techniques are based on the fact that, by applying the Hamiltonian $\mathcal{H}_a$ to one given configuration $|x\rangle$, we obtain a finite number ($6N_e + 1$) of new configurations:

$$\mathcal{H}_a|x\rangle = \left[V^a(\mathbf{x}) + \frac{3N_e}{a^2}\right]|x\rangle - \frac{1}{2a^2}\sum_{j=1}^{6N_e}|x_j\rangle, \tag{12.53}$$

where the second term in the r.h.s. comes from the fact that, according to Eq. (12.48), we have to change every electronic position by $\pm a$, for each cartesian component; in this case, each individual change will produce a new configuration $|x_j\rangle$.

Then, the fixed-node approximation can be applied to $\mathcal{H}_a$, similarly to the lattice case, see Chapter 10. Here, the fixed-node Hamiltonian $\mathcal{H}_a^{\text{FN},\gamma}$ can be defined in terms of a guiding function $\Psi_G(\mathbf{x})$ that is used to fix the nodal surface of the ground state, see Eq. (10.34). Within this approximation, the off-diagonal matrix elements $\mathcal{H}_a$ that give sign problem are multiplied by $-\gamma$, where γ is an arbitrary positive number, while the diagonal elements are changed according to:

$$V^a(\mathbf{x}) \to V^a(\mathbf{x}) + (1+\gamma)V_{\text{sf}}^a(\mathbf{x}), \tag{12.54}$$

with

$$V_{\text{sf}}^a(\mathbf{x}) = -\frac{1}{2a^2}\sum_{j:s(\mathbf{x},\mathbf{x}_j)<0}\frac{\Psi_G(\mathbf{x}_j)}{\Psi_G(\mathbf{x})}, \tag{12.55}$$

where $s(\mathbf{x}, \mathbf{x}_j) = \Psi_G(\mathbf{x}_j)\Psi_G(\mathbf{x})$. It is interesting to discuss the consequences of this approach for $a \to 0$. Since the condition $\Psi_G(\mathbf{x}_j)\Psi_G(\mathbf{x}) < 0$ is verified only very close (i.e., at distance $\approx a$) to the nodal surface where $\Psi_G(\mathbf{x}) = 0$, the matrix elements of $\mathcal{H}_a^{\text{FN},\gamma}$ will almost always coincide with the matrix elements of $\mathcal{H}_a$. For the same reason, the potential $V_{\text{sf}}^a(\mathbf{x})$ is non-zero only close to the nodal surface. Therefore, we obtain that for $a \to 0$, the fixed-node Hamiltonian will differ from the regularized one by a boundary condition at the nodal surface, where the fixed-node ground state of $\mathcal{H}_a^{\mathrm{FN},\gamma}$ vanishes. Indeed, the potential $V_{\mathrm{sf}}^a(\mathbf{x})$ diverges for $a \to 0$ and wave functions that do not vanish in the nodal surface will have an infinite (positive) expectation value of $\mathcal{H}_a^{\mathrm{FN},\gamma}$ for $a \to 0$. Therefore, in this limit we clearly see that the fixed-node approximation with the lattice regularization is equivalent to a boundary condition and provides the best variational energy compatible with the nodes of the guiding function. Notice that this fact is obtained without using the so-called tilling theorem (Ceperley, 1991), since the fixed-node Monte Carlo sampling on the lattice is always ergodic for $\gamma > 0$ and $a > 0$ and the ground state of $\mathcal{H}_a^{\mathrm{FN},\gamma}$ is defined in all nodal pockets. In particular, the simulation with $\gamma = 1$ is the most efficient one, while the case with $\gamma = 0$ is not ergodic because the nodal pockets cannot be crossed.

Then, the fixed-node approach can be implemented within the Green's function Monte Carlo algorithm; however, since the spectrum of the Hamiltonian is not bounded from above, it is necessary to consider the continuous-time limit, which has been described in section 8.4. In this way, we obtain the approximate ground state, which fulfills the fixed-node constraint given by the guiding function $\Psi_G(\mathbf{x})$, in analogy to the standard diffusion Monte Carlo framework (Casula et al., 2005). We remark that the conventional diffusion Monte Carlo technique has a time-step error in $\Delta\tau$, while the present formulation yields a lattice-space error; both approaches share the same upper-bound property and converge to the same projected fixed-node energy, in the limit $\Delta\tau \to 0$ and $a \to 0$, respectively.

12.6 An Improved Scheme for the Lattice Regularization

Unfortunately, although it is possible to obtain a remarkably smooth convergence with $a \to 0$, this simple approach is not very useful in practice, as the error is often large, implying that an exceedingly small value of a is necessary to obtain accurate results. In order to improve the convergence for $a \to 0$, we should consider an alternative approach, in which the Laplacian is still discretized by using Eq. (12.48) and a different version of the potential operator is considered, i.e., $V(\mathbf{x}) \to V_{\mathrm{zv}}^a(\mathbf{x})$. In practice, inspired from the fixed-node approximation on a lattice, we can modify the potential operator by requiring that the local energy of the original Hamiltonian coincides with that of the regularized one:

$$\frac{\langle x|\mathcal{H}_a|\Psi_G\rangle}{\langle x|\Psi_G\rangle} = \frac{\langle x|\mathcal{H}|\Psi_G\rangle}{\langle x|\Psi_G\rangle}. \tag{12.56}$$

This equality is assumed to be fulfilled for the best guiding function $|\Psi_G\rangle$ (i.e., the one corresponding to the lowest variational energy). This is a useful condition because it satisfies the zero-variance property, also in the presence of the lattice regularization: if the guiding function coincides with an exact eigenstate, then the local energy of $\mathcal{H}_a$ is equal to the exact eigenvalue and the corresponding fixed-node energy will be exact, regardless on the magnitude of a. Then, Eq. (12.56) implies a unique definition of $V_{\text{zv}}^a(\mathbf{x})$. Indeed, if we assume that the Laplacian is discretized by the replacement $\nabla^2 \to \nabla_a^2$, then $V_{\text{zv}}^a(\mathbf{x})$ is given by:

$$V_{\text{zv}}^a(\mathbf{x}) = V(\mathbf{x}) + \frac{\left(\nabla_a^2 - \nabla^2\right)\Psi_G(\mathbf{x})}{2\Psi_G(\mathbf{x})}, \tag{12.57}$$

which gives that $V_{\text{zv}}^a(\mathbf{x}) \to V(\mathbf{x})$ for $a \to 0$. This choice has been originally proposed by Casula et al. (2005). There, the simulations were very stable and the limit $a \to 0$ was extracted in an easy way; unfortunately, the calculations are not generally well defined for fermions, as the Hamiltonian $\mathcal{H}_a$ does not always have a finite ground-state energy for $a > 0$. Indeed, the regularized potential $V_{\text{zv}}^a(\mathbf{x})$ can assume, close to the nodal surface of the guiding function $\Psi_G(\mathbf{x})$, unbounded negative values, which cannot be compensated by the lattice-regularized kinetic energy (whose matrix elements are always finite). Nevertheless, a simulation with a limited amount of computer resources may produce reasonable results, since it is very difficult, with a statistical method, to select the configurations on the nodal surface causing the instability. However, the possibility of having instabilities cannot be excluded *a priori* and it is important to find a solution to this possible catastrophe. A similar scenario is often reported also in the standard diffusion Monte Carlo technique (Foulkes et al., 2001; Umrigar et al., 1993), especially when pseudo-potentials are used. In the latter case, for a given $\Delta\tau$, a general remedy to this catastrophe does not exist yet, although it is widely believed that, for small enough $\Delta\tau$ simulations should be stable with appropriate cutoffs. Within the lattice-regularized approach, the situation is more transparent and a simple remedy can be found, as we will describe in the following.

In order to define a rigorously stable and robust method that works for *any* finite value of a, we need to modify the electron-ion interaction. In particular, we notice that the regularization of Eq. (12.57) just corresponds to the modification of the electron-ion potential by replacing in Eq. (12.45):

$$v_{\text{ei}}(\mathbf{r}_i) \to v_{\text{zv},i}^a(\mathbf{x}), \tag{12.58}$$

where

$$v_{\text{zv},i}^a(\mathbf{x}) = v_{\text{ei}}(\mathbf{r}_i) + \frac{\left(\nabla_{i,a}^2 - \nabla_i^2\right)\Psi_G(\mathbf{x})}{2\Psi_G(\mathbf{x})}. \tag{12.59}$$

Then, with a good guiding function, which satisfies the electron-ion cusp conditions (as discussed in section 12.2), the above replacement cancels out most of the singularities of the attractive potential; however, we can still have unbounded negative values on the nodal surface of the guiding function $\Psi_G(\mathbf{x})$. A safe possibility to protect the simulation from *all* possible instabilities coming from the nodal surface of the guiding function is given by the following modification of the electron-ion potential for each electron i:

$$v^a_{\max,i}(\mathbf{x}) = \text{Max}\left[v^a_{zv,i}(\mathbf{x}), v^a_{ei}(\mathbf{r}_i)\right], \tag{12.60}$$

where $v^a_{ei}(\mathbf{r}_i)$ is defined in Eq. (12.49). Within this formulation, the regularized potential is protected from the possible instabilities close to the nodal surface and guarantees the existence of a finite ground-state energy for any $a > 0$. Since the condition $v^a_{ei}(\mathbf{r}_i) > v^a_{zv,i}(\mathbf{x})$ is often satisfied, this choice can be improved by applying the condition (12.60) only when electrons are very close to the nodal surface. For this purpose, we apply the (single-electron) condition (12.60) only when the electrons cross the nodal surface with the discretized Laplacian:

$$v^a_{\text{opt},i}(\mathbf{x}) = \begin{cases} v^a_{\max,i}(\mathbf{x}) & \text{if } \Psi_G(\mathbf{x})\Psi_G(\mathbf{x}_j) < 0 \text{ for at least one } \mathbf{x}_j, \\ v^a_{zv,i}(\mathbf{x}) & \text{otherwise;} \end{cases} \tag{12.61}$$

here, $\mathbf{x}_j$ indicates the possible configurations, generated by the discretized Laplacian $\nabla^2_{a,i}$ for the electron i. Notice that, for finite values of the lattice spacing a, the above condition could miss the cases where a double sign change occurs. However, this circumstance has a negligible effect for small values of a. In Fig. 12.2, we report the results of the previously defined schemes for the case of 18 equally spaced Hydrogens placed in a one-dimensional chain (where nearest-neighbor atoms are at distance $R = 2a.u.$). Here, the guiding function is given by an optimized Jastrow factor (with a $3s2p1d$ basis) applied to a Slater determinant that is obtained from a density-functional calculation with a $4s2p1d$ basis. The most simple approach, defined by Eq. (12.57) with no further regularization, already gives quite accurate results. By imposing the modification of Eq. (12.60), we obtain much higher energies, suggesting that this kind of scheme is too strict. Finally, very accurate and smooth energies are obtained within the final choice of Eq. (12.61), indicating that this approach gives the best possible approximation for the lattice-regularized diffusion Monte Carlo method.

This scheme for the lattice regularization is perfectly size consistent, whenever the guiding function is also size consistent. In fact, let us consider a wave function $\Psi_G(\mathbf{x})$ that factorizes at large distances:

$$\Psi_G(\mathbf{x}) \approx \Psi^A_G(\mathbf{x}_A)\Psi^B_G(\mathbf{x}_B), \tag{12.62}$$

where we assume that the relevant electronic configurations $\mathbf{x}$ are obtained by N_A (N_B) electrons corresponding to a configuration $\mathbf{x}_A$ ($\mathbf{x}_B$) within the region A (B).

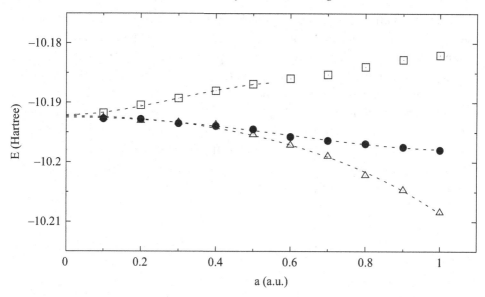

Figure 12.2 Results for different regularization schemes for the total energy of 18 Hydrogens (i.e., 18 electrons) placed in a one-dimensional chain at distance $R = 2a.u.$. The guiding function used here is obtained by applying an optimized Jastrow factor (with a $3s2p1d$ basis) on a Slater determinant obtained by a density-functional calculation with a $4s2p1d$ basis. The simple approximation of Eq. (12.57) (empty triangles), the one given by Eq. (12.60) (empty squares), and the one given by Eq. (12.61) (full circles) are reported. The $a \to 0$ extrapolated energy is $E = -10.1925(2)$ Hartree.

Then, the regularized potential of Eq. (12.59) will act independently in the two regions:

$$v_{zv,i}^a(\mathbf{x}) = \begin{cases} v_{ei}(\mathbf{r}_i) + \dfrac{\left(\nabla_{i,a}^2 - \nabla_i^2\right) \Psi_G^A(\mathbf{x}_A)}{2\Psi_G^A(\mathbf{x}_A)} & \text{if } \mathbf{r}_i \in A, \\[1em] v_{ei}(\mathbf{r}_i) + \dfrac{\left(\nabla_{i,a}^2 - \nabla_i^2\right) \Psi_G^B(\mathbf{x}_B)}{2\Psi_G^B(\mathbf{x}_B)} & \text{if } \mathbf{r}_i \in B, \end{cases} \qquad (12.63)$$

which lead to:

$$v_{max,i}^a(\mathbf{x}) = v_{max,A,i}^a(\mathbf{x}_A) + v_{max,B,i}^a(\mathbf{x}_B). \qquad (12.64)$$

This fact explains the reason why we have done a regularization that considers each single-electron contribution independently, see Eq. (12.60). By contrast, a regularization that takes $V_{max}^a(\mathbf{x}) = \text{Max}\left[V_{zv}^a(\mathbf{x}), V^a(\mathbf{x})\right]$, where $V^a(\mathbf{x})$ is given by Eq. (12.50), is not size consistent, as the global relation implied by the Max operation unavoidably couples the two regions A and B, even when they are at large distance.

Finally, the lattice-regularized approach can be naturally extended to the use of pseudo-potentials, which are usually considered to remove the large energy scales coming from the core electrons. The pseudo-potentials have terms that simply add to $v_{ei}(\mathbf{r}_i)$ with a non-local contribution, which has a natural description by means of lattice operators. Indeed, they can be safely considered as further hopping terms to be included in the regularized Laplacian ∇_a^2. Therefore, it is straightforward to apply the fixed-node scheme described before, maintaining its upper bound property for the energy. Moreover, the use of soft pseudo-potentials (that cancel the electron-ion singularity) leads to a smooth local potential that does not need to be regularized, so that the regularization of Eq. (12.49) does not have to be applied.

Appendix

Pseudo-Random Numbers Generated
by Computers

In most cases, the random numbers that are produced by computers are not really random but *pseudo-random*, which means that they are generated in a predictable way by using precise deterministic algorithms. Indeed, even though there are algorithms that are purely random (based upon some atmospheric noise in the computer), in many extents it is much better to have codes that determine reproducible calculations. Indeed, debugging is facilitated by the ability to run the same sequence of random numbers. Moreover, random-number generators that are based upon some thermal noise in the computer are usually much slower than deterministic algorithms, thus giving another reason to avoid them. All routines that produce pseudo-random numbers are just complicated algorithms that condense into few hundreds lines a lot of clever tricks of number theory. The output of these algorithms looks random but it is not. Indeed, when we run them twice (under the same initial conditions), they always give the same output. In practice, these algorithms generate a sequence of numbers $r_0, r_1, r_2, \ldots$ in which, at each iteration, r_k is given by some complicated function of the preceding ones. The simplest example is given by linear congruential generators, which are of the form:

$$r_k = f(r_{k-1}). \tag{A.1}$$

A simple choice (that should not be used anymore) is given by:

$$i_k = (ai_{k-1} + b) \mod M, \tag{A.2}$$

where a and b are suitable integers (the quality of the generator crucially depends upon them) and $M = 2^{31} - 1$ (for 32-bit generators). A real number in $[0, 1)$ is then obtained by taking $r_k = i_k/M$. Starting from a given *seed* (here, the initial value for i_0), the full chain of pseudo-random numbers is iteratively generated. Here, the main problem is that the sequence repeats identically after (at most) $2^{31} - 1$ iterations. In this sense, the lagged Fibonacci generators give much better results:

$$i_k = i_{k-p} \bigotimes i_{k-q} \quad \mod M, \tag{A.3}$$

where $\bigotimes$ indicates a generic binary operation, and p and q are fixed integers. Depending on the chosen binary operation and the integers p and q, we can achieve generators with incredibly long periodicities, e.g., as large as 10^{400} or even longer.

When dealing with pseudo-random numbers, we must keep in mind that subsequent numbers are always correlated, since they come from a deterministic procedure. However, the important point is that they must look nearly random when the algorithm is not known, at least on a "coarse grained" level. More precisely, the conditional probability $\omega(i_k | i_{k-1})$ should be as flat as possible, indicating that the probability to have a given i_k should be independent on the fact of having obtained i_{k-1} at the previous step. This is clearly false for an algorithm like Eq. (A.2), where the outcome at the step $k - 1$ completely determines the outcome at the next step. However, once we consider a coarse graining of the interval $[0, 1)$, a constant $\omega(i_k | i_{k-1})$ may emerge. Moreover, for the lagged Fibonacci generators,

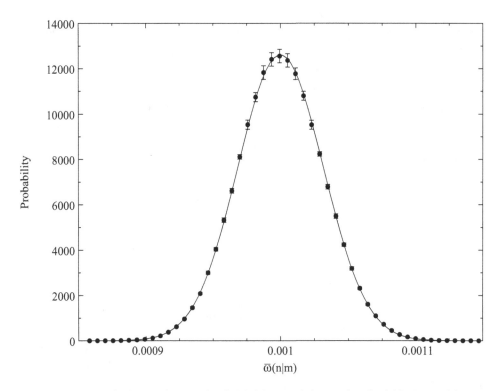

Figure A.1 The interval $[0, 1)$ is divided into sub-intervals of width $1/L$, with $L = 1000$. The conditional probability $\overline{\omega}(n|m)$ that two subsequent numbers fall in the sub-intervals $[(n-1)/L, n/L]$ and $[(m-1)/L, m/L]$, respectively, is evaluated and averaged over $M = 10^9$ samples. The distribution function of $\overline{\omega}(n|m)$ is shown.

whenever the period is much longer than $2^{31} - 1 \approx 2 \times 10^9$, the same i_{k-1} will give rise to several different values for i_k along the sequence, leading to a practically uniform probability. In order to exemplify this concept, we divide the interval $[0, 1)$ into L small sub-intervals of width $1/L$, generate M pseudo-random numbers, and calculate the number of times $N(n|m)$ that a number falls in $[(n-1)/L, n/L]$ when the previous one was in $[(m-1)/L, m/L]$. Then, the estimated conditional probability is obtained as $\overline{\omega}(n|m) = N(n|m)/\sum_n N(n|m)$, which is also a pseudo-random number. For a good generator, the probability distribution of $\overline{\omega}(n|m)$ is a Gaussian with mean value equal to $1/L$ and variance $1/M$, as expected from the central limit theorem (the mean of $\overline{\omega}(n|m)$ is $1/L$ and its variance is $1/L + O(1/L^2)$; the number of samplings for each m is $O(M/L)$, on average). In Fig. A.1, we report the case with $L = 1000$ and $M = 10^9$, where the distribution function of $\overline{\omega}(n|m)$ is perfectly fitted by a Gaussian.

In general, random-number generators must satisfy two main requirements (Knuth, 1997):

1. To have a long period. Indeed, every deterministic generator must eventually loop and the goal is to make the period as long as possible.
2. To have a good statistical quality. In this respect, the output of the generator should be practically indistinguishable from the one of a true random-number generator and it should not exhibit any correlation along the sequence of numbers.

A poor quality of the generator can ruin the results of Monte Carlo applications (Ferrenberg et al., 1992); for this reason, it is very important that generators are able to pass the set of empirical tests, as for example the battery of *Diehard Tests*, developed by George Marsaglia, or the *TestU01 Library*, introduced by Pierre L'Ecuyer and Richard Simard.

References

Allen, M. P., and Tildesley, D. J. 1987. *Computer Simulation of Liquids*. Oxford University Press.

Anderson, H. 1986. Metropolis, Monte Carlo and the MANIAC. *Los Alamos Science*, **14**, 96–108.

Anderson, J. B. 1975. Random walk simulation of the Schrödinger equation: He_3^+. *J. Chem. Phys.*, **63**, 1499–1503.

Anderson, J. B. 1976. Quantum chemistry by random walk. *J. Chem. Phys.*, **65**, 4121–4127.

Anderson, P. W. 1987. The resonating valence bond state in La_2CuO_4 and superconductivity. *Science*, **235**, 1196–1198.

Anderson, P. W., Baskaran, G., Zou, Z., and Hsu, T. 1987. Resonating-valence-bond theory of phase transitions and superconductivity in La_2CuO_4-based compounds. *Phys. Rev. Lett.*, **58**, 2790–2793.

Aoki, H., Tsuji, N., Eckstein, M., Kollar, M., Oka, T., and Werner, P. 2014. Non-equilibrium dynamical mean-field theory and its applications. *Rev. Mod. Phys.*, **86**, 779–837.

Arovas, D., Schrieffer, J. R., and Wilczek, F. 1984. Fractional statistics and the quantum Hall effect. *Phys. Rev. Lett.*, **53**, 722–723.

Bajdich, M., Mitas, L., Drobny, G., Wagner, L. K., and Schmidt, K. E. 2006. Pfaffian pairing wave functions in electronic-structure quantum Monte Carlo simulations. *Phys. Rev. Lett.*, **96**, 130201.

Bajdich, M., Mitas, L., Wagner, L. K., and Schmidt, K. E. 2008. Pfaffian pairing and backflow wave functions for electronic-structure quantum Monte Carlo methods. *Phys. Rev. B*, **77**, 115112.

Baldereschi, A. 1973. Mean-value point in the Brillouin zone. *Phys. Rev. B*, **7**, 5212–5215.

Bardeen, J., Cooper, L. N., and Schrieffer, J. R. 1957. Theory of superconductivity. *Phys. Rev.*, **108**, 1175–1204.

Baroni, S., and Moroni, S. 1998. Reptation quantum Monte Carlo. *arXiv:cond-mat/9808213*.

Bartlett, R. J. 1981. Many-body perturbation theory and coupled cluster theory for electron correlation in molecules. *Ann. Rev. Phys. Chem.*, **32**, 359–401.

Baskaran, G., and Anderson, P. W. 1988. Gauge theory of high-temperature superconductors and strongly correlated Fermi systems. *Phys. Rev. B*, **37**, 580–583.

Bethe, H. 1931. Zur Theorie der metalle. I. eigenwerte und eigenfunktionen der linearen atomkette. *Z. Phys.*, **71**, 205–226.

Bouchaud, J. P., Georges, A., and Lhuillier, C. 1988. Pair wave functions for strongly correlated fermions and their determinantal representation. *J. Phys. (Paris)*, **49**, 553–559.

Calandra Buonaura, M., and Sorella, S. 1998. Numerical study of the two-dimensional Heisenberg model using a Green function Monte Carlo technique with a fixed number of walkers. *Phys. Rev. B*, **57**, 11446–11456.

Capello, M., Becca, F., Fabrizio, M., Sorella, S., and Tosatti, E. 2005. Variational description of Mott insulators. *Phys. Rev. Lett.*, **94**, 026406.

Capello, M., Becca, F., Yunoki, S., and Sorella, S. 2006. Unconventional metal-insulator transition in two dimensions. *Phys. Rev. B*, **73**, 245116.

Capello, M., Becca, F., Fabrizio, M., and Sorella, S. 2007. Superfluid to Mott-insulator transition in Bose-Hubbard models. *Phys. Rev. Lett.*, **99**, 056402.

Capello, M., Becca, F., Fabrizio, M., and Sorella, S. 2008. Mott transition in bosonic systems: insights from the variational approach. *Phys. Rev. B*, **77**, 144517.

Capriotti, L., Becca, F., Parola, A., and Sorella, S. 2001. Resonating valence bond wave functions for strongly frustrated spin systems. *Phys. Rev. Lett.*, **87**, 097201.

Carleo, G., Becca, F., Schiro, M., and Fabrizio, M. 2012. Localization and glassy dynamics of many-body quantum system. *Sci. Rep.*, **2**, 243.

Carleo, G., Becca, F., Sanchez-Palencia, L., Sorella, S., and Fabrizio, M. 2014. Light-cone effect and supersonic correlations in one- and two-dimensional bosonic superfluids. *Phys. Rev. A*, **89**, 031602.

Casula, M., Filippi, C., and Sorella, S. 2005. Diffusion Monte Carlo method with lattice regularization. *Phys. Rev. Lett.*, **95**, 100201.

Ceperley, D., Chester, G. V., and Kalos, M.H. 1977. Monte Carlo simulation of a many-fermion study. *Phys. Rev. B*, **16**, 3081–3099.

Ceperley, D. M. 1991. Fermion nodes. *J. Stat. Phys.*, **63**, 1237–1267.

Ceperley, D. M., and Alder, B. J. 1980. Ground state of the electron gas by a stochastic method. *Phys. Rev. Lett.*, **45**, 566–569.

Corboz, P., Rice, T. M., and Troyer, M. 2014. Competing states in the $t-J$ model: uniform d-wave state versus stripe state. *Phys. Rev. Lett.*, **113**, 046402.

Daley, A. J., Kollath, C., Schollwöck, U., and Vidal, G. 2004. Time-dependent density-matrix renormalization-group using adaptive effective Hilbert spaces. *J. Stat. Mech.: Theor. Exp.*, P04005.

Dirac, P. A. M. 1930. Note on exchange phenomena in the Thomas atom. *Math. Proc. Cambridge Philos. Soc.*, **26**, 376–385.

Dongarra, J., and Sullivan, F. 2000. Guest editors' introduction: the top 10 algorithms. *Comput. Sci Eng.*, **2**, 22–23.

Dovesi, R., Orlando, R., Erba, A., Zicovich-Wilson, C. M., Civalleri, B., Casassa, S., Maschio, L., Ferrabone, M., De La Pierre, M., D'Arco, P., Noël, Y., Causà, M., Rérat, M., and Kirtman, B. 2014. CRYSTAL14: a program for the ab initio investigation of crystalline solids. *Int. J. Quantum Chem.*, **114**, 1287–1317.

Eichenberger, D., and Baeriswyl, D. 2007. Superconductivity and antiferromagnetism in the two-dimensional Hubbard model: a variational study. *Phys. Rev. B*, **76**, 180504.

Eisert, J., Cramer, M., and Plenio, M. B. 2010. Area laws for the entanglement entropy. *Rev. Mod. Phys.*, **82**, 277–306.

Faddeev, L. D., and Takhtajan, L. A. 1981. What is the spin of a spin wave? *Phys. Lett. A*, **85**, 375–377.

Fahy, S., and Hamann, D. R. 1991. Diffusive behavior of states in the Hubbard-Stratonovich transformation. *Phys. Rev. B*, **43**, 765–779.

Fano, G., Ortolani, F., and Colombo, E. 1986. Configuration-interaction calculations on the fractional quantum Hall effect. *Phys. Rev. B*, **34**, 2670–2680.

Fazekas, P. 1999. *Lecture Notes on Electron Correlation and Magnetism*. World Scientific.

Fazekas, P., and Anderson, P. W. 1974. On the ground state properties of the anisotropic triangular antiferromagnet. *Phil. Mag.*, **30**, 423–440.

Ferrenberg, A. M., Landau, D. P., and Wong, Y. J. 1992. Monte Carlo simulations: Hidden errors from "good" random number generators. *Phys. Rev. Lett.*, **69**, 3382–3384.

Fetter, A. L., and Walecka, J. D. 2003. *Quantum Theory of Many-Particle Systems*. Dover Publications Inc.

Feynman, R. P. 1954. Atomic theory of the two-fluid model of liquid Helium. *Phys. Rev.*, **94**, 262–277.

Feynman, R. P., and Cohen, M. 1956. Energy spectrum of the excitations in liquid Helium. *Phys. Rev.*, **102**, 1189–1204.

Filippi, C., and Umrigar, C. 1996. Multiconfiguration wave functions for quantum Monte Carlo calculations of first-row diatomic molecules. *J. Chem. Phys.*, **105**, 213–226.

Foulkes, W. M. C., Mitas, L., Needs, R. J., and Rajagopal, G. 2001. Quantum Monte Carlo simulations of solids. *Rev. Mod. Phys.*, **73**, 33–83.

Gnedenko, B. V. 2014. *The Theory of Probability*. Martino Fine Books.

Gnedenko, B. V., and Kolmogorov, A. N. 1954. *Limit Distributions for Sums of Independent Random Variables*. Addison-Wesley.

Gros, C. 1988. Superconductivity in correlated wave functions. *Phys. Rev. B*, **38**, 931–934.

Gros, C. 1989. Physics of projected wave functions. *Ann. Phys.*, **189**, 53–88.

Gros, C., Joynt, R., and Rice, T. M. 1987. Antiferromagnetic correlations in almost-localized Fermi liquids. *Phys. Rev. B*, **36**, 381–393.

Gubernatis, J., Kawashima, N., and Werner, P. 2016. *Quantum Monte Carlo methods: algorithms for lattice models*. Cambridge University Press.

Gutzwiller, M. C. 1963. Effect of correlation on the ferromagnetism of transition metals. *Phys. Rev. Lett.*, **10**, 159–162.

Haldane, F. D. M. 1983. Fractional quantization of the Hall effect: a hierarchy of incompressible quantum fluid states. *Phys. Rev. Lett.*, **51**, 605–608.

Haldane, F. D. M. 1988. Exact Jastrow-Gutzwiller resonating-valence-bond ground state of the spin-1/2 antiferromagnetic Heisenberg chain with $1/r^2$ exchange. *Phys. Rev. Lett.*, **60**, 635–638.

Haldane, F. D. M. 1991. "Spinon gas" description of the $S = 1/2$ Heisenberg chain with inverse-square exchange: exact spectrum and thermodynamics. *Phys. Rev. Lett.*, **66**, 1529–1532.

Haldane, F. D. M., and Rezayi, E. H. 1985. Periodic Laughlin-Jastrow wave functions for the fractional quantized Hall effect. *Phys. Rev. B*, **60**, 2529–2531.

Harju, A., Barbiellini, B., Siljamäki, S., Nieminen, R. M., and Ortiz, G. 1997. Stochastic gradient approximation: an efficient method to optimize many-body wave functions. *Phys. Rev. Lett.*, **79**, 1173–1177.

Hastings, W. K. 1970. Monte Carlo sampling methods using Markov chains and their applications. *Biometrika*, **57**, 97–109.

Heitler, W., and London, F. 1927. Wechselwirkung neutraler atome und homöopolare bindung nach der quantenmechanik. *Z. Phys.*, **44**, 455–472.

Hirsch, J. E. 1985. Two-dimensional Hubbard model: numerical simulation study. *Phys. Rev. B*, **31**, 4403–4419.

Hu, W.-J., Becca, F., and Sorella, S. 2012. Absence of static stripes in the two-dimensional $t–J$ model determined using an accurate and systematic quantum Monte Carlo approach. *Phys. Rev. B*, **85**, 081110.

Hubbard, J. 1959. Calculation of partition functions. *Phys. Rev. Lett.*, **3**, 77–78.

Hubbard, J. 1963. Electron correlations in narrow energy bands. *Proc. Royal Soc. of London*, **276**, 238–257.

Ido, K., Ohgoe, T., and Imada, M. 2015. Time-dependent many-variable variational Monte Carlo method for non-equilibrium strongly correlated electron systems. *Phys. Rev. B*, **92**, 245106.

Imada, M., Fujimori, A., and Tokura, Y. 1998. Metal-insulator transitions. *Rev. Mod. Phys.*, **70**, 1039–1263.

Iqbal, Y., Becca, F., and Poilblanc, D. 2011. Projected wave function study of Z_2 spin liquids on the kagome lattice for the spin-1/2 quantum Heisenberg antiferromagnet. *Phys. Rev. B*, **84**, 020407.

Jain, J. K. 2012. *Composite Fermions*. Cambridge University Press.

Jastrow, R. 1955. Many-body problem with strong forces. *Phys. Rev.*, **98**, 1479–1484.

Kalos, M. H., Levesque, D., and Verlet, L. 1974. Helium at zero temperature with hard-sphere and other forces. *Phys. Rev. A*, **9**, 2178–2195.

Kanamori, J. 1963. Electron correlation and ferromagnetism of transition metals. *Prog. Theor. Phys.*, **30**, 275–289.

Kaneko, R., Tocchio, L. F., Valentí, R., Becca, F., and Gros, C. 2016. Spontaneous symmetry breaking in correlated wave functions. *Phys. Rev. B*, **93**, 125127.

Kim, J., Esler, K. P., McMinis, J., Morales, M. A., Clark, B. K., Shulenburger, L., and D. M., Ceperley. 2012. Hybrid algorithms in quantum Monte Carlo. *J. Phys.: Conf. Ser.*, **402**, 012008.

Kivelson, S. A., Rokhsar, D. S., and Sethna, J. P. 1987. Topology of the resonating valence-bond state: solitons and high-T_c superconductivity. *Phys. Rev. B*, **35**, 8865–8868.

Knuth, D. 1997. *The Art of Computer Programming*. Addison-Wesley.

Krauth, W. 2006. *Statistical Mechanics: Algorithms and Computations*. Oxford University Press.

Krauth, W., Caffarel, M., and Bouchaud, J. 1992. Gutzwiller wave function for a model of strongly interacting bosons. *Phys. Rev. B*, **45**, 3137–3140.

Kwon, Y., Ceperley, D. M., and Martin, R. M. 1993. Effects of three-body and backflow correlations in the two-dimensional electron gas. *Phys. Rev. B*, **48**, 12037–12046.

Kwon, Y., Ceperley, D. M., and Martin, R. M. 1998. Effects of backflow correlation in the three-dimensional electron gas: quantum Monte Carlo study. *Phys. Rev. B*, **58**, 6800–6806.

Laughlin, R. B. 1983. Anomalous quantum Hall effect: an incompressible quantum fluid with fractionally charged excitations. *Phys. Rev. Lett.*, **50**, 1395–1398.

Lee, P. A., Nagaosa, N., and Wen, X.-G. 2006. Doping a Mott insulator: physics of high-temperature superconductivity. *Rev. Mod. Phys.*, **78**, 17–85.

Liang, S., Doucot, B., and Anderson, P. W. 1988. Some new variational resonating-valence-bond-type wave functions for the spin-1/2 antiferromagnetic Heisenberg model on a square lattice. *Phys. Rev. Lett.*, **61**, 365–368.

Lieb, E. H., and Wu, F. Y. 1968. Absence of Mott transition in an exact solution of the short-range, one-band model in one dimension. *Phys. Rev. Lett.*, **20**, 1445–1448.

Lin, C., Zong, F. H., and Ceperley, D. M. 2001. Twist-averaged boundary conditions in continuum quantum Monte Carlo algorithms. *Phys. Rev. E*, **64**, 016702.

Loh, E. Y., Gubernatis, J. E., Scalettar, R. T., White, S. R., Scalapino, D. J., and Sugar, R. L. 1990. Sign problem in the numerical simulation of many-electron systems. *Phys. Rev. B*, **41**, 9301–9307.

Marchi, M., Azadi, S., Casula, M., and Sorella, S. 2009. Resonating valence bond wave function with molecular orbitals: Application to first-row molecules. *J. Chem. Phys.*, **131**, 154116.

Marshall, W. 1955. Antiferromagnetism. *Proc. R. Soc. London Ser.*, **A 232**, 48–68.

Martin, R. M. 2004. *Electronic Structure: Basic Theory and Practical Methods*. Cambridge University Press.

Martin, R.M., Reining, L., and Ceperley, D.M. 2016. *Interacting Electrons: Theory and Computational Approaches*. Cambridge University Press.

Mazzola, G., and Sorella, S. 2017. Accelerating *ab initio* molecular dynamics and probing the weak dispersive forces in dense liquid Hydrogen. *Phys. Rev. Lett.*, **118**, 015703.

McMillan, W. L. 1965. Ground State of Liquid He^4. *Phys. Rev.*, **138**, A442–A451.

Metropolis, N. 1987. The beginning of the Monte Carlo method. *Los Alamos Science*, **15**, 125–130.

Metropolis, N., and Ulam, S. 1949. The Monte Carlo method. *J. Am. Stat. Ass.*, **44**, 335–341.

Metropolis, N., Rosenbluth, A., Rosenbluth, M., Teller, A., and Teller, E. 1957. Equations of state calculations by fast computing machines. *J. Chem. Phys.*, **21**, 1087–1092.

Meyer, C. D. 2000. *Matrix Analysis and Applied Linear Algebra*. SIAM.

Moore, G., and Read, N. 1991. Nonabelions in the fractional quantum Hall effect. *Nucl. Phys. B*, **360**, 362–396.

Moroni, S., and Baroni, S. 1999. Reptation quantum Monte Carlo: A method for unbiased ground-state averages and imaginary-time correlations. *Phys. Rev. Lett.*, **82**, 4745–4748.

Moskowitz, J. W., Schmidt, K. E., Lee, M. A., and Kalos, M. H. 1982. A new look at correlation energy in atomic and molecular systems. II: the application of the Green's function Monte Carlo method to LiH. *J. Chem. Phys.*, **77**, 349–355.

Mott, N. F. 1949. The basis of the electron theory of metals, with special reference to the transition metals. *Proc. Phys. Soc. (London)*, **62**, 416–422.

Mott, N. F. 1990. *Metal-Insulator Transitions*. Taylor and Francis.

Nagaoka, Y. 1966. Ferromagnetism in a narrow, almost half-filled s band. *Phys. Rev.*, **147**, 392–405.

Needs, R. J., Towler, M. D., Drummond, N. D., and López Ríos, P. 2010. Continuum variational and diffusion quantum Monte Carlo calculations. *J. Phys.: Condens. Matter*, **22**, 023201.

Nelson, E. 1966. Derivation of the Schrödinger equation from Newtonian mechanics. *Phys. Rev.*, **150**, 1079–1085.

Neuscamman, E. 2012. Size consistency error in the antisymmetric geminal power wave function can be completely removed. *Phys. Rev. Lett.*, **109**, 203001.

Neuscamman, E., Umrigar, C. J., and Chan, G.K.-L. 2012. Optimizing large parameter sets in variational quantum Monte Carlo. *Phys. Rev. B*, **85**, 045103.

Nightingale, M. P., and Melik-Alaverdian, V. 2001. Optimization of ground- and excited-state wave functions and van der Waals clusters. *Phys. Rev. Lett.*, **87**, 043401.

Norris, J. R. 1997. *Markov Chains*. Cambridge Series in Statistical and Probabilistic Mathematics.

Nozieres, P. 1964. *Theory of Interacting Fermi Systems*. New York: W.A. Benjamin.

Ortiz, G., Ceperley, D. M., and Martin, R. M. 1993. New stochastic method for systems with broken time-reversal symmetry: 2D fermions in a magnetic field. *Phys. Rev. Lett.*, **71**, 2777–2780.

Oshikawa, M., and Senthil, T. 2006. Fractionalization, topological order, and quasiparticle statistics. *Phys. Rev. Lett.*, **96**, 060601.

Pandharipande, V. R., and Itoh, N. 1973. Effective mass of ^{3}He in liquid ^{4}He. *Phys. Rev. A*, **8**, 2564–2566.

Parisi, G. 1984. Prolegomena to any future computer evaluation of the QCD mass spectrum. Pages 531–541 of: *Progress in Gauge Field Theory*. NATO ASI Series, vol. 115. Springer.

Parisi, G., and Wu, Y.-S. 1981. Perturbation theory without gauge fixing. *Sci. Sinica*, **24**, 483–496.

Pauling, L. 1960. *The Nature of the Chemical Bond*. Cornell University Press.

Pierleoni, C., and Ceperley, D. M. 2005. Computational methods in coupled electronion Monte Carlo simulations. *ChemPhysChem*, **6**, 1872–1878.

Pitaevskii, L., and Stringari, S. 1991. Uncertainly principle, quantum fluctuations, and broken symmetries. *J. Low Temp. Phys.*, **85**, 377–388.

Press, W. H., Teukolsky, S. A., Vetterling, W. T., and Flannery, B. P. 2007. *Numerical Recipes 3rd edition: The Art of Scientific Computing*. Cambridge University Press.

Read, N., and Chakraborty, B. 1989. Statistics of the excitations of the resonating-valence-bond state. *Phys. Rev. B*, **40**, 7133–7140.

Read, N., and Rezayi, E. 1999. Beyond paired quantum Hall states: Parafermions and incompressible states in the first excited Landau level. *Phys. Rev. B*, **59**, 8084–8092.

Reger, J. D., and Young, A. P. 1988. Monte Carlo simulations of the spin-1/2 Heisenberg antiferromagnet on a square lattice. *Phys. Rev. B*, **37**, 5978–5981.

Reynolds, P. J., Ceperley, D. M., Alder, B. J., and Lester, Jr. 1982. Fixednode quantum Monte Carlo for molecules. *J. Chem. Phys.*, **77**, 5593–5603.

Ring, P., and Schuck, P. 2004. *The Nuclear Many-body Problem*. Springer-Verlag.

Rokhsar, D. S., and Kotliar, B. G. 1991. Gutzwiller projection for bosons. *Phys. Rev. B*, **44**, 10328–10332.

Schmidt, K. E., and Pandharipande, V. R. 1979. New variational wave function for liquid ^{3}He. *Phys. Rev. B*, **19**, 2504–2519.

Schollwöck, U. 2005. The density-matrix renormalization group. *Rev. Mod. Phys.*, **77**, 259–315.

Schollwöck, U. 2011. The density-matrix renormalization group in the age of matrix product states. *Ann. Phys.*, **326**, 96–192.

Schrieffer, J. R. 1964. *Theory of Superconductivity*. New York: W.A. Benjamin.

Senthil, T., and Fisher, M. P. A. 2000. Z_2 gauge theory of electron fractionalization in strongly correlated systems. *Phys. Rev. B*, **62**, 7850–7881.

Shastry, B. S. 1988. Exact solution of an $S = 1/2$ Heisenberg antiferromagnetic chain with long-ranged interactions. *Phys. Rev. Lett.*, **60**, 639–642.

Slater, J. C. 1930. The electronic structure of metals. *Rev. Mod. Phys.*, **6**, 209–280.

Sorella, S. 1998. Green function Monte Carlo with stochastic reconfiguration. *Phys. Rev. Lett.*, **80**, 4558–4561.

Sorella, S. 2001. Generalized Lanczos algorithm for variational quantum Monte Carlo. *Phys. Rev. B*, **64**, 024512.

Sorella, S. 2002. Effective Hamiltonian approach for strongly correlated lattice models. *arXiv:cond-mat/0201388*.

Sorella, S. 2005. Wave function optimization in the variational Monte Carlo method. *Phys. Rev. B*, **71**, 241103.

Sorella, S., Baroni, S., Car, R., and Parrinello, M. 1989. A novel technique for the simulation of interacting fermion systems. *Europhys. Lett.*, **8**, 663–668.

Sorella, S., Martins, G. B., Becca, F., Gazza, C., Capriotti, L., Parola, A., and Dagotto, E. 2002. Superconductivity in the two-dimensional $t-J$ model. *Phys. Rev. Lett.*, **88**, 117002.

Sorella, S., Casula, M., and Rocca, D. 2007. Weak binding between two aromatic rings: Feeling the van der Waals attraction by quantum Monte Carlo methods. *J. Chem. Phys.*, **127**, 014105.

Sorella, S., Devaux, N., Dagrada, M., Mazzola, G., and Casula, M. 2015. Geminal embedding scheme for optimal atomic basis set construction in correlated calculations. *J. Chem. Phys.*, **143**, 244112.

Stewart, G. R. 1984. Heavy-fermion systems. *Rev. Mod. Phys.*, **56**, 755–787.

Stratonovich, R. L. 1957. A method for the computation of quantum distribution functions. *Doklady Akad. Nauk S.S.S.R.*, **115**, 1097–1100.

Sutherland, B. 1971. Exact results for a quantum many-body problem in one dimension. *Phys. Rev. A*, **4**, 2019–2021.

Sutherland, B. 1975. Model for a multicomponent quantum system. *Phys. Rev. B*, **12**, 3795–3805.

Suzuki, M. 1976a. Generalized Trotter's formula and systematic approximants of exponential operators and inner derivations with applications to many-body problems. *Commun. Math. Phys.*, **51**, 183–190.

Suzuki, M. 1976b. Relationship between d-dimensional quantal spin systems and $(d + 1)$-dimensional Ising systems. *Prog. Theor. Phys.*, **56**, 1454–1469.

Szabo, A., and Ostlund, N. S. 1996. *Modern Quantum Chemistry: Introduction to Advanced Electronic Structure Theory*. Dover Publications Inc.

Tasaki, H. 1998. From Nagaoka's ferromagnetism to flat-band ferromagnetism and beyond. *Prog. Theor. Phys.*, **99**, 489–548.

ten Haaf, D. F. B., van Bemmel, H. J. M., van Leeuwen, J. M. J., van Saarloos, W., and Ceperley, D. M. 1995. Proof of upper bound in fixed-node Monte Carlo for lattice fermions. *Phys. Rev. B*, **51**, 13039–13045.

Tocchio, L. F., Becca, F., Parola, A., and Sorella, S. 2008. Role of backflow correlations for the nonmagnetic phase of the t–t' Hubbard model. *Phys. Rev. B*, **78**, 041101.

Tocchio, L. F., Becca, F., and Gros, C. 2011. Backflow correlations in the Hubbard model: An efficient tool for the study of the metal-insulator transition and the large-U limit. *Phys. Rev. B*, **83**, 195138.

Toulouse, J., and Umrigar, C. J. 2007. Optimization of quantum Monte Carlo wave functions by energy minimization. *J. Chem. Phys.*, **126**, 084102.

Trivedi, N., and Ceperley, D. M. 1989. Green-function Monte Carlo study of quantum antiferromagnets. *Phys. Rev. B*, **40**, 2737–2740.

Trivedi, N., and Ceperley, D. M. 1990. Ground-state correlations of quantum antiferromagnets: a Green-function Monte Carlo study. *Phys. Rev. B*, **41**, 4552–4569.

Trotter, H. F. 1959. On the product of semi-groups of operators. *Proc. Am. Math. Soc.*, **10**, 545–551.

Tuckerman, M. E. 2010. *Statistical Mechanics: Theory and Molecular Simulation*. Oxford University Press.

Umrigar, C. J., and Filippi, C. 2005. Energy and variance optimization of many-body wave functions. *Phys. Rev. Lett.*, **94**, 150201.

Umrigar, C. J., Wilson, K. G., and Wilkins, J. W. 1988. Optimized trial wave functions for quantum Monte Carlo calculations. *Phys. Rev. Lett.*, **60**, 1719–1722.

Umrigar, C. J., Nightingale, M. P., and Runge, K. J. 1993. A diffusion Monte Carlo algorithm with very small timestep errors. *J. Chem. Phys.*, **99**, 2865–2890.

Umrigar, C. J., Toulouse, J., Filippi, C., Sorella, S., and Hennig, R. G. 2007. Alleviation of the fermion-sign problem by optimization of many-body wave functions. *Phys. Rev. Lett.*, **98**, 110201.

Wagner, L. K., Bajdich, M., and Mitas, L. 2009. QWalk: A quantum Monte Carlo program for electronic structure. *J. Comp. Phys.*, **228**, 3390–3404.

Wen, X.-G. 1991. Topological orders and Chern-Simons theory in strongly correlated quantum liquid. *Int. J. Mod. Phys. B*, **5**, 1641–1648.

Wen, X.-G., and Niu, Q. 1990. Ground-state degeneracy of the fractional quantum Hall states in the presence of a random potential and on high-genus Riemann surfaces. *Phys. Rev. B*, **41**, 9377–9396.

White, S. R. 1992. Density matrix formulation for quantum renormalization groups. *Phys. Rev. Lett.*, **69**, 2863–2866.

White, S. R, and Feiguin, A. 2004. Real-time evolution using the density matrix renormalization group. *Phys. Rev. Lett.*, **93**, 076401.

White, S. R., and Scalapino, D.J. 1998. Density matrix renormalization group study of the striped phase in the 2D $t-J$ model. *Phys. Rev. Lett.*, **80**, 1272–1275.

Wigner, E., and Seitz, F. 1934. On the constitution of metallic Sodium. II. *Phys. Rev.*, **46**, 509–524.

Yokoyama, H., and Shiba, H. 1987a. Variational Monte Carlo studies of Hubbard model. I. *J. Phys. Soc. Jpn.*, **56**, 1490–1506.

Yokoyama, H., and Shiba, H. 1987b. Variational Monte Carlo studies of Hubbard model. II. *J. Phys. Soc. Jpn.*, **56**, 3582–3592.

Yokoyama, H., and Shiba, H. 1990. Variational Monte Carlo studies of Hubbard model. III. Intersite correlation effects. *J. Phys. Soc. Jpn.*, **59**, 3669–3686.

Young, P. 2012. Everything you wanted to know about data analysis and fitting but were afraid to ask. *arXiv:1210.3781*.

Zhang, F. C., and Rice, T. M. 1988. Effective Hamiltonian for the superconducting Cu oxides. *Phys. Rev. B*, **37**, 3759–3761.

Zhang, S., Carlson, J., and Gubernatis, J. E. 1995. Constrained path quantum Monte Carlo method for fermion ground states. *Phys. Rev. Lett.*, **74**, 3652–3655.

Index

Acceptance-rejection method, 62
Anti-symmetric geminal power
    Constrained optimization, 246
    Definition, 237
Area law, 31
Atomic orbitals, 238
    Contracted, 241
    Gaussian type, 240
    Periodic systems, 241
    Slater type, 240
    Twisted boundary conditions, 241
Auxiliary-field quantum Monte Carlo
    Attractive Hubbard model, 228
    Backward propagation, 221
    Forward propagation, 221
    Pseudo-partition function, 220
    Repulsive Hubbard model, 229
    Sequential updates, 223
    Sign problem, 221

Backflow correlations, 21, 122
Bayes formula, 43
BCS Hamiltonian, 27, 115, 125
BCS wave function, 26, 125
Block analysis, 83
Boltzmann distribution, 86, 90
Bootstrap method, 78
Bosonic wave function
    Condensed state, 110
    Fast update of the Jastrow factor, 111
    Permanent, 111
Box-Muller trick, 65
Brownian motion, 86

Central limit theorem, 53
Changing random variables, 47
Characteristic function, 45
Chebyshev's inequality, 48
Cholesky decomposition, 222
Classical mapping, 33

Configuration interaction, 20
Constrained-path quantum Monte Carlo, 219
Continuous time, 179, 196
Correlated sampling, 59
Correlation energy, 243
Correlation time, 73
Covariance, 45
Cumulative probability, 43
Cusp conditions, 239

Delayed updates, 121, 227
Direct sampling, 60
Distribution function
    Bernulli, 46
    Binomial, 46
    Gaussian, 46
    Poisson, 46
    Uniform, 46

Energy derivatives, 134
Errorbars
    Block analysis, 83
    Bootstrap method, 78
    Error propagation, 77
    Jackknife method, 81

Fermionic wave function
    Determinant, 112
    Fast update of determinants, 116
    Fast update of Pfaffians, 126
    Green's functions, 119, 128
    Pfaffian, 123
Fixed-node approximation, 204, 209, 255
Fixed-node energy
    Expectation value, 212
    Ground-state value, 211
    Upper-bound property, 211
Fixed-node error, 212
Fixed-node Hamiltonian, 210, 255
Fixed-phase approximation, 204

Fokker-Planck equation, 90
  Approach to equilibrium, 92
  Schrödinger equation, 91

Gram-Schmidt orthogonalization, 222
Green's function Monte Carlo
  Accumulated weight, 173, 176, 184
  Branching, 181
  Continuous time, 179
  Forward walking, 176, 186
  Guiding function, 173
  Importance sampling, 173
  Master equation, 169
  Mixed average, 177
  Sign problem, 168
  Single walker technique, 170
Gutzwiller factor, 17
Gutzwiller projector, 18, 107
Gutzwiller wave function, 17

Haldane-Shastry wave function, 23
Hartree-Fock wave function, 4, 15
Heisenberg model, 10
Heitler-London wave function, 3
Hellmann-Feynman theorem, 212
Hessian matrix, 148
Hubbard model (bosonic), 12
Hubbard model (fermionic), 7
Hubbard-Stratonovich transformation, 218

Imaginary-time propagation, 35, 202, 205, 214
Importance sampling, 61

Jackknife method, 81
Jastrow factor
  One-body (electron-ion on the continuum), 239
  Two-body (electron-electron on the continuum), 239
  Two-body (electron-electron on the lattice), 19, 107
Jastrow wave function, 18, 107
Jastrow-Slater wave function, 20, 107, 237

Langevin dynamics
  Accelerated dynamics, 96
  Detailed balance, 92
  Discretization, 87
  Discretization error, 95
  First-order equations, 85, 140
  Master equation, 90
  Second-order equations, 86
Large deviations, 55
Lattice regularization, 235
  Electron-ion potential, 253
  Laplacian, 253
  Size consistency, 258
  Zero-variance property, 256

Laughlin wave function, 25, 237
Light cone, 163
Linear method, 132, 147, 248
  "Strong" zero-variance property, 149
  Generalized eigenvalue equation, 149
Local energy, 105, 172, 175, 191
Local estimator, 104

Markov chains
  Auxiliary-field quantum Monte Carlo, 223
  Definition, 66
  Detailed balance, 70
  Ergodicity, 70
  Green's function Monte Carlo, 169, 200
  Master equation, 67
  Periodicity, 70
  Reducibility, 70
  Reptation Monte Carlo, 191
  Stationary solution, 70
Mean value, 44
Metropolis-Hastings algorithm, 74
  Acceptance probability, 74
  Trial probability, 74
Molecular orbitals, 243
Mott insulator, 9
Mott transition, 10
Multideterminant wave function, 20

Newton-Raphson optimization, 132
Nodal surface, 202

Optimization
  Linear method, 147
  Reweighting technique, 132
  Steepest descent, 137
  Stochastic reconfiguration, 139

Pairing (geminal) function, 243
Particle-hole transformation, 115, 229
Path integral Monte Carlo, 190, 220
Pauli surface, 202
Perron-Frobenius theorem, 72, 199, 210
Pfaffian, 123
Power method, 34
Probability
  Conditional, 42
  Definition (frequentist), 41
  Joint, 42
  Marginal, 42
  Reproducibility of the experiments, 40, 52
Probability density, 43
Projection technique, 34, 167, 189, 214, 253
Pseudo-random numbers
  Definition, 261
  Fibonacci (lagged) generators, 261
  Periodicity, 261

Quantum quench, 161

Random process, 66
Random variable, 42
Real-time propagation, 156
Reptation Monte Carlo
    Bounce algorithm, 195
    Continuous time, 196
    Importance sampling, 190
    Master equation, 195
    Pseudo-partition function, 189
    Sign problem, 190
    Simple update, 191
Resonating-valence bond wave function, 5,
    29, 237, 244
Reweighting technique, 59

Sampling, 56
    Continuous distribution, 64
    Discrete distribution, 62
Sign problem, 199
Size consistency, 244
Size extensivity, 31
Slater determinant, 4
Sparse Hamiltonian matrix, 35
Standard deviation, 45
Steepest descent, 88, 132, 137, 248
Stochastic reconfiguration, 132, 139, 248
    Covariance property, 142
    Preconditioning, 143
    Projection, 146
    Regularization, 143, 147
    Scale invariant regularization, 144
    Signal to noise ratio, 144

Thermalization time, 73, 172, 176, 193
Time-dependent Hartree-Fock, 156
Time-dependent variational Monte Carlo
    Continuous-time limit, 159
    Energy conservation, 160
    Equations, 158
    Norm conservation, 160
    Stationary action, 160
Trotter approximation, 216
Trotter error, 180, 196, 216

Vandermonde determinant, 22
Variance, 45
Variational Hamiltonian, 209
Variational Monte Carlo
    Expectation value, 104
    Local basis set, 108
    Local energy, 105
    Local estimator, 104
    Zero-variance property, 105
Variational principle
    Accuracy on the correlations, 14
    Accuracy on the energy, 13
    Generalities, 13

Wave function metric, 139
Weak law of large numbers, 52
White noise, 85
Wick theorem, 119, 128, 224

Zero-variance property, 105

Printed in the United States
by Baker & Taylor Publisher Services